AF361840

Silver-Based Antimicrobials

Silver-Based Antimicrobials

Editor

Raymond J. Turner

MDPI • Basel • Beijing • Wuhan • Barcelona • Belgrade • Manchester • Tokyo • Cluj • Tianjin

Editor
Raymond J. Turner
University of Calgary
Canada

Editorial Office
MDPI
St. Alban-Anlage 66
4052 Basel, Switzerland

This is a reprint of articles from the Special Issue published online in the open access journal *Antibiotics* (ISSN 2079-6382) (available at: https://www.mdpi.com/journal/antibiotics/special_issues/silver_antimicrobials).

For citation purposes, cite each article independently as indicated on the article page online and as indicated below:

LastName, A.A.; LastName, B.B.; LastName, C.C. Article Title. *Journal Name* **Year**, *Volume Number*, Page Range.

ISBN 978-3-03943-891-4 (Hbk)
ISBN 978-3-03943-892-1 (PDF)

Contents

About the Editor

Raymond J. Turner Ph.D. (Professor of Microbiology and Biochemistry), received a BSc in Biochemistry/Chemistry and Ph.D. in Physical Biochemistry. His research interests are bacterial resistance mechanisms, molecular microbiology, and bioenergetics. He has been at the Department of Biological Sciences, University of Calgary, since 1998, where he lectures in courses of Introductory Biology and Biochemistry, Biomembranes, Molecular and Biochemical Advanced Techniques, Environmental Chemistry, and Biochemical Toxicology. He has held visiting professorships at the University of Bologna and the University of Verona, Italy. His service highlights include Associate Department Head, Dean's Advisory Committee, and Vice-President of Research Advisory Group. To date, he has trained 46 graduate students and 21 PDFs. He was awarded the Scientist Award: International Association of Advanced Materials (2020); University Teaching Award in graduate supervision (2017); and University Award of Research Excellence (2013). During his career, he has contributed to 263 publications with >12,000 citations, 33 patents/licenses, and 369 conference posters/talks/seminars. He has an h-index of 54 (GS) or 44 (ISI).

 *antibiotics*

Editorial

Is Silver the Ultimate Antimicrobial Bullet?

Raymond J Turner

Department of Biological Sciences, University of Calgary, Calgary, Alberta T2N 1N4, Canada;
turnerr@ucalgary.ca; Tel.: +1-403-220-4308

Received: 10 December 2018; Accepted: 17 December 2018; Published: 19 December 2018

The use of metal compounds as antimicrobial agents has been around since antiquity, only to be replaced by the introduction of organic antibiotics and antiseptics in the mid-20th century. The discovery of penicillin by Alexander Fleming in 1928 began the era of antibiotics. Unfortunately, this time is rapidly coming to an end, as antibiotic resistance is now the norm for most pathogen strains. We now accept that we have entered the Antibiotic Resistance Era, where the World Health Organization considers antibiotic resistance one of the biggest threats to global health, food security and development today. Their 2017 report confirms the world is running out of useful antibiotics [1]. Since the turn of the century, interest into alternatives to antibiotics has seen an explosion of attention into inorganic antimicrobial agents including metal-based antimicrobials [2].

Beyond its malleable and aesthetic qualities, silver has been used since antiquity to control infection. For example, ancient mariners would toss silver coins into the drinking water barrel on ships to prevent fouling. Nowadays, silver and silver nanoparticles (AgNPs) are widely used in healthcare, food industry, cosmetic industry, coatings to surface materials and in textiles. Most of these applications are targeting for infection control or treatment, however, in textiles the antimicrobial properties are exploited for odor control.

Considerable efforts have been made towards understanding the molecular mechanism(s) of action of silver [3,4]. The rules for efficacy of metals as antimicrobials are poorly understood but may follow some fundamental chemical rules (discussed in [5]). In the case of silver, there are likely multiple targets, both direct and indirect, leading various cellular systems to be affected [6]. Regardless of the efficacy, bacteria can develop resistance to metal-based antimicrobials [7], with a silver resistance determinant identified as early as 1975, primarily through an efflux mechanism as well as others (reviewed in [8]).

With a few exceptions, the articles of this special issue of 'Silver-based antimicrobials' focus on AgNPs or nanomaterials, which reflects field-wide research trends. Different AgNP synthesis methods or formulations that are in combination with other antimicrobials are of interest. The various methods, either biological or chemical-physical, produce different types of AgNPs. The articles here and the review of patents from Sim et al., [9] reflects an explosion of such exploratory activity towards potential industrial and health care applications. It is becoming clear that the different formulations of AgNPs that lead to differences in their size, shape, structure and their release of Ag atoms lead to very different antimicrobial activities. This body of work suggests the possibility of tuning silver's antimicrobial activity towards specific strains. Research to date suggests AgNPs to be very effective antimicrobial silver bullets.

As we research the mechanisms of toxicity and resistance of silver, as well as how to prepare novel AgNP formulations, we must keep in mind how we intend to use silver in order to preserve its efficacy. It is prudent to consider stewardship and sustainability at the start before misuse runs rampant. However, given the present overuse of silver in textiles, are we already too late?

References

1. WHO World health organization Antibiotic resistance fact sheet. 2018. Available online: https://www.who.int/news-room/fact-sheets/detail/antibiotic-resistance (accessed on 12 August 2018).
2. Turner, R.J. Metal-based antimicrobial strategies. *Microbial. Biotechnol.* **2017**, *10*, 1062–1065. [CrossRef] [PubMed]
3. Maillard, J.-Y.; Haremann, P. Silver as an antimicrobial: facts and gaps in knowledge. *Crit. Rev. Microbiol.* **2013**, *39*, 373–383. [CrossRef] [PubMed]
4. Rizzello, L.; Pompa, P.O. Nanosilver-based antibacterial drugs and devices: Mechanisms methodological drawbacks and guidelines. *Chem. Soc. Rev.* **2014**, *43*, 1501–1518. [CrossRef] [PubMed]
5. Lemire, J.; Harrison, J.J.; Turner, R.J. Antimicrobial activity of metals: Mechanisms, molecular targets and applications. *Nat. Rev. Microbiol.* **2013**, *11*, 371–384. [CrossRef] [PubMed]
6. Gugala, N.; Lemire, J.A.; Chatfield-Reed, K.; Yan, Y.; Chua, G.; Turner, R.J. Using a chemical genetic screen to enhance our understanding of the antibacterial properties of silver. *Genes* **2018**, *9*, 344. [CrossRef] [PubMed]
7. Harrison, J.J.; Ceri, H.; Turner, R.J. Multimetal resistance and tolerance in microbial biofilms. *Nat. Rev. Microbiol.* **2007**, *5*, 928–938. [CrossRef] [PubMed]
8. Silver, S. Bacterial silver resistance: molecular biology and uses and missuses of silver compounds. *FEMS Microbiol. Rev.* **2003**, *27*, 341–353. [CrossRef]
9. Sim, S.; Barnard, R.T.; Blaskovich, M.A.T.; Ziora, Z.M. Antimicrobial silver in medicinal and consumer applications: A patent review of the past decade (2007–2017). *Antibiotics.* **2018**, *7*, 93. [CrossRef] [PubMed]

 *antibiotics*

Review

Silver and Antibiotic, New Facts to an Old Story

Frédéric Barras [1,2,*], Laurent Aussel [1] and Benjamin Ezraty [1,*]

[1] Laboratoire de Chimie Bactérienne (LCB), Institut de Microbiologie de la Méditerranée (IMM),
    Aix Marseille Université, Centre National de la Recherche Scientifique (CNRS), 13009 Marseille, France;
    aussel@imm.cnrs.fr
[2] Département de Microbiologie, Institut Pasteur, 75015 Paris, France
* Correspondence: fbarras@pasteur.fr (F.B.); ezraty@imm.cnrs.fr (B.E.)

Received: 2 August 2018; Accepted: 21 August 2018; Published: 22 August 2018

Abstract: The therapeutic arsenal against bacterial infections is rapidly shrinking, as drug resistance spreads and pharmaceutical industry are struggling to produce new antibiotics. In this review we cover the efficacy of silver as an antibacterial agent. In particular we recall experimental evidences pointing to the multiple targets of silver, including DNA, proteins and small molecules, and we review the arguments for and against the hypothesis that silver acts by enhancing oxidative stress. We also review the recent use of silver as an adjuvant for antibiotics. Specifically, we discuss the state of our current understanding on the potentiating action of silver ions on aminoglycoside antibiotics.

Keywords: silver; antibiotics; adjuvant; combinatorial; metal; ROS

1. Introduction

The antibacterial effect of silver ions (Ag^+) has been known for centuries as ancient Greek used silver for stomach pains or wound healing. According to Mijnendonckx et al. [1], "silver was perhaps the most important antimicrobial compound before the introduction of antibiotics". Currently, it is used on surfaces in hospitals to reduce nosocomial disease. It is also widely used in water cleaning systems such as hospital hot water circuits, swimming pool and potable water delivery systems. And as Simon Silver repeatedly recalled it, silver can even be found within Japanese Jintan pills meant "to cure from nausea, vomiting, hangover, bad breath and sunstroke among others" [1]. More recently, a series of initiatives aimed at fighting multidrug resistant bacteria elected combinatorial strategies as a way of potentiating drug efficiency [2]. Likewise, silver ions were identified as a highly efficient potent of antibiotics of different classes [3,4]. Last, silver nanoparticles (AgNP) rank currently among the most widely commercialised nanomaterial used in medical, bactericidal and electrical products [5]. Despite this old and broad use, the mechanism underlying antimicrobial activity of silver ions is not fully understood.

In this Review, we will first describe the multiple cases of silver ions being used as biocides. We will then give a broad overview of the many situations wherein combining silver and antibiotics yielded to enhanced antibacterial efficiency. Last, the molecular mechanism allowing silver to potentiate aminoglycoside toxicity will be discussed. Strategies based upon silver nanostructures, as well as their synthesis, toxicity and efficiency, will not be covered in the present review and interested readers can find examples of such studies in References [6–8].

2. Molecular Basis of Silver Toxicity toward Microbes

Silver antibacterial activity has been studied for a long period of time [9]. Silver ions were proposed to target macromolecules and their associated alteration was predicted to be the cause of silver mediated toxicity (Figure 1). Yet, although some consistent trend emerged from this bulk of studies, several discrepancies remained.

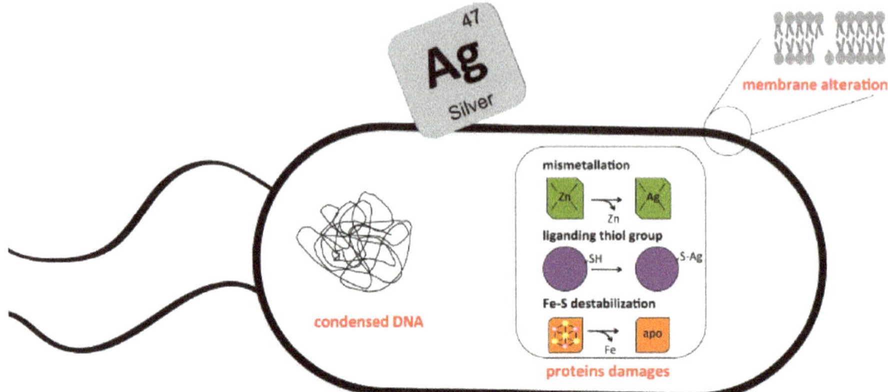

Figure 1. Pleiotropic molecular basis of antimicrobial effects of silver. Silver targets different macromolecules in bacteria. Here are depicted modifications observed in silver-treated bacteria such as DNA condensation, membrane alteration and protein damages. In this latter case, several situations were reported wherein silver ions interacted with thiol group, destabilised Fe-S clusters or substituted to metals in metalloproteins.

2.1. Silver Ions Target DNA

Silver ions are strong nucleic acids binders and form several complexes with DNA or RNA. They interact preferentially with bases rather than the negatively charged backbone of DNA. Thermodynamic experiments showed that silver ions formed homo-base pairs with a higher affinity with guanine, which could potentially lead to pyrimidine dimerization. At a high concentration, silver ions were observed to interact with adenine [10]. Microscopy analysis of silver treated bacteria showed a dense electron-light region assigned as condensed DNA in the centre of the cells. While all of these in vitro observations support the hypothesis that silver could lead to DNA modification prone to mutation or replication inhibition, actual contribution of DNA-silver adducts formation to silver antimicrobial toxicity remains to be assessed in vivo.

2.2. Silver Ions Target Proteins

A silver target on which everybody agrees is the sulfhydryl group, which results in the formation of S-silver bond [11]. Sulfhydryl groups belong to lateral chains of Cys residues. Cys residue frequently served as ligand for metal and/or cofactors in metalloproteins, including those forming respiratory chains. Accordingly, silver ions were found to alter respiration of *E. coli* [9,12] and it was thought that proton motive force (PMF) collapse due to respiration inhibition constituted the basis of silver toxicity [13]. However, subsequent work revealed that silver had additional targets besides respiration [14]. For instance, in *Vibrio cholerae*, proton leakage, which could be a consequence of PMF collapse, was observed even in the absence of the NADH-ubiquinone oxidoreductase [13]. This suggested that silver had multiple protein targets in the membrane. Recently, Xu and Imlay investigated the toxicity of different soft metals in *E. coli* and identified Fe-S cluster containing proteins as primary targets of silver [15]. Importantly, NADH dehydrogenase I activity, a main component of the aerobic respiratory electron transfer chain, was untouched by silver treatment. Instead dehydratases like fumarase A appeared as preferred targets. The 4Fe4S cluster from fumarase was degraded to 3Fe4S cluster that could be reactivated by exogenous Fe^{2+} under reducing conditions. The reason for the apparent specificity of silver ions for Fe-S cluster from dehydratases likely stems from the exposed nature of their solvent and the lability of the catalytic Fe atom.

Other candidate targets include thiol containing cytoplasmic proteins. For instance, OxyR, the H_2O_2-sensing transcriptional activator, was reported to be inactivated in silver-exposed *E. coli* strains [3]. The authors argued that silver antagonises disulphide bond formation within OxyR monomer, which is required for activating transcription.

Last, one should keep in mind that the high thiophilicity of silver ions could allow them to substitute for any SH-liganded metal. For instance, it is conceivable that silver acts upon Fe-S cluster by substituting for labile Fe atom, which would apply to the dehydratase situation depicted above. Alternatively, silver could substitute for zinc ions in, for instance, zinc-finger proteins. Overall, these substitutions could lead to massive protein mis-metallation, loss of function and associated defects. It is noteworthy that cytosolic dense granules were observed in silver treated *E. coli* cells; such granules were interpreted as being constituted of misfolded protein aggregates [3].

2.3. Silver Mediated Membrane Alteration

Transmission electronic microscopy (TEM) observation of silver treated *E. coli* revealed morphological and structural changes of the cell envelope. Moreover, use of propidium iodine showed an enhanced permeability of the cell envelope [3]. In a separate study, TEM revealed an enlargement of the periplasmic space in *E. coli*, suggesting the shrinking of the inner membrane and its detachment from the cell wall. Interestingly a gram-positive bacterium, *Staphylococcus aureus*, which exhibits a thicker cell envelope underwent similar morphological changes than *E. coli*, albeit to a lesser extent, suggesting a stronger resistance to silver ions [16].

2.4. Are Silver Ions Producing ROS?

There is much debate on whether silver, which is not a redox active metal, induces ROS formation, and if it is the case, how this happens. To determine whether silver ions induce ROS, Park et al. used a *soxS-lacZ* reporter strain. After exposure to silver nitrate, induction of *soxS* was observed. As *soxS* expression being under the control of SoxR, it was deduced that superoxide radicals had been produced by the presence of silver ions. However, no *soxR* control mutant was tested and the actual signal SoxR is responding to remains a matter of debate. In particular, it has been proposed that SoxR senses the ratio $NAD(P)H/NAD(P)$ [17]. Were silver ions to impair respiration, this ratio would be modified and SoxR activated without the need for superoxide production. Importantly OxyR activation was not observed, supporting the notion that no H_2O_2 accumulated in the presence of silver ions. Using 3'-p-hydroxyphenyl fluorescein (HPF), a dye, Morones et al. observed hydroxyl radicals production in silver treated *E. coli* cells [3]. Surprisingly, overproduction of superoxide dismutase (SOD), predicted to enhance hydroxyl radical production via the Fenton reaction, was found to reduce HPF-estimated hydroxyl radical. Moreover, detection of hydroxyl radicals by Morones et al. [3] somehow did not fit with the lack of H_2O_2 enhanced production and lack of OxyR induction reported by Park et al. [18]. If ROS were instrumental in conveying silver toxicity, a prediction is that anaerobically grown cultures should be less sensitive to silver ions. This issue was investigated in several studies but unfortunately conflicting observations were reported and it is so far impossible to draw a firm conclusion from the literature (see [1]). Last, a very recent transcriptomic analysis of *E. coli* exposed to silver ions failed to identify anti-ROS defence genes induction, while dysregulation of silver transport and detoxification (*copA, cueO, mgtA, nhaR*), stress response genes (*dnaK, dnaJ, pspA*, oxidoreductase genes), methionine biosynthesis (*metA, metR*), membrane homeostasis (*fadL*), and cell wall integrity (*lpxA, arnA, ycfS, ycbB*) were identified [19]. Hence, experimental evidences for silver ions to induce ROS production remain scant and open to discussion.

On the other hand, if one admits that silver ions are perturbing iron homeostasis as well as destabilizing Fe-S clusters, it seems quite likely that eventually this will indirectly lead to ROS production (Figure 2). Indeed, because Fe-S proteins are central to respiration, this latter is expected to be perturbed and this could provoke electron leakage and associated ROS production. Also, destabilization of Fe-S clusters is expected to release free iron, which should fuel in the Fenton reaction.

Last, silver ions by binding to thiols will preclude endogenous anti-ROS defences such as free cysteine and glutathione, two compounds with ROS-scavenging properties.

Hence it seems indeed a safe prediction that silver ions will favour ROS production, yet the causal chain linking silver, a non-redox soft metal, and ROS production remains to be established and described in precise molecular terms.

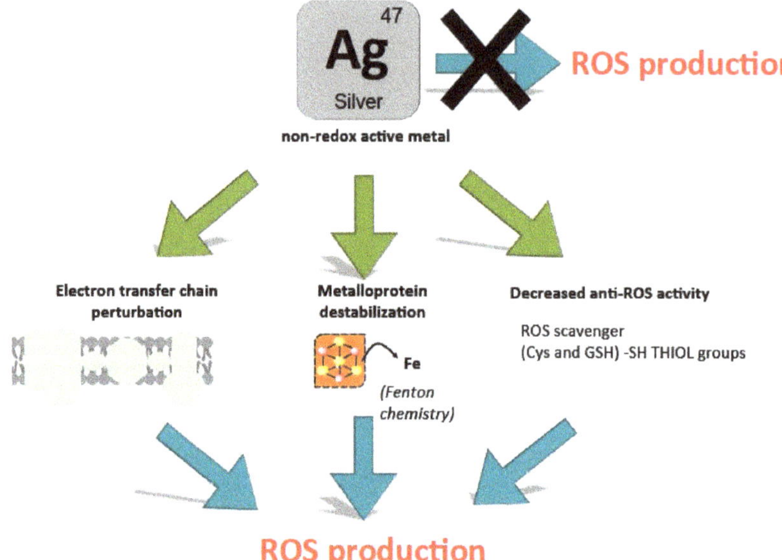

Figure 2. Searching for the causal link between silver ions and ROS production. Silver is a non-redox active metal that cannot directly produce ROS. Some experimental evidences however pointed to the enhanced production of ROS in the presence of silver ions. Depicted here are possible indirect ways silver ions could participate to ROS production: Perturbation of respiratory electron transfer chain, Fenton chemistry following destabilization of Fe-S clusters, or displacement of iron, inhibition of anti-ROS defences by thiol-silver bond formation.

3. Silver Enhances Antibacterial Activity of Antibiotics

In 2007, Morones et al. investigated the capacity of silver ions to synergise antibiotics [3]. They reported that silver potentiates bactericidal antibiotics both in laboratory growth conditions and animal models. The three major classes of bactericidal antibiotics in *E. coli* were tested, i.e., ß-lactams (ampicillin), which target cell-wall synthesis, quinolones (ofloxacin), which target DNA replication and repair, aminoglycosides (gentamicin) that are ribosome binders known to cause protein mistranslation. All of these drugs were tested at a concentration close or inferior to the MIC values, and in the presence of sublethal concentrations of silver. In all of these cases, a significantly enhanced antimicrobial activity was observed. A more precise analysis revealed that the highest synergistic effect was found when combining gentamicin and silver as viability dropped 2 logs. In the case of ampicilin and ofloxacin, presence of silver decreased viability 1 log at the maximum. After showing that mice tolerated the silver concentration used (3–6 mg/kg), the authors reported that silver potentiated both the gentamicin activity in a urinary tract infection mouse model, and the vancomycin activity in a mouse peritonitis infection mouse model.

The potentiating activity of silver on antibiotic toxicity in *E. coli* K12 was further investigated by Herisse et al. [4]. An extended set of bactericidal and bacteriostatic antibiotics including tetracycline and chloramphenicol were tested [4]. According to changes in MIC values, silver was found to be most potent with aminoglycosides (gentamicin, kanamycin, tobramycin, streptomycin) as MIC value decreased by more than 10-fold. A reduction in the MIC value of 2-fold was noted with spectinomycin, a bacteriostatic antibiotic related to aminoglycoside and also with tetracycline. Moreover, they reported a slight potentiating effect (less than 20%) when silver was used in conjunction with quinolone (nalidixic acid and norfloxacin) or with chloramphenicol [4].

Another study showed that silver enhances the toxicity of the selenazol drug ebselen, a competitive inhibitor of bacterial thioredoxin reductase activity against clinically multidrug-resistant Gram-negative bacteria (*Klebsiella pneumoniae*, *Acinetobacter baumannii*, *Pseudomnas aeruginosa*, *Enterobacter cloacae*, *Escherichia coli*) [20]. Potentiating effects were observed both in laboratory growth conditions and mice peritonitis model (6 mg/kg). Similarly, Wan and collaborators in a study on AgNP showed that ionic nitrate silver acts synergistically with polymixin B and rifampicin to combat carbapenem-resistant *A. baumannii* obtained from clinical patients. Interestingly, AgNP and AgNO$_3$ showed the same potentiating effect with both antibiotics, but cytotoxicity of AgNP was lower than that of AgNO$_3$ [21]. Silver was also reported to potentiate polymixin B and a series of antimicrobial peptides to combat gram-negative bacteria [22].

Many antibiotics that are effective against planktonic cells turned out to be ineffective against biofilms. Combination of silver with tobramycin combated biofilm of *E. coli* and *Pseudomonas aeruginosa* as a 3-fold enhancement of antimicrobial efficiency was observed [23]. A similar potentiating effect of silver (6 mg/kg) with gentamicin was noted in combating biofilm formed on a catheter located into a mouse model [3].

Last, silver made antibiotics effective against resistant bacteria. Indeed, silver was able to sensitise *E. coli* to the Gram-positive-specific antibiotic vancomycin and the highly tolerant anaerobic pathogen *Clostridium difficile* became sensitive to aminoglycoside [3,4]. Moreover, silver could restore antibiotic susceptibility to a tetracycline resistant *E. coli* mutant [3].

We wish to underline that the use of silver as an adjuvant might also be of interest to treat persister cells, a subpopulation of isogenic bacteria that become highly tolerant to antibiotics [3]. All these data are grouped in Table 1.

Table 1. Antibacterial activity of silver ions in combination with antibiotics.

Antibiotics		Organism	Culture Condition	Effects	References
ß-lactams	Ampicillin	*E. coli*	Laboratory medium	10-fold increase in antimicrobial activity	[3]
Quinolones	Ofloxacine, Nalidixic Acid, Norfloxacin	*E. coli*	Laboratory medium	10-fold increase in antimicrobial activity. MIC value decreased 10–25%	[3,4]
Aminoglycosides	Gentamicin	*E. coli*	Laboratory medium. Animal models	100-fold increase in antimicrobial activity. MIC value decreased more than 10-fold	[3,4]
		C. difficile	Laboratory medium	MIC value decreased 4-fold	[4]
	Tobramycin	*E. coli.,* *P. aeruginosa*	Laboratory medium	MIC value decreased 10-fold (*E. coli*). 3-fold increase in antimicrobial activity (*P. aeruginosa*)	[4,23]
	Kanamycin Streptomycin	*E. coli*	Laboratory medium	MIC value decreased more than 10-fold	[4]
Spectinomycin		*E. coli*	Laboratory medium	MIC value decreased 2-fold	[4]

Table 1. *Cont.*

Antibiotics	Organism	Culture Condition	Effects	References
Vancomycin	*E. coli*	Laboratory medium. Animal models	10-fold increase in antimicrobial activity	[3]
Chloramphenicol	*E. coli*	Laboratory medium	MIC value decreased 1.5-fold	[4]
Ebselen	*K. pneumoniae, A. baumanni, P. aeruginosa, E. cloacae, E. coli*	Laboratory medium. Animal models	10-fold increase in MIC value	[20]
Polymixin B	*E. coli*	Laboratory medium	MIC value decreased 5- to 10-fold	[21,22]
Rifampicin	*A. baumannii*	Laboratory medium	MIC value decreased 5-10 fold	[21]
Tetracycline	*E. coli* (Tet^R)	Laboratory medium	MIC value decreased 2-fold	[3]

4. Molecular Mechanism in the Aminoglycoside/Silver Synergy

Of all antibiotics tested, aminoglycosides (gentamicin, tobramycin, kanamycin, streptomycin) benefited the most from silver ions as adjuvants (Figure 3). The molecular basis of the synergistic effect between silver and aminoglycoside has been investigated in two separated studies, which we discuss below [3,4]. However, we shall first recall how aminoglycosides are predicted to kill bacteria, and in particular how they are uptaken by *E. coli*.

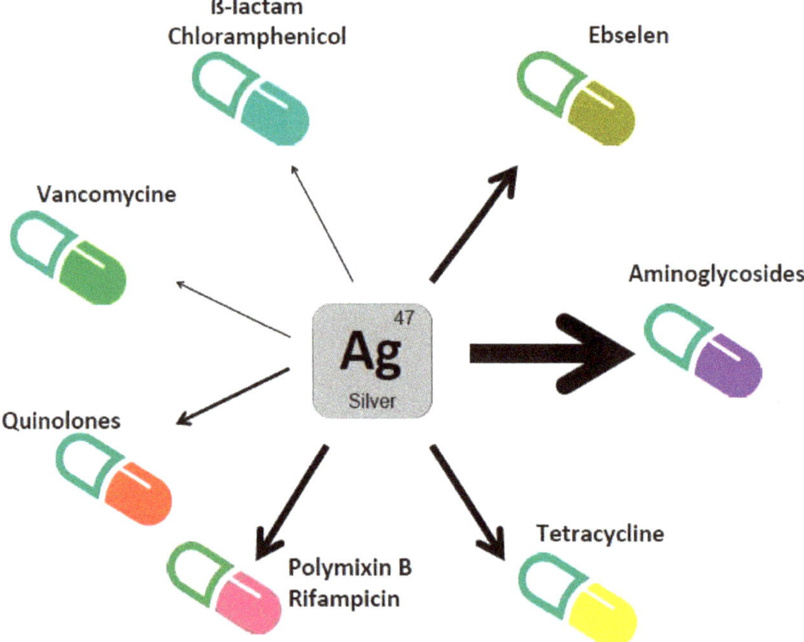

Figure 3. Silver potentiates antibiotics toxicity. The capacity of silver ions to enhance the toxicity of antibiotics from different family is represented. The size of the arrows line reflects the extent of the synergistic effect.

Aminoglycosides, first discovered in the 1940s, are the antibiotics most commonly used worldwide, due to their high efficacy and low cost [24]. Aminoglycosides are a group of bactericidal antibiotics that target the 30S ribosomal subunit and induce amino acid mis-incorporation. Aminoglycoside need to be transported through the cytoplasmic membrane to reach their target.

These transport systems are energised via proton motive force (PMF)-dependent pathways [25]. Moreover a so-called feed-forward loop model postulates the occurrence of a two-steps process: Aminoglycosides would cross quite inefficiently the cytoplasmic membrane prior to hit membrane-bound ribosome (EDP-I), resulting in aborted translated products, which would go into the membrane due to their hydrophobic characters, and destabilise further the membrane, allowing for enhanced entry of aminoglycoside (EDP-II) [26].

Herisse et al. showed that silver enhances aminoglycoside toxicity by acting independently of PMF as it by-passes the EDP-I PMF-dependent step of the aminoglycoside entry process. Silver by-passed the antagonist effect of the PMF dissipating action of the carbonyl cyanide-m-chlorophenylhydrazone (CCCP), an uncoupler H^+ ionophore [4]. Moreover, silver restored aminoglycoside uptake by strains exhibiting a reduced PMF level such as mutants lacking complex I and II (Δ*nuo* Δ*sdh*) or Fe-S cluster biosynthesis (Δ*iscUA*) [4]. In contrast, the silver-potentiating effect of aminoglycoside toxicity remained dependent on translation, the EDP-II proteins translation-dependent step [4]. Indeed, adding chloramphenicol, a bacteriostatic antibiotic inhibiting translation, prevented silver from potentiating aminoglycoside toxicity. It was proposed that silver destabilises the membrane in a protein translation-dependent pathway, allowing aminoglycoside to get access to the cytosol more efficiently. This implied that membrane disturbance induced by silver is not sufficient for massive aminoglycoside uptake and needs additional contribution from mis-localised aborted polypeptides. By acting directly on ribosomes, silver could release aborted translated products that would eventually go to the membrane and cause an EDP-II like step (Figure 4). This agrees with a proposal by Morones et al. [3] who envisioned that silver produced misfolded proteins would be directed towards the inner membrane and destabilised it. An argument supporting this view was that enhanced silver resistance of a *secG* mutant impaired in protein translocation [3]. Hence, irrespective of the origin and cause of increased level of misfolded proteins, both studies pointed out to an enhanced permeability of the cell envelope. This is consistent with morphological and structural changes observed by TEM studies of silver treated cells (see above).

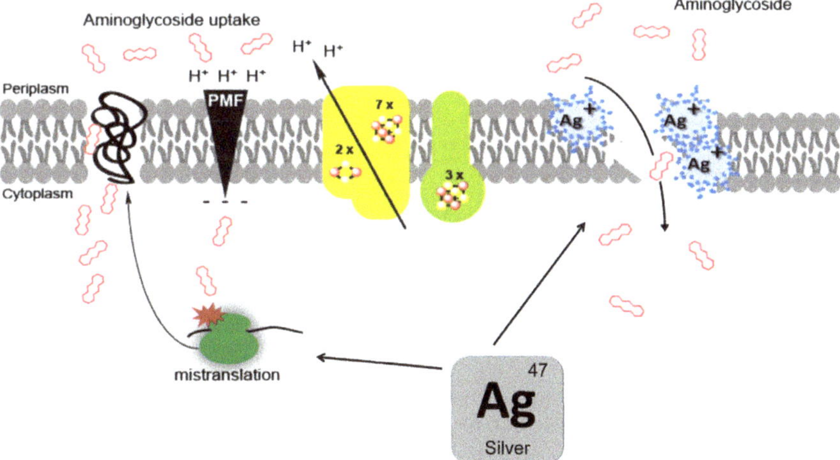

Figure 4. A molecular mechanism model for aminoglycoside and silver synergy. Silver enhances aminoglycoside toxicity by enhancing their uptake. Silver could destabilise the membrane either directly by altering intrinsic membrane proteins or indirectly by acting on ribosomes, which would produce misfolded aborted polypeptides that would eventually go to the inner membrane. Increased permeability of membrane would provoke massive aminoglycoside uptake.

In contrast, the contribution of ROS to silver toxicity was more controversial. Morones et al. postulated that silver ions enhance gentamicin toxicity via the capacity of silver to produce ROS [3]. However the enhanced production of ROS in the presence of the combination (silver+gentamicin) was not tested. Moreover, the actual production of ROS following silver addition is highly debatable as summarised above. Last, Herisse et al. directly addressed the question of the contribution of ROS to the silver potentiating effect and collected only negative evidences: (i) silver potentiated gentamicin toxicity even in anaerobic conditions; (ii) mutants altered in anti-ROS activities like the strains lacking superoxide dismutases ($\Delta sodA$ $\Delta sodB$) or the H_2O_2-stress responding master regulator ($\Delta oxyR$) exhibited similar sensitivity to silver potentiating effect as the wild type *E. coli* [4].

5. Conclusions

In this review we listed numerous cases in which silver ions were reported to exhibit efficient antibacterial activity. We also reviewed the emerging trend of using silver ions as adjuvants for potentiating antibiotic toxicity. It is compelling that after so many years, the actual reason silver kills bacteria is still eluding us. In fact, it is likely that silver ions act upon multiple different targets, from macromolecules to free amino-acid like cysteine or small molecule such as glutathion, and therefore renders it difficult, if not impossible, to trace the actual cause of death of a silver treated bacterium. Nevertheless, some pressing issues remain: Is silver destabilising protein components of respiratory chains? Does silver have any deleterious (mutagenic?) effect on genome integrity? What is the actual structural state of a silver-destabilised membrane? What is the link, if any, between silver and ROS production?

There is little doubt that new efforts should be dedicated towards the understanding of the action of silver such that this very ancient antibacterial metal can be further exploited within the context of the multiple antibiotic resistance crisis. Interest for such a potential path is reinforced by the fact that pharmacological, toxicological and pharmacokinetic modelling studies indicated that human health risks associated with silver exposure were low [27,28]. From a broader perspective, recently, we advocated the need to take into account iron in its influence on antibiotic sensitivity [29]. It is known that most metals can have antibacterial activities at high concentration, such as bismuth, cobalt, copper and cadmium, to cite a few [15,30–32]. Aiming at characterising and further exploiting their biocide activity might be a rewarding goal.

Author Contributions: Conceptualization, F.B. and B.E.; Investigation, L.A. and B.E.; Writing-Original Draft Preparation, F.B. and B.E.; Writing-Review & Editing, F.B. and B.E.; Funding Acquisition, F.B., L.A. and B.E.

Funding: This research was funded by Joint Programming Initiative on Antimicrobial Resistance (JPIAMR)/Agence Nationale de la Recherche (ANR) grant number ANR-15-JAMR-0003-02 Combinatorial, Fondation pour la Recherche Médicale (FRM), CNRS and Aix Marseille Université.

Conflicts of Interest: The authors declare no conflict of interest. The funders had no role in the design of the study; in the collection, analyses, or interpretation of data; in the writing of the manuscript, and in the decision to publish the results.

References

1. Mijnendonckx, K.; Leys, N.; Mahillon, J.; Silver, S.; Van Houdt, R. Antimicrobial silver: Uses, toxicity and potential for resistance. *BioMetals* **2013**, *26*, 609–621. [CrossRef] [PubMed]

2. Brochado, A.R.; Telzerow, A.; Bobonis, J.; Banzhaf, M.; Mateus, A.; Selkrig, J.; Huth, E.; Bassler, S.; Beas, J.Z.; Zietek, M.; et al. Species-specific activity of antibacterial drug combinations. *Nature* **2018**, *559*, 259–263. [CrossRef] [PubMed]

3. Morones-Ramirez, J.R.; Winkler, J.A.; Spina, C.S.; Collins, J.J. Silver enhances antibiotic activity against gram-negative bacteria. *Sci. Transl. Med.* **2013**, *5*, 1–11. [CrossRef] [PubMed]

4. Herisse, M.; Duverger, Y.; Martin-Verstraete, I.; Barras, F.; Ezraty, B. Silver potentiates aminoglycoside toxicity by enhancing their uptake. *Mol. Microbiol.* **2017**, *105*, 115–126. [CrossRef] [PubMed]

5. Bartłomiejczyk, T.; Lankoff, A.; Kruszewski, M.; Szumiel, I. Silver nanoparticles—Allies or adversaries? *Ann. Agric. Environ. Med.* **2013**, *20*, 48–54. [PubMed]

6. Jakobsen, V.; Viganor, L.; Blanco-Fernández, A.; Howe, O.; Devereux, M.; McKenzie, C.J.; McKee, V. Tetrameric and polymeric silver complexes of the omeprazole scaffold; synthesis, structure, in vitro and in vivo antimicrobial activities and DNA interaction. *J. Inorg. Biochem.* **2018**, *186*, 317–328. [CrossRef] [PubMed]

7. Leonhard, V.; Alasino, R.V.; Muñoz, A.; Beltramo, D.M. Silver nanoparticles with high loading capacity of amphotericin B: Characterization, bactericidal and antifungal effects. *Curr. Drug Deliv.* **2018**. [CrossRef] [PubMed]

8. De Matteis, V.; Cascione, M.; Toma, C.; Leporatti, S. Silver nanoparticles: Synthetic routes, in vitro toxicity and theranostic applications for cancer disease. *Nanomaterials* **2018**, *8*, 319. [CrossRef] [PubMed]

9. Yudkin, J. The effect of silver ions on some enzymes of Bacterium coli. *Enzymologia* **1937**, *2*, 161–170.

10. Arakawa, H.; Neault, J.F.; Tajmir-Riahi, H.A. Silver(I) complexes with DNA and RNA studied by fourier transform infrared spectroscopy and capillary electrophoresis. *Biophys. J.* **2001**, *81*, 1580–1587. [CrossRef]

11. Russell, A.D.; Hugo, W.B. Antimicrobial activity and action of silver. *Prog. Med. Chem.* **1994**, *31*, 351–370. [PubMed]

12. Rainnie, D.J.; Bragg, P.; Bragg, P.D.; Rainnie, D.J. The effect of silver ions on the respiratory chain of Escherichia coli. *Can. J. Microbiol.* **1974**, *20*, 883–889.

13. Dibrov, P.; Dzioba, J.; Gosink, K.K.; Häse, C.C.; Ha, C.C. Chemiosmotic mechanism of antimicrobial activity of Ag$^+$ in vibrio cholerae. *Antimicrob. Agents Chemother.* **2002**, *46*, 2668–2670. [CrossRef] [PubMed]

14. Schreurs, W.J.A.; Rosenberg, H. Effect of silver ions on transport and retention of phosphate by *Escherichia coli*. *J. Bacteriol.* **1982**, *152*, 7–13. [PubMed]

15. Xu, F.F.; Imlay, J.A. Silver(I), mercury(II), cadmium(II), and zinc(II) target exposed enzymic iron-sulfur clusters when they toxify Escherichia coli. *Appl. Environ. Microbiol.* **2012**, *78*, 3614–3621. [CrossRef] [PubMed]

16. Feng, Q.L.; Wu, J.; Chen, G.Q.; Cui, F.Z.; Kim, T.N.; Kim, J.O. A mechanistic study of the antibacterial effect of silver ions on *Escherichia coli* and *Staphylococcus aureus*. *J. Biomed. Mater. Res.* **2000**, *52*, 662–668. [CrossRef]

17. Gu, M.; Imlay, J.A. The SoxRS response of *Escherichia coli* is directly activated by redox-cycling drugs rather than by superoxide. *Mol. Microbiol.* **2011**, *79*, 1136–1150. [CrossRef] [PubMed]

18. Park, H.J.; Kim, J.Y.; Kim, J.; Lee, J.H.; Hahn, J.S.; Gu, M.B.; Yoon, J. Silver-ion-mediated reactive oxygen species generation affecting bactericidal activity. *Water Res.* **2009**, *43*, 1027–1032. [CrossRef] [PubMed]

19. Saulou-Bérion, C.; Gonzalez, I.; Enjalbert, B.; Audinot, J.-N.; Fourquaux, I.; Jamme, F.; Cocaign-Bousquet, M.; Mercier-Bonin, M.; Girbal, L. *Escherichia coli* under ionic silver stress: An integrative approach to explore transcriptional, physiological and biochemical responses. *PLoS ONE* **2015**, *10*, e0145748. [CrossRef] [PubMed]

20. Zou, L.; Lu, J.; Wang, J.; Ren, X.; Zhang, L.; Gao, Y.; Rottenberg, M.E.; Holmgren, A. Synergistic antibacterial effect of silver and ebselen against multidrug-resistant Gram-negative bacterial infections. *EMBO Mol. Med.* **2017**, *9*, 1165–1178. [CrossRef] [PubMed]

21. Wan, G.; Ruan, L.; Yin, Y.; Yang, T.; Ge, M.; Cheng, X. Effects of silver nanoparticles in combination with antibiotics on the resistant bacteria *Acinetobacter baumannii*. *Int. J. Nanomed.* **2016**, *11*, 3789–3800. [CrossRef] [PubMed]

22. Ruden, S.; Hilpert, K.; Berditsch, M.; Wadhwani, P.; Ulrich, A.S. Synergistic interaction between silver nanoparticles and membrane-permeabilizing antimicrobial peptides. *Antimicrob. Agents Chemother.* **2009**, *53*, 3538–3540. [CrossRef] [PubMed]

23. Kim, J.; Pitts, B.; Stewart, P.S.; Camper, A.; Yoon, J. Comparison of the antimicrobial effects of chlorine, silver ion, and tobramycin on biofilm. *Antimicrob. Agents Chemother.* **2008**, *52*, 1446–1453. [CrossRef] [PubMed]

24. Davis, B.D. Mechanism of bactericidal action of aminoglycosides. *Microbiol. Rev.* **1987**, *51*, 341–350. [PubMed]

25. Taber, H.W.; Mueller, J.P.; Miller, P.F.; Arrow, A.M.Y.S. Bacterial uptake of aminoglycoside antibiotics. *Microbiol. Rev.* **1987**, *51*, 439–457. [PubMed]

26. Hurwitz, C.; Braun, C.B.; Rosano, C.L. Role of ribosome recycling in uptake of dihydrostreptomycin by sensitive and resistant *Escherichia coli*. *BBA Sect. Nucleic Acids Protein Synth.* **1981**, *652*, 168–176. [CrossRef]

27. Lansdown, A.B.G. A Pharmacological and toxicological profile of silver as an antimicrobial agent in medical devices. *Adv. Pharmacol. Sci.* **2010**, *2010*, 1–16. [CrossRef] [PubMed]

28. Bachler, G.; von Goetz, N.; Hungerbühler, K. A physiologically based pharmacokinetic model for ionic silver and silver nanoparticles. *Int. J. Nanomed.* **2013**, *8*, 3365–3382.

29. Ezraty, B.; Barras, F. The 'liaisons dangereuses' between iron and antibiotics. *FEMS Microbiol. Rev.* **2016**, *40*, 418–435. [CrossRef] [PubMed]

30. Keogan, D.M.; Griffith, D.M. Current and potential applications of bismuth-based drugs. *Molecules* **2014**, *19*, 15258–15297. [CrossRef] [PubMed]

31. Barras, F.; Fontecave, M. Cobalt stress in *Escherichia coli* and *Salmonella enterica*: Molecular bases for toxicity and resistance. *Metallomics* **2011**, *3*, 1130–1134. [CrossRef] [PubMed]

32. Chandrangsu, P.; Rensing, C.; Helmann, J.D. Metal homeostasis and resistance in bacteria. *Nat. Rev. Microbiol.* **2017**, *15*, 338–350. [CrossRef] [PubMed]

Review

State-of-the-Art, and Perspectives of, Silver/Plasma Polymer Antibacterial Nanocomposites

Jiří Kratochvíl, Anna Kuzminova and Ondřej Kylián *

Department of Macromolecular, Faculty of Mathematics and Physics, Physics Charles University,
Prague 18000, Czech Republic; kratji@seznam.cz (J.K.); annakuzminova84@gmail.com (A.K.)
* Correspondence: ondrej.kylian@gmail.com; Tel.: +420-951-552-258

Received: 14 July 2018; Accepted: 10 August 2018; Published: 17 August 2018

Abstract: Urgent need for innovative and effective antibacterial coatings in different fields seems to have triggered the development of numerous strategies for the production of such materials. As shown in this short overview, plasma based techniques arouse considerable attention that is connected with the possibility to use these techniques for the production of advanced antibacterial Ag/plasma polymer coatings with tailor-made functional properties. In addition, the plasma-based deposition is believed to be well-suited for the production of novel multi-functional or stimuli-responsive antibacterial films.

Keywords: non-equilibrium plasma; silver; antibacterial coatings; plasma polymers; nanocomposites

---

1. Introduction

As shown in numerous recent reviews, road-maps, and "white" papers [1–7], non-equilibrium plasmas may be successfully used in various fields, such as from the automotive or aerospace industry to waste and pollution control, or from the production of advanced functional biomaterials to the design of smart textiles. The tremendous boom that plasma-based technologies experienced in the last few decades is mainly given by their unique features. These comprise a relatively low-temperature nature of the processes that employ plasma, allowing processing of heat-sensitive materials, including polymers, to treat or coat virtually any material without compromising the bulk properties of treated objects or high versatility in terms of materials that may be deposited by plasma-based techniques (e.g., metals, metal-oxides, plasma polymers). In contrast to techniques that utilize chemical synthesis, plasma processes are dry with no or limited use of solvents or potentially harmful chemical substances, making them, in many cases, a highly valuable "green" alternative to "wet" chemical methods. Enormous advancement in the use of plasma technologies was also enabled by recent progress in plasma science and technology, especially in the case of atmospheric pressure discharges, as well as by close collaborations of plasma physicists with chemists, biologists, and medical doctors. This broadened the range of possible applications of non-equilibrium plasmas that nowadays include sterilization/decontamination of surfaces [8–13], plasma medicine [14–17], agriculture [18–22], or as it will be discussed in this review, the production of antibacterial coatings [23–25].

In order to highlight the state-of-the-art plasma-based strategies for the deposition of silver-containing antibacterial coatings, the article is organized as follows: In Section 2, different approaches that may be used for the suppression of bacteria adhesion or biofilms formation will be briefly summarized, with emphasis given to Ag-containing nanocomposites. Different plasma-based strategies that were developed for the production of Ag/plasma polymer nanocomposites will be presented in Section 3. Finally, Section 4 will cover the challenges and future perspectives of the plasma deposition of antibacterial coatings.

2. Antibacterial Coatings

Preventing the bacterial colonization of surfaces is a key requirement to limit the spread of infections. Although this is important in various fields (e.g., the textile industry, food packaging production, or space missions [26–30]), most of the attention is devoted to surfaces used in medicine and healthcare services. Here, the presence of bacteria on medical instruments, tools, accessories etc. may have catastrophic consequences. Although many of the factors leading to bacterial infection can be easily avoided by appropriate hygiene procedures in the hospitals, it is impossible to completely avoid bacterial infections. For example, this is true in the case of medical implants. It is well known that planktonic bacteria present in body fluids may adhere to the implant surface in vivo, multiply, and start to form complex and highly resistant communities, such as biofilms, where bacteria can survive for extended periods of time. The biofilms may then act as persistent reservoirs for pathogens which may trigger infections. This, in turn, often necessitates re-operation and replacement of the infected implant that not only increases costs, but represents serious risk—especially for elderly patients. A promising strategy for reducing the occurrence of such undesirable events is to avoid the initial attachment of bacteria to the implant's surface. Generally, this can be done by following two strategies [31].

The first one is based on coating implants with non-fouling thin films, i.e. films that are capable of resisting protein adsorption or bacteria adhesion. The typical examples of such materials are *poly (ethylene glycol)* (PEG) derivatives or zwitterionic polymers, that have proved to be able to reduce or inhibit bio-fouling [32–36]. However, it has to be stressed that in most of the in vivo experiments, surfaces led only to the delay of biofilm formation and restricted longer-term stability and performance. Furthermore, non-fouling coatings are prone to damage during their handling.

The second way in which to combat biofilm formation is by the production of surfaces that actively kill bacteria. This may be achieved either by contact-killing, or by release-based strategies. The contact-killing of adhered bacteria is achieved either by antimicrobial compounds covalently anchoring to the material surface. Examples include quaternary ammonium compounds, cationic peptides, or enzymes [23,37,38], or the design of biomimetic nanostructured coatings with high-aspect-ratio surface topographies that exhibit biocidal efficacy [39,40]. In the case of release-based coatings, antibacterial action is connected with the gradual leaching of antibacterial agents from the coatings into the surrounding media. The antimicrobial agents may either be organic or inorganic [23,31,41]. Regardless of the antibacterial agent used, its local delivery at a specific site ensures a high local dose without exceeding the systemic toxicity. Because of this, various techniques for the production of such antibacterial coatings were developed, including plasma-based ones that utilized organic antibiotics [42,43].

However, despite the success of these approaches, silver-based materials have been receiving increasing attention since the 1960s. This was caused by ongoing reports on the development of the resistance of certain bacteria to commonly used antibiotics (e.g., [44–50]). Even though this is also the case for silver, where reports exist which indicate the occurrence of silver-resistant bacteria [51,52], the development of bacterial resistance towards Ag is highly improbable due to the multiple possible pathways that may lead to bacterial damage by silver nanoparticles (NPs). The exact mechanism of the antibacterial action of Ag NPs is still not fully understood, and is a frequent topic of discussion and controversy. Ag NPs may either induce the formation of reactive oxygen species (ROS) or may release in aqueous media, such as highly bioactive silver ions. The ROS and Ag^+ ions produced subsequently interact with DNA or the thiol groups of molecules present in the cytoplasm, cell membrane, and inner membrane of mitochondria, that may result in changes in permeability, disturbance of respiration, leakage of intracellular content, inhibition of protein synthesis and function, and cell apoptosis or necrosis [52–55].

Naturally, the silver NPs and released silver ions may interact not only with bacteria, but also with tissue cells. Fortunately, the health risks associated with systemic absorption of Ag^+ ions are rather low [56]. Nevertheless, in order to avoid any undesirable side-effects connected with silver release, its dosage has to be carefully controlled and regulated, which is one of the key challenges connected with the application of silver-based antibacterial nanocomposites.

3. Antibacterial Ag/Plasma Polymer Nanocomposites

Numerous studies have been devoted to the functionalization of surfaces (mostly polymeric ones) by antibacterial nanosilver (e.g., [57–62]). However, a limitation of the proposed methods was the poor adhesion of silver to substrates, which often led to the washing-out of Ag NPs from the substrate, which came with unacceptable pollution of the environment. Although it was demonstrated by different groups that this effect may be substantially suppressed by plasma activation/functionalization of substrate materials prior to the silver deposition (e.g., [63–68]), in many cases the stability of Ag layers still remained questionable. Because of this, alternative strategies based on the embedding of Ag NPs into a supporting matrix were suggested. From these points of view, plasma polymers are considered to be highly promising matrix materials, i.e. macromolecular solids that are created when an organic vapour or precursor passes through a plasma [69–74]. In contrast to conventional polymers, plasma polymers have a random and irregular structure with higher degrees of crosslinking, and a branched structure. Despite their complex structure, plasma polymers offer certain advantages- they can be deposited in the form of conformal, pin-hole free thin films on any substrate material with well-controlled thicknesses at a nanometre scale. Furthermore, their properties that can be tuned by a wide range of operational conditions (e.g., applied power, precursor/working gas mixture, pressure) span from soft to hard polymers, from bio-fouling to non-fouling, or from films that are stable, swelling, or dissolvable in aqueous media. The use of plasma polymers thus may not only improve the fixation on Ag NPs and avoid their release to surrounding media, but, due to the high flexibility of plasma polymers in terms of their physico-chemical and bio-adhesive properties, also tailor their antibacterial or bio-fouling performance. In the following two subsections, different approaches that were tested, with the aim of producing Ag/plasma polymer nanocomposites, will be summarized.

3.1. Layered Nanocomposites

Sandwich or multi-layer structures (see Figure 1) represent the first family of antibacterial Ag/plasma polymer nanocomposites. Such coatings typically consist of Ag NPs overcoated with a thin top layer of plasma polymer (typically from several nanometers to tens of nanometers) that fixes Ag NPs on a substrate material and acts as a diffusion barrier for leaching Ag$^+$ ions. Ag NPs may be deposited on a substrate either by immersion into a colloidal solution of Ag NPs followed by drying, salt reduction, or physical methods, such as evaporation, magnetron sputtering, or the use of gas aggregation sources (GAS) of silver NPs (see Figure 2). The physical methods of Ag NPs deposition are highly advantageous, as they limit the possible uncontrolled aggregation of Ag NPs on surfaces, ensure high purity of silver NPs, and may be easily combined with other low-pressure deposition techniques used for plasma polymer matrix deposition. In order to facilitate the adhesion of Ag NPs onto a substrate additional bottom layer (often plasma polymer film) is used as an interface between the substrate and Ag NPs.

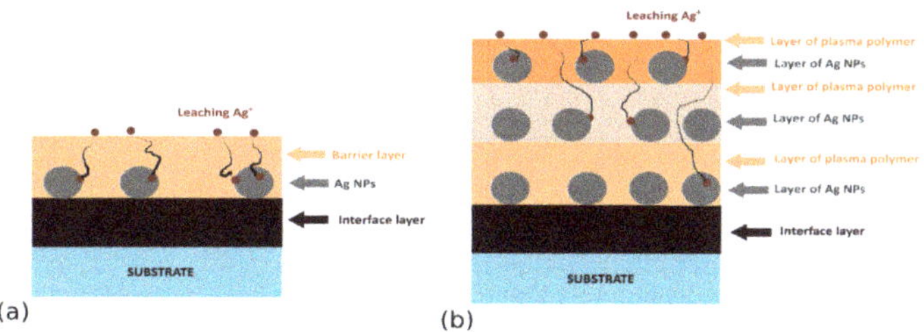

Figure 1. Common structure of (**a**) sandwich and (**b**) multi-layered Ag-based nanocomposites.

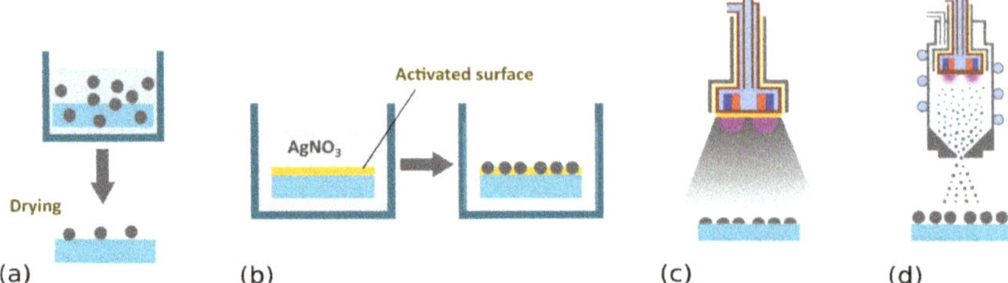

Figure 2. Schematic representation of possible strategies for coating substrates with Ag NPs. (**a**) Immersion into a colloidal solution of Ag NPs, followed by drying (e.g., [63,67,75]). (**b**) Salt reduction. Substrates are commonly exposed to silver nitrate solution. AgNO$_3$ is, in the second step, reduced by an appropriate reducing agent (e.g., sodium citrate, sodium borohydride, or even H$_2$ plasma) [76–79]. (**c**) Direct current (DC) magnetron sputtering. In this low-pressure vacuum-based plasma deposition technique, a silver target mounted onto a DC magnetron is subjected to a flux of highly energetic ions produced in the plasma bulk. This leads to the emission of silver atoms that condense on a substrate located in the deposition chamber. Depending on the deposition conditions (magnetron current, pressure, deposition time, etc.) various silver nanostructures are formed that range from individual separated nanoislands to interconnected Ag networks (e.g., [59,80–87]. (**d**) Deposition of Ag NPs by means of a gas aggregation source equipped with a silver target. Gas aggregation sources based on magnetron sputtering were introduced by Haberland et al. [88] and since then were successfully applied for the fabrication of different kinds of nanoparticles (for more information, readers should refer to the instance to monography [89] or recent review articles [90–93]). In contrast to sputter deposition, silver nanoparticles are already formed in a volume of the aggregation chamber as a result of gas-phase nucleation of sputtered atoms, which is followed by the coagulation or coalescence of growing nanoparticles. As shown by our group, the fact that NPs are formed in the gas-phase means their properties are not dependent on the substrate material, which makes it possible to independently control the size and number of deposited NPs [94,95].

In most cases, low pressure deposition systems (plasma-enhanced chemical vapor deposition or magnetron sputtering, see Figure 3a,b) were selected for the deposition of the base and top layers. The choice of overcoat material and its thickness strongly influenced the antibacterial action of the resulting coatings. For instance, the Kiel group, in their detailed study using Ag NPs that were sandwiched in between RF sputtered *polytetrafluoroethylene* (PTFE) and a layer of either plasma-sputtered PTFE, plasma-polymerized *hexamethyldisiloxane* (ppHMDSO), or SiO$_x$ films prepared by plasma polymerization of HMDSO with admixed oxygen, proved that the release of silver ions and also the antibacterial efficacy of produced coatings strongly depended on the overcoat material [96,97], and the highest Ag$^+$ release and bacteria-killing rates were observed for SiO$_x$ films. Similar conclusions were drawn by Kuzminova et al. [98], who used multi-layer structures with alternating layers of ppHMDSO or SiO$_x$ and Ag NPs prepared by the PE-CVD and GAS system. Better antibacterial behavior of an SiO$_x$-like matrix was ascribed to the different wettability of ppHMDSO and SiO$_x$—whereas ppHMDSO is hydrophobic and thus acts as a barrier for water penetration into the depth of the coatings, SiO$_x$ films are hydrophilic and more water permeable. This was confirmed by Blanchard et al. [99] who studied water penetration into ppHMDSO and SiO$_x$ thin films by means of neutron reflectometry. However, recent results of Kylian et al. [100] showed that even for highly hydrophobic C:F top layers (water contact angles of prepared coatings up to 165°), a strong antibacterial character of the coatings may still be observed when only a thin C:F layer (10 nm) is used. Such a result suggests that coating wettability is not the only parameter, and that the morphology of the coatings—especially the presence of crevices and defects in the overcoat layer which make Ag NPs accessible by water—has to be considered as well. In addition, the release of antibacterial Ag$^+$ responsible for antibacterial action of produced coatings was found to be strongly linked with the number of silver NPs in the coatings, as well as with the thickness of the top barrier layer. This allows for the fabrication of the coatings with either a burst-release of silver ions, or a slow but temporally stable leaching of Ag^+ ions.

Although the low-pressure deposition systems were traditionally used for the production of plasma polymer overcoats, the possibility to also use atmospheric pressure deposition systems was suggested by Deng et al. [101,102]. These authors used an atmospheric pressure plasma jet operated in a mixture of N_2/O_2/*tetramethyldisiloxane* (TMDSO) to produce a barrier layer deposited on top of PET fabrics coated with Ag NPs. It was demonstrated that the antibacterial effect of such prepared fabrics on *P. aeruginosa*, *S. aureus*, *C. albicans* and *E. coli* may be varied by the thickness of the top layer. Furthermore, the top layer was confirmed to enhance the stability of antibacterial effect, and no variation of antibacterial action was observed after 10 washing cycles [101].

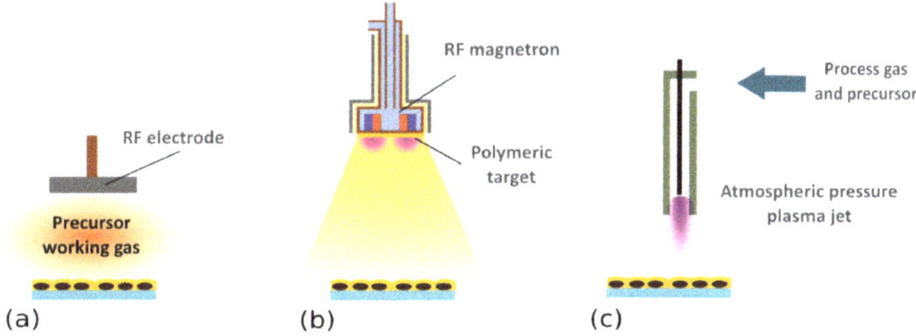

Figure 3. Schematic representation of possible plasma-based strategies for the overcoating of Ag NPs. (**a**) Plasma-enhanced chemical vapor deposition (PE-CVD). In this method, organic vapours or precursors are introduced into the plasma, where they are activated and fragmented. Formed radicals that condense on a substrate subsequently undergo free radical chain growth polymerization (for more details, please refer to the monographies [69,71,72] or reviews [2,24,74]). (**b**) RF magnetron sputtering of a polymeric target. In this case, starting material is supplied in the form of a solid-state polymeric target that is attached onto a magnetron. Atoms, molecules, and molecular fragments sputtered from the polymeric target consequently participate in plasma polymerisation, as in the PE-CVD. In contrast to sputtering of metallic targets, polymers are not conductive, meaning that RF power has to be applied. Magnetron sputtering was employed for the production of a wide range of plasma polymers, including C:H, C:H:N:O, or C:F ones (e.g., [103–109]). (**c**) Atmospheric pressure plasma jets. These systems are based on PE-CVD process, but the gaseous or liquid precursors are introduced to the plasma ignited at atmospheric pressure. Different configurations of atmospheric pressure plasma jets were demonstrated to be suitable for the deposition of various plasma polymer films (e.g., [110–118]).

3.2. Direct Embedment of Ag NPs into Plasma Polymer Matrix

As shown in the previous section, layered silver-containing structures are highly effective at killing bacteria. However, the deposition procedure involves multiple steps, which is, from a technological point of view, a limiting factor. Consequently, alternative approaches for Ag-based nanocomposite production were developed, in which Ag nanoparticles are incorporated directly into the growing plasma polymer matrix. One of the first attempts toward this direction were reported by Favia et al. [119] and Sardella et al. [120]. These authors prepared conformal coatings with silver NPs embedded in a *polyethylene oxide*-like (PEO-like) matrix, using PE-CVD from RF glow discharges fed with Ar, and *diethylglycol-dimethyl-ether* (DEGME) with simultaneous sputtering from the Ag RF electrode in an asymmetric parallel-plate configuration (see Figure 4a). Due to the reactor asymmetry, negative DC self-bias developed on the smaller Ag electrode, which led to the bombardment of the Ag target by highly energetic ions inducing silver sputtering. Emitted Ag atoms impinged upon the plasma polymer surface, which grew simultaneously by plasma polymerization. Ag atoms, due to their limited diffusion in a cross-linked matrix, subsequently formed stable metal nanoparticles through an aggregation process. The properties of the resulting nanocomposites (Ag filling factor, size of Ag NPs, and PEO-like character of the matrix) were controlled by operational parameters (power, pressure, and Ar flow). Based on the disk diffusion test that was performed with *S. epidermidis,* it was reported that the inhibition area correlated with the Ag content in the coatings. Similar approaches using simultaneous plasma polymerization and Ag sputtering was since then employed by other research groups, which utilized different precursors/working gas mixtures. Prepared nanocomposites were tested towards both Gramm-negative and Gramm-positive bacteria (for a summary of results, please refer to Table 1). Furthermore, it was found that Ag/HMDSO nanocomposites with properly-tuned silver content may not only kill bacteria, but also drastically reduce microbial adhesion [121,122].

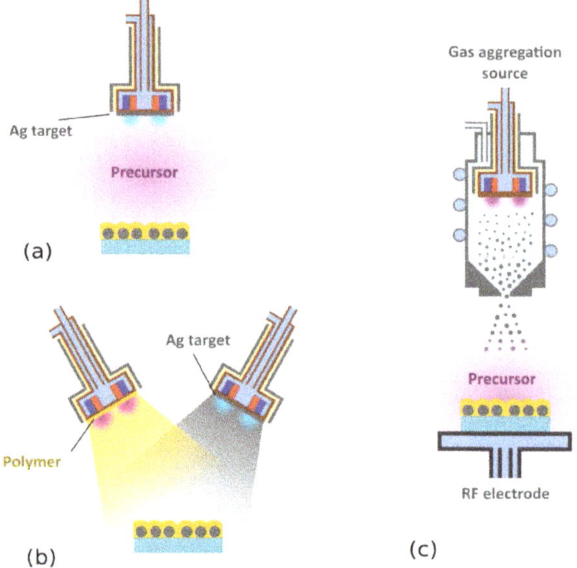

Figure 4. Different approaches for the production of Ag/plasma polymer nanocomposites: (a) Simultaneous sputtering and plasma polymerization, (b) deposition from two independent magnetrons, and (c) a combination of a gas aggregation source and plasma polymerization.

The great advantage of simultaneous sputtering and plasma polymerization is that it is a single-step process. However, as the sputtering and plasma polymerization are fully coupled, it is not possible to independently tailor the size of Ag NPs and their amount in the films and matrix properties. In order to overcome this limitation, another procedure that is based on co-sputtering from two independent magnetron sources may be used (see Figure 4b). The volume fraction of silver in a matrix material is, in this case, controlled by the power applied to the individual magnetrons. This approach, which is also applicable for the production of inorganic Ag containing antibacterial nanocomposites (e.g., Ag/silica [123], Ag/hydroxyapatite [124], or Ag/TiO$_2$ [125]), was used for the fabrication of Ag/C:F antibacterial films by Zaporojtchenko et al. [126]. It was reported that the nanocomposites produced steadily supplied silver ions for a long period of time (reported values were for 300 days) and exhibited antibacterial effects towards *S. epidermidis, S. aureus,* and *P. aeruginosa.* Moreover, an addition of a small amount of Au (1%) was found to substantially increase the release rate of silver ions (by one order of magnitude) [126]. This effect is explained by the formation of galvanically-coupled Ag and Au NPs, because in the galvanic pair, silver is more active than gold, and the presence of gold enhances Ag^+ ion formation.

Table 1. Overview of the reported antibacterial effect of Ag/plasma polymer nanocomposites.

Deposition Method	Working Gas Precursor Target	Bacteria	Result	Reference
Simultaneous sputtering and plasma polymerization	Ag target Ar/DEGME	*Staphylococcus Epidermidis*	Inhibition zone 6 mm	[120]
Simultaneous sputtering and plasma polymerization	Ag target Ar/NH$_3$/C$_2$H$_4$	*Pseudomonas aeruginosa, Staphylococcus aureus*	Bacterial growth down to 0%	[127]
Simultaneous sputtering and plasma polymerization	Ag target Ar/NH$_3$/C$_2$H$_4$ Ar/ CO$_2$/C$_2$H$_4$	*Pseudomonas aeruginosa, Staphylococcus aureus*	>7 log reduction	[128]
Simultaneous sputtering and plasma polymerization	Ag target Ar/HMDSO	*Escherichia coli, Staphylococcus aureus*	*E. Coli* > 6 log reduction *S. aureus* > 1 log	[129]
Simultaneous sputtering and plasma polymerization	Ag target Ar/HMDSO	*Saccharomyces cerevisiae*	1.9 log reduction	[130]
Simultaneous sputtering and plasma polymerization	Ag target polyaniline	*MRSA, MRSE, VRE, Escherichia coli Pseudomonas aeruginosa, Proteus mirabilis, Klebsiella pneumoniae*	Inhibition zone 20–22 mm for all tested bacteria. > 6 log reduction for *E. Coli* and MRSA	[131]
Simultaneous sputtering and plasma polymerization	Ag target C$_2$H$_2$	*Staphylococcus aureus*	99% of bacteria killed	[132]
Co-sputtering	Ag and PTFE targets	*Staphylococcus epidermidis, Staphylococcus aureus, Enterococcus faecalis, Pseudomonas aeruginosa, Escherichia coli, Klebsiella pneumoniae, Candida albicans*	7 log reduction for *S. aureus*	[126]
GAS system combined with plasma polymerization	Ag GAS Ar/n-hexane	*Escherichia coli*	Almost 3 log reduction	[133]

Another strategy that makes it possible to completely decouple the deposition of Ag NPs and the plasma polymer matrix is based on the use of gas aggregation sources [134,135]. This approach was recently used by Vaidulych et al. [133] for the deposition of Ag/a-C:H nanocomposites. In order to produce hard coatings, the substrates were placed on a RF electrode, used for a-C:H matrix deposition, perpendicularly to the direction of a beam of Ag NPs produced by the GAS system (Figure 4c). As it was shown in a previous study focusing on Cu/a-C:H nanocomposites [136], the volume fraction of metallic NPs in a-C:H matrix and connected antibacterial efficiency may be tuned either by the current applied onto the DC magnetron in the gas aggregation source, or by changing the duty cycle of RF power used for the matrix deposition. Furthermore, it was reported that antibacterial action of Ag/a-C:H coatings may be enhanced by the partial etching of nanocomposites by plasma treatment performed in the same deposition chamber resulting in partial removal of a carbonaceous layer from the outermost surface of produced nanocomposites, thus making Ag NPs more accessible for water [133].

A completely different one-step process applicable for the deposition of Ag/plasma polymer antibacterial films was proposed by Zimmerman et al. [137], Beier et al. [138], and Gerullis et al. [139]. These authors used an atmospheric pressure plasma chemical vapor deposition technique, in which a HMDSO precursor was introduced into the atmospheric pressure plasma, together with silver nitrate by a modified jet nozzle (Figure 5a). A strong antibacterial effect of nanocomposites produced in this way on *E. coli* was shown even after 10,000 washing cycles [138]. Atmospheric pressure deposition was finally also tested by Deng et al. [140], where they directly introduced Ag NPs (100 nm size) instead of silver nitrate into the feed gas (N$_2$ with admixing of O$_2$ and TMDSO) (Figure 5b). According to the reported results, prepared coatings exhibited strong antibacterial effects against Gramm-negative

E. coli, and a modest effect on Gram-positive *S. aureus.* A similar approach was recently employed by Ligouri et al. [141] for the production of plasma-polymerized *polyacrylic acid* with embedded Ag NPs. Disk diffusion tests with *E. coli* confirmed the antibacterial efficacy of fabricated coatings.

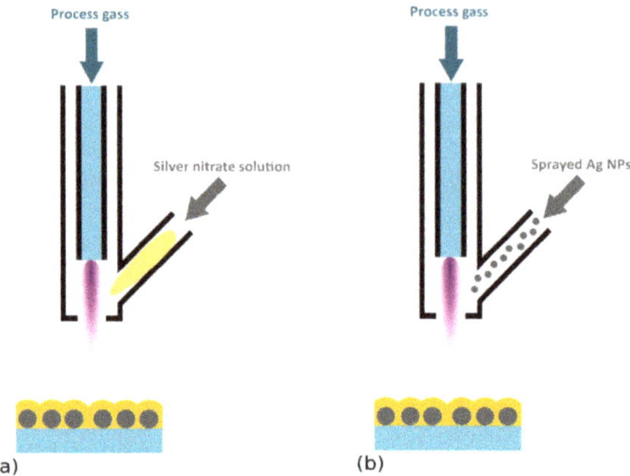

Figure 5. Schematic representation of atmospheric pressure plasma systems for the deposition of Ag/plasma polymer nanocomposites. (**a**) Plasma jet with injection of silver nitrate solution, and (**b**) plasma jet fed with suspension of Ag NPs.

4. Perspectives and Challenges

Plasma-based techniques have been shown to present itself as an interesting option for the production of antibacterial Ag-based coatings. In addition, plasma-based deposition technologies are well-suited for the production of novel antibacterial coatings with enhanced functionality. This refers not only to the possibility to produce coatings with a multi-approach character that may combine the antibacterial action of Ag NPs with the non-fouling nature of a polymeric matrix (e.g., PEO-like plasma polymers, Figure 6a), silver/plasma polymer nanocomposites with covalently immobilized bactericidal substances (Figure 6b), or surface nanotopography (Figure 6c), but also coatings with (i) a temporally non-monotonous release of silver ions, (ii) so-called multi-release coatings that employ two or more antibacterial agents, or even (iii) thin films with Ag^+ release induced by an external stimuli, i.e., coatings that were suggested to facilitate prevention of bacterial infections (e.g., [31]).

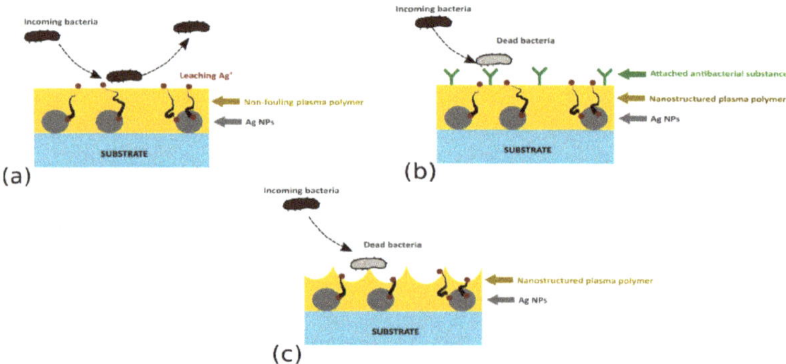

Figure 6. Ag-based multi-functional coatings with (**a**) non-fouling character, (**b**) Ag/plasma polymer nanocomposites with covalently immobilized antibacterial agents, or (**c**) a nanostructured plasma polymer top layer.

Regarding coatings with non-monotonous kinetics in the release of silver ions, two approaches are under investigation. The first one is based on the possibility to tune Ag^+ release by the thickness of the barrier top layer (Figure 7). In this case, different parts of Ag NPs containing films may be coated by plasma polymer films with different thicknesses. Zones with thin top layers will ensure burst release, whereas zones coated with thicker films will be characterized by the delayed and prolonged leaching of a smaller amount of Ag^+. The same is expected to be achieved by the second approach that is based on multi-layer coatings, or the coatings with a depth gradient in the number of embedded NPs. Such materials should fulfill the requirement of long-lasting (several months of) antibacterial action needed to prevent infections on implants.

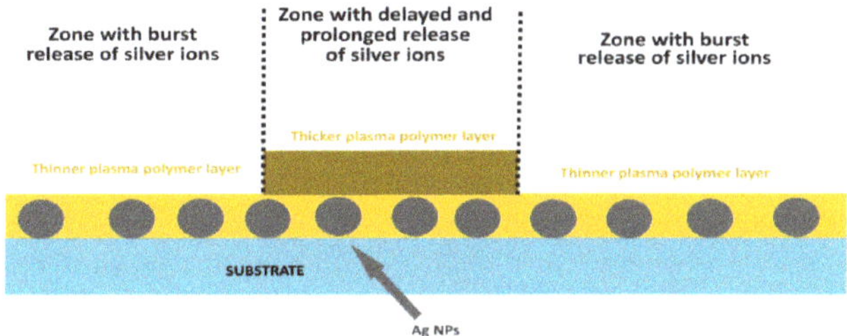

Figure 7. Scheme of Ag/plasma polymer coating with non-monotonous kinetics of release of silver ions.

In the case of multi-release coatings, different antibacterial agents (e.g., silver and copper NPs or Ag NPs and antibiotics) with different release kinetics and/or bactericidal effects can be combined. Such coatings should significantly reduce the induction of bacterial resistance and guarantee synergic antibacterial action, thus enhancing antibacterial efficiency. Possible structures under consideration include silver-containing plasma polymer films impregnated with antibiotics (Figure 8a), sandwiched structures with different metallic NPs with antibacterial character (Figure 8b), or multi-layered coatings prepared by a step-by-step deposition, in which individual layers will be loaded with different antibacterial agents (Figure 8c).

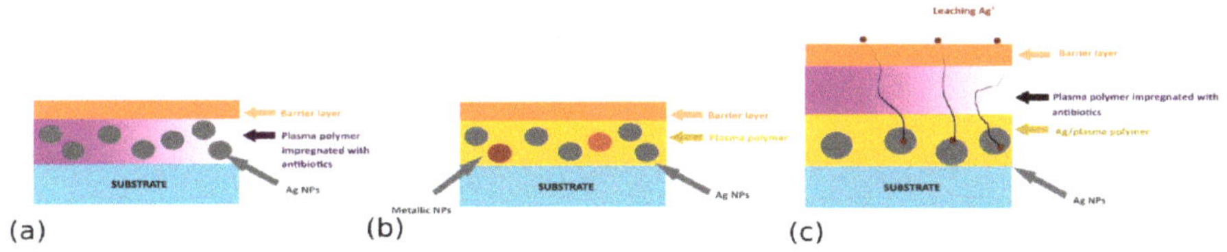

Figure 8. Different architectures suggested for multi-release coatings.

Plasma-based deposition techniques may also be utilized for producing coatings with stimuli-responsive behavior. These materials benefit from the ability of some materials to undergo volume or structural changes when exposed to a particular trigger. This property may be utilized for controlling the release of antibacterial agents, including silver ions, from the coatings "on demand". The first example of Ag/plasma polymer nanocomposites with this ability was reported by Kulaga et al. [142]. These authors coated polypropylene surgical mesh with a layer of plasma-polymerized *maleic anhydride* (MA), impregnated with silver NPs and coated by a barrier layer (plasma polymerized MA) that blocked the spontaneous release of silver from the coatings. Tailored release of silver ions was then achieved by mechanical stimulation of the coatings that led to the formation of cracks in the barrier layer. Moreover, plasma polymerization may also be used for the production of other types of stimuli-responsive plasma polymers where changes in the temperature or pH acts as a trigger.

For instance, Pan et al. [143], Spridon et al. [144], Chen et al. [145], and Moreno-Couranjou et al. [146] synthetized thermo-responsive plasma-polymerized *poly(N-isopropylacrylamide)* or *N-vinylcaprolactam* films. Muzammil et al. [147] reported on the pH-responsive coatings prepared by the plasma co-polymerization of *acrylic acid* and *octafluorocyclobutane*. Although none of the already developed stimuli-responsive plasma polymers were used for the fabrication of Ag-containing nanocomposites, this option still shows a great deal of promise.

Finally, it has to be mentioned that despite the substantial and undisputable progress in the field of antibacterial Ag/plasma polymer coatings and the large amount of suggested and tested approaches that were briefly summarized in this review, to date the use of such materials still remains very limited. This is partly due to the lack of clinical studies, as well as the not yet standardized methods applied for evaluating the antibacterial effects of produced coatings. The latter relates both to the choice of bacteria or procedures used for the quantification of antibacterial effects that makes it almost impossible to compare results reached by different groups. Because of this, there is a clear demand to propose/develop standardized, reliable, and high throughput validation methodologies for testing produced antibacterial coatings, which applies not only to the coatings produced by plasma-based methods but to other methods as well. In addition, all the experiments focused on determining the antibacterial effects of produced Ag/plasma polymer coatings were performed in vitro. It is well-known that under the more relevant in vivo conditions the performance of produced coatings may be largely altered, e.g., by the possible accumulation of dead bacteria or proteins on surfaces of produced coatings. Furthermore, other issues relate to factors which are often overlooked, such as wear resistance, long-term stability, the ability of Ag/plasma polymer nanocomposites to withstand common sterilization procedures, or, in the case of implanted materials, to integrate well with a host tissue. However, the use of plasma-based techniques—most likely in combination with other approaches—still presents itself as a vivid and auspicious option for the production of highly effective antibacterial coatings.

5. Conclusions

There has been enormous progress in silver/plasma polymer nanocomposite antibacterial coatings in the last two decades. As shown in this paper, numerous strategies have already been tested or are under consideration for the effective production of such materials. However, despite promising results, the field of Ag/plasma polymer antibacterial coatings still faces many challenges, meaning that better understanding and control of bactericide activity of the prepared coatings, as well as the development of new manufacturing procedures, are needed.

Author Contributions: Writing-Original Draft Preparation, A.K., J.K.; Writing-Review & Editing, O.K.; Visualization, J.K.

Funding: This research was funded by Grant Agency of Czech Republic grant number GAČR 16-14024S and grant GAUK 1394217 from Grant Agency of Charles University.

Conflicts of Interest: The authors declare no conflict of interest.

References

1. Poncin-Epaillard, F.; Legeay, G. Surface engineering of biomaterials with plasma techniques. *J. Biomater. Sci. Polym. Ed.* **2003**, *14*, 1005–1028. [CrossRef] [PubMed]
2. Siow, K.S.; Britcher, L.; Kumar, S.; Griesser, H.J. Plasma Methods for the Generation of Chemically Reactive Surfaces for Biomolecule Immobilization and Cell Colonization—A Review. *Plasma Process. Polym.* **2006**, *3*, 392–418. [CrossRef]
3. Pappas, D. Status and potential of atmospheric plasma processing of materials. *J. Vac. Sci. Technol. A Vacuum Surfaces Film.* **2011**, *29*. [CrossRef]
4. Bruggeman, P.J.; Kushner, M.J.; Locke, B.R.; Gardeniers, J.G.E.; Graham, W.G.; Graves, D.B.; Hofman-Caris, R.C.H.M.; Maric, D.; Reid, J.P.; Ceriani, E.; et al. Plasma–liquid interactions: a review and roadmap. *Plasma Sources Sci. Technol.* **2016**, *25*. [CrossRef]

5. Adamovich, I.; Baalrud, S.D.; Bogaerts, A.; Bruggeman, P.J.; Cappelli, M.; Colombo, V.; Czarnetzki, U.; Ebert, U.; Eden, J.G.; Favia, P.; et al. The 2017 plasma roadmap: Low temperature plasma science and technology. *J. Phys. D. Appl. Phys.* **2017**, *50*. [CrossRef]

6. Bekeschus, S.; Favia, P.; Robert, E.; von Woedtke, T. White paper on plasma for medicine and hygiene: Future in plasma health sciences. *Plasma Process. Polym.* **2018**. [CrossRef]

7. Cvelbar, U.; Walsh, J.L.; Černák, M.; de Vries, H.W.; Reuter, S.; Belmonte, T.; Corbella, C.; Miron, C.; Hojnik, N.; Jurov, A.; et al. White paper on the future of plasma science and technology in plastics and textiles. *Plasma Process. Polym.* **2018**. [CrossRef]

8. Moisan, M.; Barbeau, J.; Moreau, S.; Pelletier, J.; Tabrizian, M.; Yahia, L. Low-temperature sterilization using gas plasmas: A review of the experiments and an analysis of the inactivation mechanisms. *Int. J. Pharm.* **2001**, *226*, 1–21. [CrossRef]

9. Laroussi, M. Low Temperature Plasma-Based Sterilization: Overview and State-of-the-Art. *Plasma Process. Polym.* **2005**, *2*, 391–400. [CrossRef]

10. Rossi, F.; Kylián, O.; Hasiwa, M. Decontamination of surfaces by low pressure plasma discharges. *Plasma Process. Polym.* **2006**, *3*, 431–442. [CrossRef]

11. Von Keudell, A.; Awakowicz, P.; Benedikt, J.; Raballand, V.; Yanguas-Gil, A.; Opretzka, J.; Flötgen, C.; Reuter, R.; Byelykh, L.; Halfmann, H.; et al. Inactivation of bacteria and biomolecules by low-pressure plasma discharges. *Plasma Process. Polym.* **2010**, *7*, 327–352. [CrossRef]

12. Rossi, F.; Kylián, O.; Rauscher, H.; Hasiwa, M.; Gilliland, D. Low pressure plasma discharges for the sterilization and decontamination of surfaces. *New J. Phys.* **2009**, *11*. [CrossRef]

13. De Geyter, N.; Morent, R. nonthermal plasma sterilization of living and nonliving surfaces. *Annu. Rev. Biomed. Eng.* **2012**, *14*, 255–274. [CrossRef] [PubMed]

14. Fridman, G.; Friedman, G.; Gutsol, A.; Shekhter, A.B.; Vasilets, V.N.; Fridman, A. Applied plasma medicine. *Plasma Process. Polym.* **2008**, *5*, 503–533. [CrossRef]

15. Kong, M.G.; Kroesen, G.; Morfill, G.; Nosenko, T.; Shimizu, T.; van Dijk, J.; Zimmermann, J.L. Plasma medicine: An introductory review. *New J. Phys.* **2009**, *11*. [CrossRef]

16. Laroussi, M. Low-temperature plasmas for medicine? *IEEE Trans. Plasma Sci.* **2009**, *37*, 714–725. [CrossRef]

17. Von Woedtke, T.; Metelmann, H.-R.; Weltmann, K.-D. Clinical plasma medicine: state and perspectives of in vivo application of cold atmospheric plasma. *Contrib. Plasma Phys.* **2014**, *54*, 104–117. [CrossRef]

18. Ito, M.; Ohta, T.; Hori, M. Plasma agriculture. *J. Korean Phys. Soc.* **2012**, *60*, 937–943. [CrossRef]

19. Ambrico, P.F.; Šimek, M.; Morano, M.; De Miccolis Angelini, R.M.; Minafra, A.; Trotti, P.; Ambrico, M.; Prukner, V.; Faretra, F. Reduction of microbial contamination and improvement of germination of sweet basil (Ocimum basilicum L.) seeds via surface dielectric barrier discharge. *J. Phys. D. Appl. Phys.* **2017**, *50*. [CrossRef]

20. Ito, M.; Oh, J.-S.; Ohta, T.; Shiratani, M.; Hori, M. Current status and future prospects of agricultural applications using atmospheric-pressure plasma technologies. *Plasma Process. Polym.* **2018**, *15*. [CrossRef]

21. Šerá, B.; Šerý, M. Non-thermal plasma treatment as a new biotechnology in relation to seeds, dry fruits, and grains. *Plasma Sci. Technol.* **2018**, *20*. [CrossRef]

22. Puač, N.; Gherardi, M.; Shiratani, M. Plasma agriculture: A rapidly emerging field. *Plasma Process. Polym.* **2018**, *15*. [CrossRef]

23. Vasilev, K.; Griesser, S.S.; Griesser, H.J. Antibacterial surfaces and coatings produced by plasma techniques. *Plasma Process. Polym.* **2011**, *8*, 1010–1023. [CrossRef]

24. Sardella, E.; Palumbo, F.; Camporeale, G.; Favia, P. Non-equilibrium plasma processing for the preparation of antibacterial surfaces. *Materials* **2016**, *9*, 515. [CrossRef] [PubMed]

25. Nikiforov, A.; Deng, X.; Xiong, Q.; Cvelbar, U.; DeGeyter, N.; Morent, R.; Leys, C. Non-thermal plasma technology for the development of antimicrobial surfaces: a review. *J. Phys. D. Appl. Phys.* **2016**, *49*. [CrossRef]

26. Zille, A.; Almeida, L.; Amorim, T.; Carneiro, N.; Esteves, M.F.; Silva, C.J.; Souto, A.P. Application of nanotechnology in antimicrobial finishing of biomedical textiles. *Mater. Res. Express* **2014**, *1*. [CrossRef]

27. Zille, A.; Oliveira, F.R.; Souto, A.P. Plasma treatment in textile industry. *Plasma Process. Polym.* **2015**, *12*, 98–131. [CrossRef]

28. Balagna, C.; Perero, S.; Ferraris, S.; Miola, M.; Fucale, G.; Manfredotti, C.; Battiato, A.; Santella, D.; Vernè, E.; Vittone, E.; Ferraris, M. Antibacterial coating on polymer for space application. *Mater. Chem. Phys.* **2012**, *135*, 714–722. [CrossRef]

29. Appendini, P.; Hotchkiss, J.H. Review of antimicrobial food packaging. *Innov. Food Sci. Emerg. Technol.* **2002**, *3*, 113–126. [CrossRef]

30. Ferraris, S.; Perero, S.; Miola, M.; Vernè, E.; Rosiello, A.; Ferrazzo, V.; Valletta, G.; Sanchez, J.; Ohrlander, M.; Tjörnhammar, S.; et al. Chemical, mechanical and antibacterial properties of silver nanocluster/silica composite coated textiles for safety systems and aerospace applications. *Appl. Surf. Sci.* **2014**, *317*, 131–139. [CrossRef]

31. Cloutier, M.; Mantovani, D.; Rosei, F. Antibacterial coatings: Challenges, perspectives, and opportunities. *Trends Biotechnol.* **2015**, *33*, 637–652. [CrossRef] [PubMed]

32. Kingshott, P.; Griesser, H.J. Surfaces that resist bioadhesion. *Curr. Opin. Solid State Mater. Sci.* **1999**, *4*, 403–412. [CrossRef]

33. Li, G.; Cheng, G.; Xue, H.; Chen, S.; Zhang, F.; Jiang, S. Ultra low fouling zwitterionic polymers with a biomimetic adhesive group. *Biomaterials* **2008**, *29*, 4592–4597. [CrossRef] [PubMed]

34. Cheng, G.; Li, G.; Xue, H.; Chen, S.; Bryers, J.D.; Jiang, S. Zwitterionic carboxybetaine polymer surfaces and their resistance to long-term biofilm formation. *Biomaterials* **2009**, *30*, 5234–5240. [CrossRef] [PubMed]

35. Choukourov, A.; Gordeev, I.; Arzhakov, D.; Artemenko, A.; Kousal, J.; Kylián, O.; Slavínská, D.; Biederman, H. Does cross-link density of PEO-like plasma polymers influence their resistance to adsorption of fibrinogen? *Plasma Process. Polym.* **2012**, *9*, 48–58. [CrossRef]

36. Buxadera-Palomero, J.; Calvo, C.; Torrent-Camarero, S.; Gil, F.J.; Mas-Moruno, C.; Canal, C.; Rodríguez, D. Biofunctional polyethylene glycol coatings on titanium: An in vitro -based comparison of functionalization methods. *Colloids Surf. B Biointerfaces* **2017**, *152*, 367–375. [CrossRef] [PubMed]

37. Green, J.-B.D.; Fulghum, T.; Nordhaus, M.A. A review of immobilized antimicrobial agents and methods for testing. *Biointerphases* **2011**, *6*. [CrossRef] [PubMed]

38. Kaur, R.; Liu, S. Antibacterial surface design—Contact kill. *Prog. Surf. Sci.* **2016**, *91*, 136–153. [CrossRef]

39. Elbourne, A.; Crawford, R.J.; Ivanova, E.P. Nano-structured antimicrobial surfaces: From nature to synthetic analogues. *J. Colloid Interface Sci.* **2017**, *508*, 603–616. [CrossRef] [PubMed]

40. Tripathy, A.; Sen, P.; Su, B.; Briscoe, W.H. Natural and bioinspired nanostructured bactericidal surfaces. *Adv. Colloid Interface Sci.* **2017**, *248*, 85–104. [CrossRef] [PubMed]

41. Campoccia, D.; Montanaro, L.; Arciola, C.R. A review of the biomaterials technologies for infection-resistant surfaces. *Biomaterials* **2013**, *34*, 8533–8554. [CrossRef] [PubMed]

42. Palumbo, F.; Camporeale, G.; Yang, Y.-W.; Wu, J.-S.; Sardella, E.; Dilecce, G.; Calvano, C.D.; Quintieri, L.; Caputo, L.; Baruzzi, F.; Favia, P. Direct plasma deposition of lysozyme-embedded bio-composite thin films. *Plasma Process. Polym.* **2015**, *12*, 1302–1310. [CrossRef]

43. Kratochvíl, J.; Kahoun, D.; Štěrba, J.; Langhansová, H.; Lieskovská, J.; Fojtíková, P.; Hanuš, J.; Kousal, J.; Kylián, O.; Straňák, V. Plasma polymerized C:H:N:O thin films for controlled release of antibiotic substances. *Plasma Process. Polym.* **2018**, *15*. [CrossRef]

44. Daschner, F.; Langmaack, H.; Wiedemann, B. Antibiotic resistance in intensive care unit areas. *Infect. Control* **1983**, *4*, 382–387. [CrossRef] [PubMed]

45. Neu, H.C. The crisis in antibiotic resistance. *Science* **1992**, *257*, 1064–1073. [CrossRef] [PubMed]

46. Albrich, W.C.; Angstwurm, M.; Bader, L.; Gärtner, R. Drug resistance in intensive care units. *Infection* **1999**, *27*, S19–S23. [CrossRef] [PubMed]

47. Hanberger, H.; Diekema, D.; Fluit, A.; Jones, R.; Struelens, M.; Spencer, R.; Wolff, M. Surveillance of antibiotic resistance in European ICUs. *J. Hosp. Infect.* **2001**, *48*, 161–176. [CrossRef] [PubMed]

48. Loeffler, J.M.; Garbino, J.; Lew, D.; Harbarth, S.; Rohner, P. Antibiotic consumption, bacterial resistance and their correlation in a swiss university hospital and its adult intensive care units. *Scand. J. Infect. Dis.* **2003**, *35*, 843–850. [CrossRef] [PubMed]

49. Levy, S.B.; Marshall, B. Antibacterial resistance worldwide: Causes, challenges and responses. *Nat. Med.* **2004**, *10*, S122–S129. [CrossRef] [PubMed]

50. Hsueh, P.-R.; Chen, W.-H.; Luh, K.-T. Relationships between antimicrobial use and antimicrobial resistance in Gram-negative bacteria causing nosocomial infections from 1991–2003 at a university hospital in Taiwan. *Int. J. Antimicrob. Agents* **2005**, *26*, 463–472. [CrossRef] [PubMed]

51. Silver, S. Bacterial silver resistance: molecular biology and uses and misuses of silver compounds. *FEMS Microbiol. Rev.* **2003**, *27*, 341–353. [CrossRef]

52. Mijnendonckx, K.; Leys, N.; Mahillon, J.; Silver, S.; Van Houdt, R. Antimicrobial silver: Uses, toxicity and potential for resistance. *BioMetals* **2013**, *26*, 609–621. [CrossRef] [PubMed]

53. Chernousova, S.; Epple, M. Silver as antibacterial agent: Ion, nanoparticle, and metal. *Angew. Chem. Int. Ed.* **2013**, *52*, 1636–1653. [CrossRef] [PubMed]

54. Durán, N.; Durán, M.; de Jesus, M.B.; Seabra, A.B.; Fávaro, W.J.; Nakazato, G. Silver nanoparticles: A new view on mechanistic aspects on antimicrobial activity. *Nanomedicine* **2016**, *12*, 789–799. [CrossRef] [PubMed]

55. Wei, L.; Lu, J.; Xu, H.; Patel, A.; Chen, Z.-S.; Chen, G. Silver nanoparticles: synthesis, properties, and therapeutic applications. *Drug Discov. Today* **2015**, *20*, 595–601. [CrossRef] [PubMed]

56. Lansdown, A.B.G. A Pharmacological and toxicological profile of silver as an antimicrobial agent in medical devices. *Adv. Pharmacol. Sci.* **2010**, *2010*, 1–16. [CrossRef] [PubMed]

57. Scholz, J.; Nocke, G.; Hollstein, F.; Weissbach, A. Investigations on fabrics coated with precious metals using the magnetron sputter technique with regard to their anti-microbial properties. *Surf. Coatings Technol.* **2005**, *192*, 252–256. [CrossRef]

58. Sant, S.B.; Gill, K.S.; Burrell, R.E. Nanostructure, dissolution and morphology characteristics of microcidal silver films deposited by magnetron sputtering. *Acta Biomater.* **2007**, *3*, 341–350. [CrossRef] [PubMed]

59. Mejía, M.I.; Restrepo, G.; Marín, J.M.; Sanjines, R.; Pulgarín, C.; Mielczarski, E.; Mielczarski, J.; Kiwi, J. Magnetron-sputtered ag surfaces. New evidence for the nature of the Ag ions intervening in bacterial inactivation. *ACS Appl. Mater. Interfaces* **2010**, *2*, 230–235. [CrossRef] [PubMed]

60. Jiang, S.X.; Qin, W.F.; Guo, R.H.; Zhang, L. Surface functionalization of nanostructured silver-coated polyester fabric by magnetron sputtering. *Surf. Coat. Technol.* **2010**, *204*, 3662–3667. [CrossRef]

61. Baghriche, O.; Ruales, C.; Sanjines, R.; Pulgarin, C.; Zertal, A.; Stolitchnov, I.; Kiwi, J. Ag-surfaces sputtered by DC and pulsed DC-magnetron sputtering effective in bacterial inactivation: Testing and characterization. *Surf. Coat. Technol.* **2012**, *206*, 2410–2416. [CrossRef]

62. Baghriche, O.; Zertal, A.; Ehiasarian, A.P.; Sanjines, R.; Pulgarin, C.; Kusiak-Nejman, E.; Morawski, A.W.; Kiwi, J. Advantages of highly ionized pulse plasma magnetron sputtering (HIPIMS) of silver for improved *E. coli* inactivation. *Thin Solid Films* **2012**, *520*, 3567–3573. [CrossRef]

63. Radetić, M.; Ilić, V.; Vodnik, V.; Dimitrijević, S.; Jovančić, P.; Šaponjić, Z.; Nedeljković, J.M. Antibacterial effect of silver nanoparticles deposited on corona-treated polyester and polyamide fabrics. *Polym. Adv. Technol.* **2008**, *19*, 1816–1821. [CrossRef]

64. Kostić, M.; Radić, N.; Obradović, B.M.; Dimitrijević, S.; Kuraica, M.M.; Škundrić, P. Silver-loaded cotton/polyester fabric modified by dielectric barrier discharge treatment. *Plasma Process. Polym.* **2009**, *6*, 58–67. [CrossRef]

65. Jiang, H.Q.; Manolache, S.; Wong, A.C.L.; Denes, F.S. Plasma-enhanced deposition of silver nanoparticles onto polymer and metal surfaces for the generation of antimicrobial characteristics. *J. Appl. Polym. Sci.* **2004**, *93*, 1411–1422. [CrossRef]

66. Kramar, A.; Prysiazhnyi, V.; Dojčinović, B.; Mihajlovski, K.; Obradović, B.M.; Kuraica, M.M.; Kostić, M. Antimicrobial viscose fabric prepared by treatment in DBD and subsequent deposition of silver and copper ions-Investigation of plasma aging effect. *Surf. Coat. Technol.* **2013**, *234*, 92–99. [CrossRef]

67. Shen, T.; Liu, Y.; Zhu, Y.; Yang, D.-Q.; Sacher, E. Improved adhesion of Ag NPs to the polyethylene terephthalate surface via atmospheric plasma treatment and surface functionalization. *Appl. Surf. Sci.* **2017**, *411*, 411–418. [CrossRef]

68. Ibrahim, N.A.; Eid, B.M.; Abdel-Aziz, M.S. Effect of plasma superficial treatments on antibacterial functionalization and coloration of cellulosic fabrics. *Appl. Surf. Sci.* **2017**, *392*, 1126–1133. [CrossRef]

69. Yasuda, H. *Plasma Polymerization*; Academic Press: New York, NY, USA, 1985; ISBN 0127687602.

70. Goodman, J. The formation of thin polymer films in the gas discharge. *J. Polym. Sci.* **1960**, *44*, 551–552. [CrossRef]

71. D'Agostino, R. *Plasma Deposition, Treatment, and Etching of Polymers*; Academic Press: Orlando, FL, USA, 1990; ISBN 9780323139083.

72. Biederman, H.; Osada, Y. *Plasma Polymerization Processes*; Elsevier: New York, NY, USA, 1992; ISBN 0444887245.

73. Shi, F.F. Recent advances in polymer thin films prepared by plasma polymerization synthesis, structural characterization, properties and applications. *Surf. Coat. Technol.* **1996**, *82*, 1–15. [CrossRef]

74. Friedrich, J. Mechanisms of plasma polymerization—Reviewed from a chemical point of view. *Plasma Process. Polym.* **2011**, *8*, 783–802. [CrossRef]

75. Zille, A.; Fernandes, M.M.; Francesko, A.; Tzanov, T.; Fernandes, M.; Oliveira, F.R.; Almeida, L.; Amorim, T.; Carneiro, N.; Esteves, M.F.; Souto, A.P. Size and aging effects on antimicrobial efficiency of silver nanoparticles coated on polyamide fabrics activated by atmospheric DBD plasma. *ACS Appl. Mater. Interfaces* **2015**, *7*, 13731–13744. [CrossRef] [PubMed]

76. Vasilev, K.; Sah, V.; Anselme, K.; Ndi, C.; Mateescu, M.; Dollmann, B.; Martinek, P.; Ys, H.; Ploux, L.; Griesser, H.J. Tunable antibacterial coatings that support mammalian cell growth. *Nano. Lett.* **2010**, *10*, 202–207. [CrossRef] [PubMed]

77. Ploux, L.; Mateescu, M.; Anselme, K.; Vasilev, K. antibacterial properties of silver-loaded plasma polymer coatings. *J. Nanomater.* **2012**, *2012*, 1–9. [CrossRef]

78. Kumar, V.; Jolivalt, C.; Pulpytel, J.; Jafari, R.; Arefi-Khonsari, F. Development of silver nanoparticle loaded antibacterial polymer mesh using plasma polymerization process. *J. Biomed. Mater. Res. A* **2013**, *101A*, 1121–1132. [CrossRef] [PubMed]

79. Fahmy, A.; Friedrich, J.; Poncin-Epaillard, F.; Debarnot, D. Plasma polymerized allyl alcohol/O_2 thin films embedded with silver nanoparticles. *Thin Solid Films* **2016**, *616*, 339–347. [CrossRef]

80. Maréchal, N.; Quesnel, E.; Pauleau, Y. Silver thin films deposited by magnetron sputtering. *Thin Solid Films* **1994**, *241*, 34–38. [CrossRef]

81. Charton, C.; Fahland, M. Growth of Ag films on PET deposited by magnetron sputtering. *Vacuum* **2002**, *68*, 65–73. [CrossRef]

82. Asanithi, P.; Chaiyakun, S.; Limsuwan, P. Growth of silver nanoparticles by DC magnetron sputtering. *J. Nanomater.* **2012**, *2012*, 1–8. [CrossRef]

83. Siegel, J.; Polívková, M.; Kasálková, N.; Kolská, Z.; Švorčík, V. Properties of silver nanostructure-coated PTFE and its biocompatibility. *Nanoscale Res. Lett.* **2013**, *8*. [CrossRef] [PubMed]

84. Šubr, M.; Kuzminova, A.; Kylián, O.; Procházka, M. Surface-enhanced Raman scattering (SERS) of riboflavin on nanostructured Ag surfaces: The role of excitation wavelength, plasmon resonance and molecular resonance. *Spectrochim. Acta Part A Mol. Biomol. Spectrosc.* **2018**, *197*, 202–207. [CrossRef] [PubMed]

85. Shahidi, S.; Ghoranneviss, M. Plasma sputtering for fabrication of antibacterial and ultraviolet protective fabric. *Cloth. Text. Res. J.* **2016**, *34*, 37–47. [CrossRef]

86. Hanuš, J.; Libenská, H.; Khalakhan, I.; Kuzminova, A.; Kylián, O.; Biederman, H. Localized surface plasmon resonance tuning via nanostructured gradient Ag surfaces. *Mater. Lett.* **2017**, *192*, 119–122. [CrossRef]

87. Wang, L.; Li, L.; Chen, W.-D. Investigation of the properties of silver thin films deposited by DC magnetron sputtering. *Surf. Rev. Lett.* **2017**, *24*. [CrossRef]

88. Haberland, H.; Karrais, M.; Mall, M.; Thurner, Y. Thin films from energetic cluster impact: A feasibility study. *J. Vac. Sci. Technol. A Vacuum Surfaces Film.* **1992**, *10*, 3266–3271. [CrossRef]

89. Huttel, Y. *Gas-Phase Synthesis of Nanoparticles*; Wiley-VCH: Weinheim, Germany, 2017; ISBN 9783527340606.

90. Binns, C. Nanoclusters deposited on surfaces. *Surf. Sci. Rep.* **2001**, *44*, 1–49. [CrossRef]

91. Wegner, K.; Piseri, P.; Tafreshi, H.V.; Milani, P. Cluster beam deposition: a tool for nanoscale science and technology. *J. Phys. D. Appl. Phys.* **2006**, *39*. [CrossRef]

92. Popok, V.N.; Barke, I.; Campbell, E.E.B.; Meiwes-Broer, K.-H. Cluster–surface interaction: From soft landing to implantation. *Surf. Sci. Rep.* **2011**, *66*, 347–377. [CrossRef]

93. Grammatikopoulos, P.; Steinhauer, S.; Vernieres, J.; Singh, V.; Sowwan, M. Nanoparticle design by gas-phase synthesis. *Adv. Phys. X* **2016**, *1*, 81–100. [CrossRef]

94. Kratochvíl, J.; Kuzminova, A.; Kylián, O.; Biederman, H. Comparison of magnetron sputtering and gas aggregation nanoparticle source used for fabrication of silver nanoparticle films. *Surf. Coat. Technol.* **2015**, *275*, 296–302. [CrossRef]

95. Petr, M.; Kylián, O.; Hanuš, J.; Kuzminova, A.; Vaidulych, M.; Khalakhan, I.; Choukourov, A.; Slavínská, D.; Biederman, H. Surfaces with roughness gradient and invariant surface chemistry produced by means of gas aggregation source and magnetron sputtering. *Plasma Process. Polym.* **2016**, *13*, 663–671. [CrossRef]

96. Alissawi, N.; Zaporojtchenko, V.; Strunskus, T.; Hrkac, T.; Kocabas, I.; Erkartal, B.; Chakravadhanula, V.S.K.; Kienle, L.; Grundmeier, G.; Garbe-Schönberg, D.; et al. Tuning of the ion release properties of silver nanoparticles buried under a hydrophobic polymer barrier. *J. Nanopart. Res.* **2012**, *14*, 928. [CrossRef]

97. Alissawi, N.; Peter, T.; Strunskus, T.; Ebbert, C.; Grundmeier, G.; Faupel, F. Plasma-polymerized HMDSO coatings to adjust the silver ion release properties of Ag/polymer nanocomposites. *J. Nanopart. Res.* **2013**, *15*, 2080. [CrossRef]

98. Kuzminova, A.; Beranová, J.; Polonskyi, O.; Shelemin, A.; Kylián, O.; Choukourov, A.; Slavínská, D.; Biederman, H. Antibacterial nanocomposite coatings produced by means of gas aggregation source of silver nanoparticles. *Surf. Coat. Technol.* **2016**, *294*, 225–230. [CrossRef]

99. Blanchard, N.E.; Naik, V.V.; Geue, T.; Kahle, O.; Hegemann, D.; Heuberger, M. Response of plasma-polymerized hexamethyldisiloxane films to aqueous environments. *Langmuir* **2015**, *31*, 12944–12953. [CrossRef] [PubMed]

100. Kylián, O.; Kratochvíl, J.; Petr, M.; Kuzminova, A.; Slavínská, D.; Biederman, H.; Beranová, J. Ag/C:F Antibacterial and hydrophobic nanocomposite coatings. *Funct. Mater. Lett.* **2017**, *10*. [CrossRef]

101. Deng, X.; Yu Nikiforov, A.; Coenye, T.; Cools, P.; Aziz, G.; Morent, R.; De Geyter, N.; Leys, C. Antimicrobial nano-silver non-woven polyethylene terephthalate fabric via an atmospheric pressure plasma deposition process. *Sci. Rep.* **2015**, *5*. [CrossRef] [PubMed]

102. Deng, X.; Nikiforov, A.; Vujosevic, D.; Vuksanovic, V.; Mugoša, B.; Cvelbar, U.; De Geyter, N.; Morent, R.; Leys, C. Antibacterial activity of nano-silver non-woven fabric prepared by atmospheric pressure plasma deposition. *Mater. Lett.* **2015**, *149*, 95–99. [CrossRef]

103. Morrison, D.T.; Robertson, T.R.F. sputtering of plastics. *Thin Solid Films* **1973**, *15*, 87–101. [CrossRef]

104. Holland, L.; Biederman, H.; Ojha, S. Sputtered and plasma polymerized fluorocarbon films. *Thin Solid Films* **1976**, *35*, 19–21. [CrossRef]

105. Biederman, H.; Ojha, S.M.; Holland, L. The properties of fluorocarbon films prepared by RF sputtering and plasma polymerization in inert and active gas. *Thin Solid Films* **1977**, *41*, 329–339. [CrossRef]

106. Biederman, H. RF sputtering of polymers and its potential application. *Vacuum* **2000**, *59*, 594–599. [CrossRef]

107. Wang, W.-C. Ultrathin fluoropolymer films deposited on a polyimide (kapton®) surface by RF magnetron sputtering of poly(tetrafluoroethylene). *Plasma Process. Polym.* **2007**, *4*, 88–97. [CrossRef]

108. Kylián, O.; Hanuš, J.; Choukourov, A.; Kousal, J.; Slavínská, D.; Biederman, H. Deposition of amino-rich thin films by RF magnetron sputtering of nylon. *J. Phys. D. Appl. Phys.* **2009**, *42*. [CrossRef]

109. Drábik, M.; Polonskyi, O.; Kylián, O.; Čechvala, J.; Artemenko, A.; Gordeev, I.; Choukourov, A.; Slavínská, D.; Matolínová, I.; Biederman, H. Super-hydrophobic coatings prepared by RF magnetron sputtering of PTFE. *Plasma Process. Polym.* **2010**, *7*, 544–551. [CrossRef]

110. Schäfer, J.; Foest, R.; Quade, A.; Ohl, A.; Weltmann, K.-D. Local deposition of SiO x plasma polymer films by a miniaturized atmospheric pressure plasma jet (APPJ). *J. Phys. D. Appl. Phys.* **2008**, *41*. [CrossRef]

111. Lommatzsch, U.; Ihde, J. Plasma polymerization of HMDSO with an atmospheric pressure plasma jet for corrosion protection of aluminum and low-adhesion surfaces. *Plasma Process. Polym.* **2009**, *6*, 642–648. [CrossRef]

112. Vogelsang, A.; Ohl, A.; Foest, R.; Schröder, K.; Weltmann, K.-D. Hydrophobic coatings deposited with an atmospheric pressure microplasma jet. *J. Phys. D. Appl. Phys.* **2010**, *43*. [CrossRef]

113. Chen, G.; Zhou, M.; Zhang, Z.; Lv, G.; Massey, S.; Smith, W.; Tatoulian, M. Acrylic acid polymer coatings on silk fibers by room-temperature APGD plasma jets. *Plasma Process. Polym.* **2011**, *8*, 701–708. [CrossRef]

114. Belmonte, T.; Henrion, G.; Gries, T. Nonequilibrium atmospheric plasma deposition. *J. Therm. Spray Technol.* **2011**, *20*, 744–759. [CrossRef]

115. Yim, J.H.; Rodriguez-Santiago, V.; Williams, A.A.; Gougousi, T.; Pappas, D.D.; Hirvonen, J.K. Atmospheric pressure plasma enhanced chemical vapor deposition of hydrophobic coatings using fluorine-based liquid precursors. *Surf. Coat. Technol.* **2013**, *234*, 21–32. [CrossRef]

116. Gordeev, I.; Šimek, M.; Prukner, V.; Artemenko, A.; Kousal, J.; Nikitin, D.; Choukourov, A.; Biederman, H. Deposition of poly (ethylene oxide)-like plasma polymers on inner surfaces of cavities by means of atmospheric-pressure SDBD-based jet. *Plasma Process. Polym.* **2016**, *13*, 823–833. [CrossRef]

117. Stallard, C.P.; Solar, P.; Biederman, H.; Dowling, D.P. Deposition of non-fouling PEO-like coatings using a low temperature atmospheric pressure plasma jet. *Plasma Process. Polym.* **2016**, *13*, 241–252. [CrossRef]

118. Ricci Castro, A.H.; Kodaira, F.V.P.; Prysiazhnyi, V.; Mota, R.P.; Kostov, K.G. Deposition of thin films using argon/acetylene atmospheric pressure plasma jet. *Surf. Coat. Technol.* **2017**, *312*, 13–18. [CrossRef]

119. Favia, P.; Vulpio, M.; Marino, R.; D'Agostino, R.; Mota, R.P.; Catalano, M. Plasma-deposition of ag-containing polyethyleneoxide-like coatings. *Plasmas Polym.* **2000**, *5*, 1–14. [CrossRef]

120. Sardella, E.; Favia, P.; Gristina, R.; Nardulli, M.; D'Agostino, R. Plasma-Aided Micro-and Nanopatterning Processes for Biomedical Applications. *Plasma Process. Polym.* **2006**, *3*, 456–469. [CrossRef]

121. Guillemot, G.; Despax, B.; Raynaud, P.; Zanna, S.; Marcus, P.; Schmitz, P.; Mercier-Bonin, M. Plasma deposition of silver nanoparticles onto stainless steel for the prevention of fungal biofilms: A case study on saccharomyces cerevisiae. *Plasma Process. Polym.* **2008**, *5*, 228–238. [CrossRef]

122. Saulou, C.; Despax, B.; Raynaud, P.; Zanna, S.; Marcus, P.; Mercier-Bonin, M. Plasma deposition of organosilicon polymer thin films with embedded nanosilver for prevention of microbial adhesion. *Appl. Surf. Sci.* **2009**, *256*, S35–S39. [CrossRef]

123. Ferraris, M.; Perero, S.; Miola, M.; Ferraris, S.; Verné, E.; Morgiel, J. Silver nanocluster-silica composite coatings with antibacterial properties. *Mater. Chem. Phys.* **2010**, *120*, 123–126. [CrossRef]

124. Chen, W.; Liu, Y.; Courtney, H.S.; Bettenga, M.; Agrawal, C.M.; Bumgardner, J.D.; Ong, J.L. In vitro anti-bacterial and biological properties of magnetron co-sputtered silver-containing hydroxyapatite coating. *Biomaterials* **2006**, *27*, 5512–5517. [CrossRef] [PubMed]

125. Xiong, J.; Ghori, M.Z.; Henkel, B.; Strunskus, T.; Schürmann, U.; Deng, M.; Kienle, L.; Faupel, F. Tuning silver ion release properties in reactively sputtered Ag/TiOx nanocomposites. *Appl. Phys. A* **2017**, *123*, 470. [CrossRef]

126. Zaporojtchenko, V.; Podschun, R.; Schürmann, U.; Kulkarni, A.; Faupel, F. Physico-chemical and antimicrobial properties of co-sputtered Ag-Au/PTFE nanocomposite coatings. *Nanotechnology* **2006**, *17*, 4904–4908. [CrossRef]

127. Lischer, S.; Korner, E.; Balazs, D.J.; Shen, D.; Wick, P.; Grieder, K.; Haas, D.; Heuberger, M.; Hegemann, D. Antibacterial burst-release from minimal Ag-containing plasma polymer coatings. *J. R. Soc. Interface* **2011**, *8*, 1019–1030. [CrossRef] [PubMed]

128. Körner, E.; Hanselmann, B.; Cierniak, P.; Hegemann, D. Tailor-made silver release properties of silver-containing functional plasma polymer coatings adjusted through a macroscopic kinetics approach. *Plasma Chem. Plasma Process.* **2012**, *32*, 619–627. [CrossRef]

129. Allion-Maurer, A.; Saulou-Bérion, C.; Briandet, R.; Zanna, S.; Lebleu, N.; Marcus, P.; Raynaud, P.; Despax, B.; Mercier-Bonin, M. Plasma-deposited nanocomposite polymer-silver coating against *Escherichia coli* and *Staphylococcus aureus*: Antibacterial properties and ageing. *Surf. Coat. Technol.* **2015**, *281*, 1–10. [CrossRef]

130. Saulou, C.; Despax, B.; Raynaud, P.; Zanna, S.; Seyeux, A.; Marcus, P.; Audinot, J.N.; Mercier-Bonin, M. Plasma-mediated nanosilver-organosilicon composite films deposited on stainless steel: Synthesis, surface characterization, and evaluation of anti-adhesive and anti-microbial properties on the model yeast saccharomyces cerevisiae. *Plasma Process. Polym.* **2012**, *9*, 324–338. [CrossRef]

131. Agarwala, M.; Barman, T.; Gogoi, D.; Choudhury, B.; Pal, A.R.; Yadav, R.N.S. Highly effective antibiofilm coating of silver-polymer nanocomposite on polymeric medical devices deposited by one step plasma process. *J. Biomed. Mater. Res. Part B Appl. Biomater.* **2014**, *102*, 1223–1235. [CrossRef] [PubMed]

132. Juknius, T.; Ružauskas, M.; Tamulevičius, T.; Šiugždinienė, R.; Juknienė, I.; Vasiliauskas, A.; Jurkevičiūtė, A.; Tamulevičius, S. Antimicrobial properties of diamond-like carbon/silver nanocomposite thin films deposited on textiles: Towards smart bandages. *Materials* **2016**, *9*, 371. [CrossRef] [PubMed]

133. Vaidulych, M.; Hanuš, J.; Steinhartová, T.; Kylián, O.; Choukourov, A.; Beranová, J.; Khalakhan, I.; Biederman, H. Deposition of Ag/a-C:H nanocomposite films with Ag surface enrichment. *Plasma Process. Polym.* **2017**, *14*. [CrossRef]

134. Polonskyi, O.; Solař, P.; Kylián, O.; Drábik, M.; Artemenko, A.; Kousal, J.; Hanuš, J.; Pešička, J.; Matolínová, I.; Kolíbalová, E.; et al. Nanocomposite metal/plasma polymer films prepared by means of gas aggregation cluster source. *Thin Solid Films* **2012**, *520*, 4155–4162. [CrossRef]

135. Peter, T.; Rehders, S.; Schürmann, U.; Strunskus, T.; Zaporojtchenko, V.; Faupel, F. High rate deposition system for metal-cluster/SiOxCyHz –polymer nanocomposite thin films. *J. Nanopart. Res.* **2013**, *15*. [CrossRef]

136. Hanuš, J.; Steinhartová, T.; Kylián, O.; Kousal, J.; Malinský, P.; Choukourov, A.; Macková, A.; Biederman, H. Deposition of Cu/a-C:H Nanocomposite Films. *Plasma Process. Polym.* **2016**, *13*, 879–887. [CrossRef]

137. Zimmermann, R.; Pfuch, A.; Horn, K.; Weisser, J.; Heft, A.; Röder, M.; Linke, R.; Schnabelrauch, M.; Schimanski, A. An approach to create silver containing antibacterial coatings by use of atmospheric pressure plasma chemical vapour deposition (APCVD) and combustion chemical vapour deposition (CCVD) in an economic way. *Plasma Process. Polym.* **2011**, *8*, 295–304. [CrossRef]
138. Beier, O.; Pfuch, A.; Horn, K.; Weisser, J.; Schnabelrauch, M.; Schimanski, A. Low temperature deposition of antibacterially active silicon oxide layers containing silver nanoparticles, prepared by atmospheric pressure plasma chemical vapor deposition. *Plasma Process. Polym.* **2013**, *10*, 77–87. [CrossRef]
139. Gerullis, S.; Pfuch, A.; Spange, S.; Kettner, F.; Plaschkies, K.; Küzün, B.; Kosmachev, P.V.; Volokitin, G.G.; Grünler, B. Thin antimicrobial silver, copper or zinc containing SiOx films on wood polymer composites (WPC) applied by atmospheric pressure plasma chemical vapour deposition (APCVD) and sol–gel technology. *Eur. J. Wood Wood Prod.* **2018**, *76*, 229–241. [CrossRef]
140. Deng, X.; Leys, C.; Vujosevic, D.; Vuksanovic, V.; Cvelbar, U.; De Geyter, N.; Morent, R.; Nikiforov, A. Engineering of Composite Organosilicon Thin Films with Embedded Silver Nanoparticles via Atmospheric Pressure Plasma Process for Antibacterial Activity. *Plasma Process. Polym.* **2014**, *11*, 921–930. [CrossRef]
141. Liguori, A.; Traldi, E.; Toccaceli, E.; Laurita, R.; Pollicino, A.; Focarete, M.L.; Colombo, V.; Gherardi, M. Co-deposition of plasma-polymerized polyacrylic acid and silver nanoparticles for the production of nanocomposite coatings using a non-equilibrium atmospheric pressure plasma jet. *Plasma Process. Polym.* **2016**, *13*, 623–632. [CrossRef]
142. Kulaga, E.; Ploux, L.; Balan, L.; Schrodj, G.; Roucoules, V. Mechanically responsive antibacterial plasma polymer coatings for textile biomaterials. *Plasma Process. Polym.* **2014**, *11*, 63–79. [CrossRef]
143. Pan, Y.V.; Wesley, R.A.; Luginbuhl, R.; Denton, D.D.; Ratner, B.D. Plasma polymerized N-Isopropylacrylamide: Synthesis and characterization of a smart thermally responsive coating. *Biomacromolecules* **2001**, *2*, 32–36. [CrossRef] [PubMed]
144. Spridon, D.; Curecheriu, L.; Dobromir, M.; Dumitrascu, N. Synthesis of poly (N-isopropylacrylamide) under atmospheric pressure plasma conditions. *J. Appl. Polym. Sci.* **2012**, *124*, 2377–2382. [CrossRef]
145. Chen, Y.; Tang, X.L.; Chen, B.T.; Qiu, G. Low temperature plasma vapor treatment of thermo-sensitive poly(*N*-isopropylacrylamide) and its application. *Appl. Surf. Sci.* **2013**, *268*, 332–336. [CrossRef]
146. Moreno-Couranjou, M.; Palumbo, F.; Sardella, E.; Frache, G.; Favia, P.; Choquet, P. Plasma deposition of thermo-responsive thin films from n-vinylcaprolactam. *Plasma Process. Polym.* **2014**, *11*, 816–821. [CrossRef]
147. Muzammil, I.; Li, Y.; Lei, M. Tunable wettability and pH-responsiveness of plasma copolymers of acrylic acid and octafluorocyclobutane. *Plasma Process. Polym.* **2017**, *14*. [CrossRef]

 *antibiotics*

Review

Silver Camphor Imine Complexes: Novel Antibacterial Compounds from Old Medicines

Jorge H. Leitão [1,*], Silvia A. Sousa [1], Silvestre A. Leite [1,2] and Maria Fernanda N. N. Carvalho [2,*]

[1] IBB—Institute for Bioengineering and Biosciences, Department of Bioengineering, Instituto Superior Técnico, Universidade de Lisboa. Av Rovisco Pais, 1049-001 Lisboa, Portugal; sousasilvia@tecnico.ulisboa.pt (S.A.S.); silvestre.leite@tecnico.ulisboa.pt (S.A.L.)

[2] Centro de Química Estrutural, Instituto Superior Técnico, Universidade de Lisboa. Av Rovisco Pais, 1049-001 Lisboa, Portugal

* Correspondence: jorgeleitao@tecnico.ulisboa.pt (J.H.L.); fcarvalho@tecnico.ulisboa.pt (M.F.N.N.C.); Tel.: +351-218417688 (J.H.L.)

Received: 12 June 2018; Accepted: 24 July 2018; Published: 26 July 2018

Abstract: The emergence of bacterial resistance to available antimicrobials has prompted the search for novel antibacterial compounds to overcome this public health problem. Metal-based complexes have been much less explored than organic compounds as antimicrobials, leading to investigations of the antimicrobial properties of selected complexes in which silver may occupy the frontline due to its use as medicine since ancient times. Like silver, camphor has also long been used for medicinal purposes. However, in both cases, limited information exists concerning the mechanisms of their antimicrobial action. This work reviews the present knowledge of the antimicrobial properties of camphor-derived silver complexes, focusing on recent research on the synthesis and antimicrobial properties of complexes based on silver and camphor imines. Selected examples of the structure and antimicrobial activity relationships of ligands studied so far are presented, showing the potential of silver camphorimine complexes as novel antimicrobials.

Keywords: silver; camphor derivatives; silver camphorimine complexes; antimicrobial activity

---

1. Introduction

The discovery and use of antimicrobials is a landmark of modern medicine. Most of the antimicrobials currently in use were developed last century, between the 1940s and 1960s [1,2]. These "magic bullets" have since then saved millions of human lives. However, bacterial resistance to multiple antimicrobials has increased worldwide over the last decades, mainly due to their misuse and abuse. Antibiotic resistance poses a serious threat to infection treatment, and thus a significant pressure on health systems, and is considered a major menace to human health [3,4]. A particular group of bacteria, referred to as the ESKAPE group (comprising *Enterococcus faecium*, *Staphyloccus aureus*, *Klebsiella pneumoniae*, *Acinetobacter baumanni*, *Pseudomonas aeruginosa*, and *Enterobacter* species) has emerged worldwide, with an increasing prevalence in hospitals and resistance to antibiotics [5,6].

Despite this emergence of resistance, the number of new antimicrobials reaching the market has been declining since the 1990s [7]. Multiple factors, including low investment return and regulatory requirements, are the major reasons [8] for the exodus of pharmaceutical and biotechnological companies from antimicrobial research and development.

Due to the new and emergent risks posed by bacterial resistance, the World Health Organization (WHO, February 2017) [9] has appealed for the investment of publicly funded agencies and the private sector in the research and development of new antibiotics. Motivated by the urgent need for new and effectively active antimicrobials, we have initiated the investigation of the antimicrobial activity of silver camphorimine complexes. Silver and camphor are medicines empirically used since ancient times, although the mechanisms underlying their antimicrobial activities remain to be fully understood.

2. The Antimicrobial Properties of Silver

Silver (Ag^0) vessels were used in ancient times not only for aesthetic reasons but because they preserved the quality of water. Although microbes were unknown, the consequences of microbial contaminations were well known, and silver was a great help to control them. Aware of that, Hippocrates (the father of medicine, 5th to 4th centuries BC) prescribed silver preparations against infections. Later, the Romans extended the medicinal use to silver nitrate ($AgNO_3$, Figure 1), and during the Medieval Period, it continued to be used for the treatment of skin wounds and ulcers [10,11]. Silver sulfadiazine (Figure 1) came later and is still in use to treat infections associated with burns [12,13].

Silver nitrate

Silver sulfadiazine

Figure 1. Examples of silver compounds with antibacterial properties.

Nowadays, silver metal and silver nanoparticles are used for medical device coatings and the reduction of bacterial adhesion to the surfaces of implants as gels or films [14,15]. Silver composites are used as antimicrobials against Gram-positive and/or Gram-negative bacteria [16–18]. Therapeutic and antiseptic applications of silver and silver derivatives actually extend to the control of surgical infections [19], a domain of increasing concern.

Despite the recognized antimicrobial activity of silver salts (e.g., silver nitrate, silver sulfadiazine) and silver particles, the pharmacological uses of silver derivatives have been restricted mostly to external uses in creams and dressings due to toxicity concerns. Recent studies showed that silver nitrate and silver nanoparticles have no significant toxicity, although $AgNO_3$ accumulates more than silver nanoparticles in the organs of rats [20].

Notwithstanding years of use, only recently has some information been gathered on the molecular mechanisms underlying the antimicrobial properties of silver. Proteins have been pointed out as the major targets of silver ions, which can react with thiol groups, inactivating membrane-associated enzymes (e.g., those involved in electron transfer and energy generation) as the Na^+-translocating NADH:ubiquinone oxidoreductase [21]. For instance, the addition of $AgNO_3$ to *Escherichia coli* has been shown to lead to proton motive force collapse and subsequent cell death [22].

Using reflectance Fourier transform infrared (ATR-FTIR) spectroscopy, Ansari et al. [23] investigated changes in *E. coli* lipopolysaccharide (LPS) and L-α-phosphatidyl-ethanolamine (PE) upon exposure to silver nanoparticles. These authors showed that the LPS *O*-antigen was involved in the interaction with silver nanoparticles through hydrogen bonding. In addition, the silver nanoparticles induced the break of the phosphodiester bond of PE, forming phosphate monoesters and resulting in a highly disordered alkyl chain, most probably causing the destruction of the membrane and cell leaking.

Dibrov et al. [24] reported that low concentrations of silver ions induce massive proton leakage and loss of cell viability in *Vibrio cholerae*, suggesting that the antimicrobial activity of Ag^+ results from its unspecific action on membrane proteins and/or the Ag^+-modified phospholipid bilayer. Consistently, the increase in d/cis ratios of unsaturated membrane fatty acids were reported upon exposure to silver species [25], most probably affecting membrane fluidity and culminating with membrane integrity loss [23,26]. A proteomics study carried out with *Pseudomonas aeruginosa* revealed that treatment with silver nanoparticles led to the identification of 59 proteins related to membrane functions and intracellular oxygen reactive species generation, and 5 silver-binding proteins were found by proteomics [27]. Feng and colleagues [28] showed that, upon exposure to silver ions, *E. coli* and *S. aureus* lost the ability to replicate DNA, suggesting that the nucleoid is another bacterial structure targeted by silver ions.

A few studies have also used electron microscopy techniques to inspect the effects of silver nanoparticles on bacterial cell morphology. An example is the work by Huang et al. [29], who used transmission electron microscopy (TEM) and scanning electron microscopy (SEM) to study the antimicrobial mechanisms of catechol-functional chitosan silver nanoparticles on the Gram-positive *S. aureus* and the Gram-negative *E. coli*. These authors suggest that the silver nanoparticles killed *S. aureus* through disruption of the cell wall and the consequent membrane damage and cytoplasmic content leak out. In the case of the Gram-negative *E. coli*, the antimicrobial mechanisms involved the adsorption of the silver nanoparticles to the surface of surface bacterial cells, interaction with the outer membrane, and damage of its permeability. This change in permeability was suggested to allow the silver ions to enter into the cytoplasm, interfering with cellular functioning.

Silver nanoparticles have also been reported as active against a wide range of multidrug-resistant (MDR) bacteria, including the Gram-positive *S. aureus, S. epidermidis, Streptococcus mutans, Enterococcus faecalis*, and *Bacillus subtilis*, and the Gram-negative *E. coli, V. cholerae, P. aeruginosa, Klebsiella pneumoniae*, and *Salmonella typhi* [30].

3. The Biological Properties of Camphor

(1*R*)-(+)-camphor (Figure 2) is a natural product (also produced by synthetic means) with very ancient applications as an insect repellent, muscular relaxant, and anesthetic [31,32] and is currently used as a cough suppressant and decongestant according to conditions established by the Food and Drug Administration. Recent studies show that camphor derivatives may also have relevant antimicrobial [33], antiviral [34,35], and/or cytotoxic properties [36], being also used as photoinitiators [37] or neuro-blocking agents [38]. However, in contrast with some knowledge of silver antibacterial mechanisms, such knowledge is inexistent in the case of camphor or camphor derivatives.

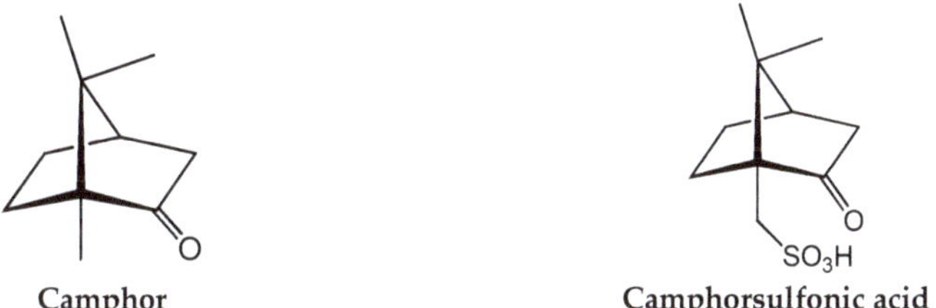

Figure 2. Molecular structures of camphor and camphorsulfonic acid.

The wide range of pharmacological applications of camphor has fostered studies on the properties of camphor derivatives and camphor complexes. Camphor derivatives can be obtained by the introduction of suitable substituents on the camphor molecule (e.g., imine or carboxylic groups), keeping the camphor skeleton intact. Camphor imines, camphor carboxylates, camphor sulfonimines, and camphor sulfonamides (derived from camphorsulfonic acid, Figure 2) are among the camphor derivatives with suitable coordinating properties, some of which (e.g., camphor sulfonamides) display therapeutic properties [39].

Such as for silver and silver derivatives, not much information is available on the molecular mechanisms underlying the antimicrobial activity of camphor derivatives. Recent computer simulations of interactions with the viral surface hemagglutinin (HA) glycoprotein suggest that camphor derivatives inhibit HA activity by binding to hydrophobic sites of the protein [40]. Apparently, camphor imines are able to effectively block conformational rearrangements of HA, required for membrane fusion during virus entry in host cells.

4. Silver Camphorimine Complexes

In complexes, the characteristics of the metal and the ligands combine in a cooperative way to generate compounds with distinct and frequently improved properties relative to the precursor species. Complexes have a wide variety of applications in catalysis, medicine, and the materials industry. Furthermore, properties of complexes can be tuned towards specific applications by varying the metal and ligands. The catalytic properties of silver complexes have been explored to promote organic transformations and bio-conjugation [41]. In the last decade, the evaluation of the biological properties of silver complexes has garnered considerable attention, in particular, fostering their use as antimicrobials or anticancer agents [1,4,42–47]. The therapeutic uses of silver complexes include silver sulfadiazine for treatment of infections associated with burns.

The core of the silver camphor imine complexes, the biological properties of which we have been studying, is formed by silver nitrate ($AgNO_3$) that binds one or two camphor ligands per metal atom. Data obtained by X-rays show that nitrate acts as a bidentate ligand and that the camphor ligand binds the silver atom through the imine nitrogen atom. The complexes arrange according to monomeric (type 1, Figure 3) or polymeric (type 2, Figure 3) structures. The polymeric arrangement was typically found in imine and bicamphor imine complexes of general formula $[Ag(NO_3)L]_n$ [33], while the monomeric arrangement was found in camphor sulphonylimine ($[Ag(NO_3)L_2]$) [43] complexes (Figure 3).

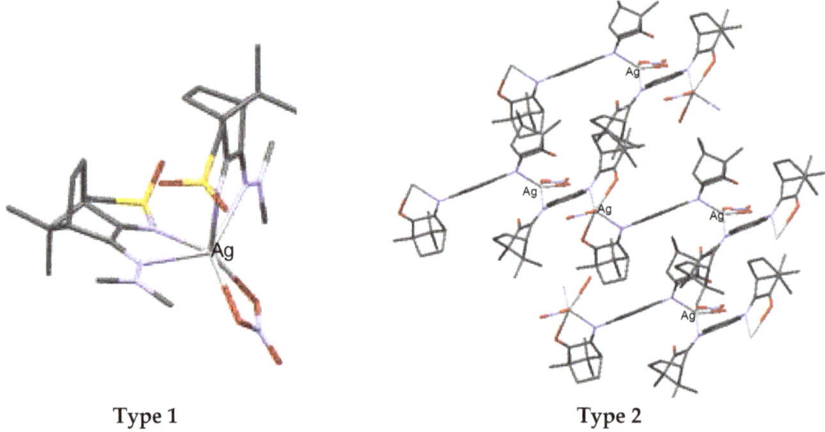

Type 1   **Type 2**

Figure 3. Structural arrangements corroborated by X-ray analysis in Ag(I)-camphor-derived complexes. The synthesis of the Ag(I) camphor complexes includes the preparation of suitable camphor compounds to be used as ligands, since just a few imine functionalized camphor derivatives are commercially available (e.g., camphor oxime and camphor sulphonylimine). (1R)-(+)-camphor is the starting material for the camphor compounds of type A1–A5 (Figure 4).

(1S)-(+)-10-camphorsulfonic acid is the precursor for the tricyclic camphor sulfonylimine derivatives of types B1 and B2 (Figure 5) within a process that involves the sequential formation of chloride and amine derivatives prior to ring closure and oxidation to obtain 3-*oxo*-camphorsulfonylimine, which is then condensed with amines or hydrazines to afford camphor sulfonylimines of types B1 and B2 (Figure 5).

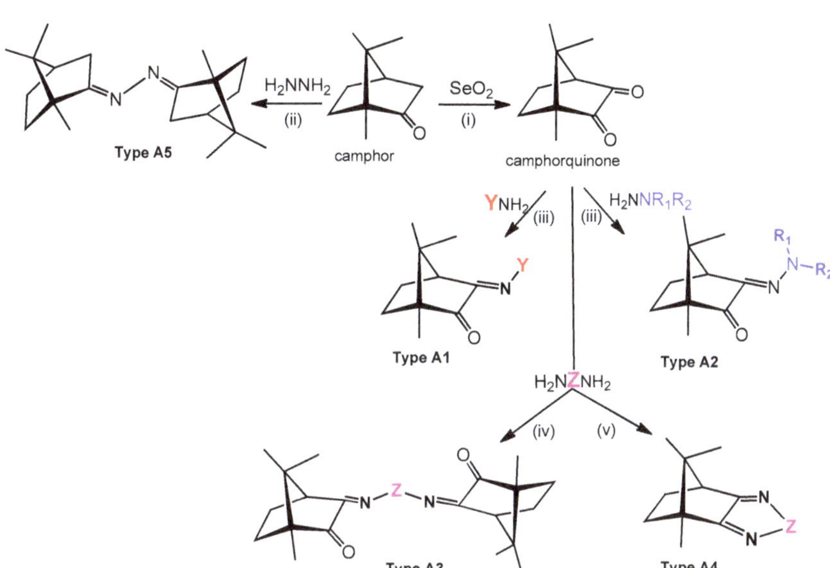

Figure 4. Camphor derivatives of the imine type: Y-aromatic or aliphatic; R_1 = H, Me, Ph; R_2 = H, Me, Ph, among others; Z = phenyl; biphenyl, linear alkane. For preparative details see references (i)—[48]; (ii)—[42]; (iii)—[49,50]; (iv)—[42,51]; (v)—[42,52].

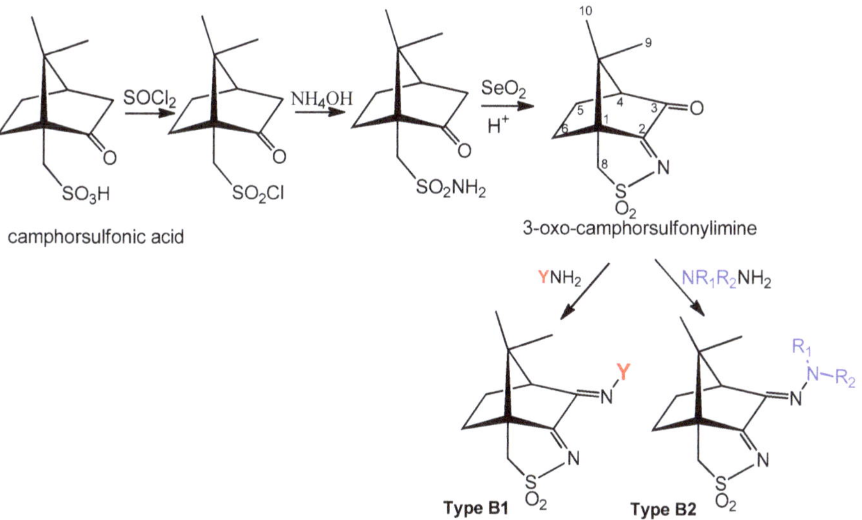

Figure 5. Camphorsulfonic acid derivatives of the camphorsulfonimine type.

A wide variety of amines (YNH_2) and hydrazines ($NR_1R_2NNH_2$) can be used to tune the electronic and steric characteristics of the imine group at position 3 (see Figure 5 for labelling) of the camphor skeleton. Thus, the design and synthesis of camphor Ag(I) complexes can be driven to control electronic and steric parameters towards the aimed objectives.

The camphor imine complexes, the antimicrobial properties of which were studied by us [42], were obtained in acetonitrile by the reaction of silver nitrate with the suitable camphor ligand

(L) under metal-to-ligand ratios directed to mononuclear (Equation (1)) or polymeric complexes (Equation (2)). Strict control of the experimental conditions (exposure to light and reaction time) is necessary depending on the electronic characteristics of the camphor ligands to inhibit Ag(I) reduction, the formation of silver nanoparticles (AgNPs) (Equation (3)), and products of oxidation of the camphor derivatives (e.g., camphorquinone).

$$AgNO_3 + 2\,L \rightarrow [Ag(N\,O_3)(L)_2] \tag{1}$$

$$AgNO_3 + L \rightarrow [Ag(N\,O_3)(L)] \tag{2}$$

$$AgNO_3 + 2\,L \rightarrow AgNPs + camphorquinone + other\ products \tag{3}$$

Studies carried out so far evidence that all silver camphor imine complexes display biological activity that, according to the structural and electronic characteristics of the ligands, favors antibacterial, antifungal, and/or anticancer activity [33–35]. In some cases, antibacterial, antifungal, and cytotoxic activity combine in the same complex [42,44]. Such performance is highly relevant to cancer treatments because opportunistic fungi and bacteria might develop when body defenses diminish due to the use of anticancer drugs.

5. Antibacterial Activity

To date, several silver camphor complexes have been synthesized and characterized, and their antibacterial activity towards Gram-positive (*S. aureus*) and Gram-negative bacteria (*E. coli, P. aeruginosa, Burkholderia contaminans*) has been screened using the Kirby–Bauer disk diffusion method and quantified based on the determination of minimal inhibitory concentrations (MIC) by microdilution assay standard methods [45]. *E. coli, S. aureus*, and *P. aeruginosa* strains were chosen because these species belong to the ESKAPE group. *B. contaminans* is a member of the *Burkholderia cepacia* complex (Bcc), a group of Gram-negative bacteria capable of causing life-threatening respiratory infections, of particular severity among cystic fibrosis patients [46,47,53,54]. Bcc bacteria are intrinsically resistant to multiple antimicrobials, rendering their eradication difficult to achieve [55]. Photographs in Figure 6 illustrate results from disk diffusion and MIC assays, obtained with complex **1** (see Table 1) for *E. coli* ATCC25922 and *S. aureus* Newman.

E. coli ATCC25922

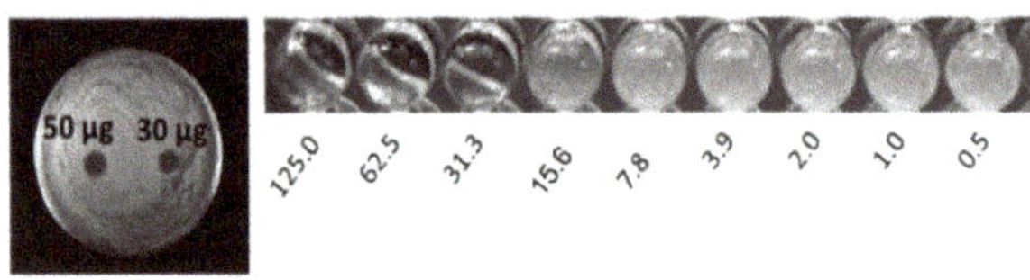

S. aureus Newman

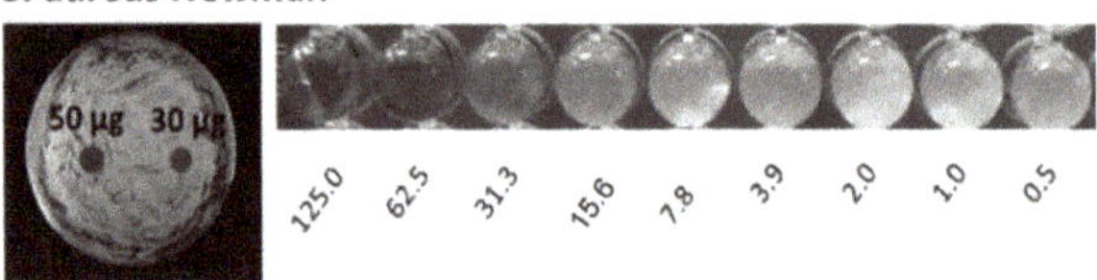

Figure 6. Photographs illustrating results from the assessment of complex **1** antibacterial activity towards *E. coli* ATCC25922 or *S. aureus* Newman by the disk diffusion method and MIC determination by broth microdilution assays. Circular growth inhibition zones are visible in plates spotted with 50 or 30 µg of the complex. The numbers below the wells indicate the respective complex **1** final concentration, in µg/mL. Photographs (not at the same scale) taken after 24 h of incubation at 37 °C.

Table 1. Minimal inhibitory concentration (MIC) values calculated for complexes [Ag(NO$_3$)L], in µg/mL, towards the bacterial strains *Staphylococcus aureus* Newman, *E scherichia coli* ATCC25922, *Pseudomonas aeruginosa* 477, and *Burkholderia contaminans* IST408. Data taken from previous publications [42–44].

Ligand (L)	Complex	*S. aureus*	*E. coli*	*P. aeruginosa*	*B. contaminans*
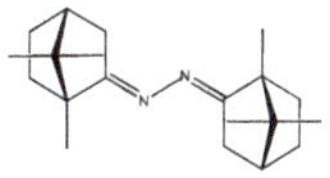	1	66 ± 5	50 ± 1	56 ± 4	79 ± 4
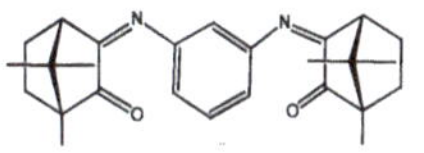	2	183 ± 3	65 ± 2	121 ± 2	144 ± 1
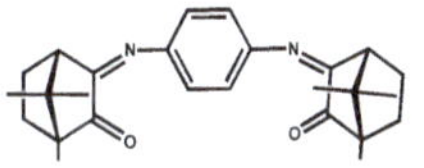	3	73 ± 2	20 ± 1	19 ± 4	36 ± 3
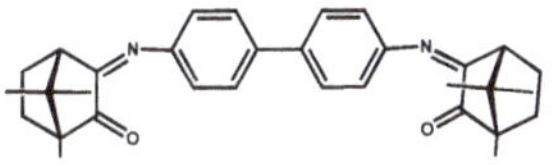	4	<100	<100	86 ± 7	<100
AgNO$_3$ (Control)	-	73	47	39	74

MIC values of a selection of silver camphor imine complexes towards *E. coli, S. aureus, P. aeruginosa,* and *B. contaminans* are shown in Table 1. The examples aim to highlight the profound effects of the presence, position, and number of aryl groups in the antibacterial activity of the silver camphor imine complexes. These selected examples also evidence that structural differences in the complexes due to *para* (complex **3**) or *meta* (complex **2**) substituents in the aromatic ring drive significant variations in the MIC values for the strains tested, not only between Gram-positive and Gram-negative strains but also within the Gram-negative strains. From the set of bicamphor complexes in Table 1, [Ag(NO$_3$)(OC$_{10}$H$_{14}$N)$_2$(*p*-C$_6$H$_4$)] (**3**, ligand of type A3, Figure 4) displays the highest antibacterial activity against Gram-negative strains, followed by complex **1** with a type A5 ligand (Figure 4). In both cases, electron delocalization throughout the camphor ligand may not be innocent since it affects the electron density at the silver center and consequently the electron transfer processes in which the silver ion may be involved. The MIC values for complexes **1**, **3**, and **4** show that one aromatic group between the two camphor moieties (**3**) increased the antimicrobial activity towards the Gram-negative strains, while two sequential aromatic groups in between the camphor moieties (**4**) resulted in a general loss of antibacterial activity (Table 1). Comparing the biological activity of complexes **2** and **3** (selected to illustrate the effects of the geometry of the ligand in the MIC values according to a relevant effect of geometry and/or electron delocalization on activity), a marked decrease in activity is observed upon replacement of a *para* (**3**) by a *meta* (**2**)-substituted aromatic spacer (Table 1).

A step forward in this work considers the design of new camphor ligands and the synthesis of silver and eventually other metal complexes to investigate the effects of electron density, electron delocalization, distinct geometries, and substituents at the camphor skeleton on the antibacterial activity. The identification of the complexes' bacterial targets is expected to enable the design of suitable ligands to tailor complexes with enhanced antimicrobial activity and allow rationalization of the mechanisms.

A major concern when developing novel antimicrobials is the emergence of resistant strains. Preliminary unpublished work from our team indicates that the frequency of spontaneous emergence of resistance for *E. coli* ATCC25922, *P. aeruginosa* 477, *B. contaminans* IST408, and *S. aureus* Newman is lower than 4×10^{-10}. These data were obtained after spreading approximately 10^9 CFUs of each strain onto the surface of Mueller-Hinton solid medium containing twice the estimated MIC concentration of complex **3** and enumerating the CFUs after five days of incubation at 37 °C. Another issue that still needs to be addressed is the toxicity of the silver camphorimine complexes to humans and animals when envisaging their use in human and veterinary medicine.

6. Conclusions

Camphor and silver derivatives have been used since ancient times as medicines. Although their combination in complexes and the search for their antimicrobial activities only started recently, there is no application of silver camphor imine complexes in medicine, pharmacy, or industry. The results obtained until now show that some silver camphor imine complexes combine antimicrobial activity against bacteria and fungi with cytotoxic activity [42–44]. This feature is highly relevant since opportunistic bacteria and fungi usually develop during cancer treatments due to the immunosuppressive effects of anticancer drugs.

The insights already made into the antimicrobial properties of some silver camphor imine complexes show they have moderate to high antibacterial activity, depending on the characteristics of the ligands. Such information fosters further enhancement of the ligands to optimize the antibacterial properties of the camphor imine complexes. Meanwhile, synthetic strategies to synthesize, characterize, and evaluate the antimicrobial properties of silver camphor imine complexes were developed and described in the present minireview. Detailed knowledge of the bacterial targets of these compounds is still missing, and therefore, future work will focus on the unveiling of the molecular targets and mechanisms underlying their antimicrobial activity. To address these issues, studies will focus on the use of transcriptomic approaches to gain clues regarding the bacterial gene expression responses to exposure to inhibitory concentrations of silver camphor imine complexes. It will be also necessary to use more classical biochemical approaches to identify the bacterial targets of these complexes. The comprehensive knowledge of the molecular details of the antimicrobial activity of silver camphor imine complexes is expected to enable the exploitation of structure and activity relationships to tailor complexes with enhanced antimicrobial activity.

Author Contributions: Conceptualization, J.H.L., M.F.N.N.C.; Writing-Original Draft Preparation, J.H.L., M.F.N.N.C., S.A.S., S.A.L.; Writing-Review & Editing, J.H.L., M.F.N.N.C., S.A.S.; Funding Acquisition, J.H.L., M.F.N.N.C.

Funding: This research was funded by from FCT-Fundação para a Ciência e a Tecnologia (Portugal) through projects UID/QUI/00100/2013, UID/BIO/04565/2013, Programa Operacional Regional de Lisboa 2020 (Project N. 007317) and the NMR Network (IST-UTL Node) for facilities.

Acknowledgments: S.A.S. acknowledges a postdoctoral grant from FCT (SFRH/BPD/102006/2014).

Conflicts of Interest: The authors declare no conflict of interest.

References

1. Labischinski, H. New Antibiotics. *Int. J. Med. Microbiol.* **2001**, *291*, 317–318. [CrossRef] [PubMed]
2. Spellberg, B.; Powers, J.H.; Brass, E.P.; Miller, L.G.; Edwards, J.E. Trends in Antimicrobial Drug Development: Implications for the Future. *Clin. Infect. Dis.* **2004**, *38*, 1279–1286. [CrossRef] [PubMed]
3. Institute of Medicine. *Microbial Threats to Health*; Smolinski, M.S., Hamburg, M.A., Lederberg, J., Eds.; The National Academies Press: Washington, DC, USA, 2003.
4. Johnston, B.L.; Conly, J.M. Bioterrorism in 2001: How ready are we? *Can. J. Infect. Dis.* **2001**, *12*, 77–80. [CrossRef] [PubMed]
5. Rice, L.B. Federal funding for the study of antimicrobial resistance in nosocomial pathogens: No ESKAPE. *J. Infect. Dis.* **2008**, *197*, 1079–1081. [CrossRef] [PubMed]
6. Pendleton, J.N.; Gorman, S.P.; Gilmore, B.F. Clinical relevance of the ESKAPE pathogens. *Expert Rev. Anti-Infect. Ther.* **2013**, *11*, 297–308. [CrossRef] [PubMed]
7. Theuretzbacher, U. Future antibiotics scenarios: Is the tide starting to turn? *Int. J. Antimicrob. Agents* **2009**, *34*, 15–20. [CrossRef] [PubMed]
8. Payne, D.J.; Gwynn, M.N.; Holmes, D.J.; Pompliano, D.L. Drugs for bad bugs: Confronting the challenges of antibacterial discovery. *Nat. Rev. Drug Discov.* **2007**, *6*, 29–40. [CrossRef] [PubMed]

9. WHO Publishes List of Bacteria for which New Antibiotics Are Urgently Needed. Available online: http://www.who.int/en/news-room/detail/27-02-2017-who-publishes-list-of-bacteria-for-which-new-antibiotics-are-urgently-needed (accessed on 1 February 2018).

10. Azócar, M.I.; Gómez, G.; Levín, P.; Paez, M.; Muñoz, H.; Dinamarca, N. Review: Antibacterial behavior of carboxylate silver(I) complexes. *J. Coord. Chem.* **2014**, *67*, 3840–3853. [CrossRef]

11. Klasen, H.J. Historical review of the use of silver in the treatment of burns. I. Early uses. *Burns* **2000**, *26*, 117–130. [CrossRef]

12. Moyer, C.A.; Brentano, L.; Gravens, D.L.; Margraf, H.W.; Monafo, W.W. Treatment of large human burns with 0.5% silver nitrate solution. *Arch. Surg.* **1965**, *90*, 812–867. [CrossRef] [PubMed]

13. Kremer, E.; Facchin, G.; Estévez, E.; Alborés, P.; Baran, E.J.; Ellena, J.; Torre, M.H. Copper complexes with heterocyclic sulfonamides: Synthesis, spectroscopic characterization, microbiological and SOD-like activities: Crystal structure of [Cu(sulfisoxazole)$_2$(H$_2$O)$_4$]·2H$_2$O. *J. Inorg. Biochem.* **2006**, *100*, 1167–1175. [CrossRef] [PubMed]

14. Brunetto, P.S.; Slenters, T.V.; Fromm, K.M. In vitro biocompatibility of new silver(I) coordination compound coated-surfaces for dental implant applications. *Materials (Basel)* **2011**, *4*, 355–367. [CrossRef] [PubMed]

15. Maillard, J.-Y.; Hartemann, P. Silver as an antimicrobial: Facts and gaps in knowledge. *Crit. Rev. Microbiol.* **2013**, *39*, 373–383. [CrossRef] [PubMed]

16. Bormio Nunes, J.H.; de Paiva, R.E.F.; Cuin, A.; Lustri, W.R.; Corbi, P.P. Silver complexes with sulfathiazole and sulfamethoxazole: Synthesis, spectroscopic characterization, crystal structure and antibacterial assays. *Polyhedron* **2015**, *85*, 437–444. [CrossRef]

17. McCann, M.; Curran, R.; Ben-Shoshan, M.; McKee, V.; Devereux, M.; Kavanagh, K.; Kellett, A. Synthesis, structure and biological activity of silver(I) complexes of substituted imidazoles. *Polyhedron* **2013**, *56*, 180–188. [CrossRef]

18. Atiyeh, B.S.; Costagliola, M.; Hayek, S.N.; Dibo, S.A. Effect of silver on burn wound infection control and healing: Review of the literature. *Burns* **2007**, *33*, 139–148. [CrossRef] [PubMed]

19. Alexander, J.W. History of the medical use of silver. *Surg. Infect. (Larchmt)* **2009**, *10*, 289–292. [CrossRef] [PubMed]

20. Qin, G.; Tang, S.; Li, S.; Lu, H.; Wang, Y.; Zhao, P.; Li, B.; Zhang, J.; Peng, L. Toxicological evaluation of silver nanoparticles and silver nitrate in rats following 28 days of repeated oral exposure. *Environ. Toxicol.* **2017**, *32*, 609–618. [CrossRef] [PubMed]

21. Semeykina, A.L.; Skulachev, V.P. Submicromolar Ag$^+$ increases passive Na$^+$ permeability and inhibits the respiration-supported formation of Na$^+$ gradient in *Bacillus* FTU vesicles. *FEBS Lett.* **1990**, *269*, 69–72. [CrossRef]

22. Schreurs, W.J.; Rosenberg, H. Effect of silver ions on transport and retention of phosphate by *Escherichia coli*. *J. Bacteriol.* **1982**, *152*, 7–13. [PubMed]

23. Ansari, M.A.; Khan, H.M.; Khan, A.A.; Ahmad, M.K.; Mahdi, A.A.; Pal, R.; Cameotra, S.S. Interaction of silver nanoparticles with *Escherichia coli* and their cell envelope biomolecules. *J. Basic Microbiol.* **2014**, *54*, 905–915. [CrossRef] [PubMed]

24. Dibrov, P.; Dzioba, J.; Gosink, K.K.; Häse, C.C. Chemiosmotic mechanism of antimicrobial activity of Ag(+) in *Vibrio cholerae*. *Antimicrob. Agents Chemother.* **2002**, *46*, 2668–2670. [CrossRef] [PubMed]

25. Hachicho, N.; Hoffmann, P.; Ahlert, K.; Heipieper, H.J. Effect of silver nanoparticles and silver ions on growth and adaptive response mechanisms of *Pseudomonas putida* mt-2. *FEMS Microbiol. Lett.* **2014**, *355*, 71–77. [CrossRef] [PubMed]

26. Li, W.-R.; Xie, X.-B.; Shi, Q.-S.; Zeng, H.-Y.; OU-Yang, Y.-S.; Chen, Y.-B. Antibacterial activity and mechanism of silver nanoparticles on *Escherichia coli*. *Appl. Microbiol. Biotechnol.* **2010**, *85*, 1115–1122. [CrossRef] [PubMed]

27. Yan, X.; He, B.; Liu, L.; Qu, G.; Shi, J.; Hu, L.; Jiang, G. Antibacterial mechanism of silver nanoparticles in *Pseudomonas aeruginosa*: Proteomics approach. *Metallomics* **2018**, *10*, 557–564. [CrossRef] [PubMed]

28. Feng, Q.L.; Wu, J.; Chen, G.Q.; Cui, F.Z.; Kim, T.N.; Kim, J.O. A mechanistic study of the antibacterial effect of silver ions on *Escherichia coli* and *Staphylococcus aureus*. *J. Biomed. Mater. Res.* **2000**, *52*, 662–668. [CrossRef]

29. Huang, X.; Bao, X.; Liu, Y.; Wang, Z.; Hu, Q. Catechol-functional chitosan/silver nanoparticle composite as a highly effective antibacterial agent with species-specific mechanisms. *Sci. Rep.* **2017**, *7*, 1860. [CrossRef] [PubMed]

30. Rai, M.K.; Deshmukh, S.D.; Ingle, A.P.; Gade, A.K. Silver nanoparticles: The powerful nanoweapon against multidrug-resistant bacteria. *J. Appl. Microbiol.* **2012**, *112*, 841–852. [CrossRef] [PubMed]

31. Chen, W.; Vermaak, I.; Viljoen, A. Camphor—A fumigant during the black death and a coveted fragrant wood in ancient Egypt and Babylon—A review. *Molecules* **2013**, *18*, 5434–5454. [CrossRef] [PubMed]

32. Xu, H.; Blair, N.T.; Clapham, D.E. Camphor activates and strongly desensitizes the transient receptor potential vanilloid subtype 1 channel in a vanilloid-independent mechanism. *J. Neurosci.* **2005**, *25*, 8924–8937. [CrossRef] [PubMed]

33. Sokolova, A.S.; Yarovaya, O.I.; Korchagina, D.V.; Zarubaev, V.V.; Tretiak, T.S.; Anfimov, P.M.; Kiselev, O.I.; Salakhutdinov, N.F. Camphor-based symmetric diimines as inhibitors of influenza virus reproduction. *Bioorg. Med. Chem.* **2014**, *22*, 2141–2148. [CrossRef] [PubMed]

34. Justino, G.C.; Pinheiro, P.F.; Roseiro, A.P.S.; Knittel, A.S.O.; Gonçalves, J.; Justino, M.C.; Carvalho, M.F.N.N. Camphor-based CCR5 blocker lead compounds—A computational and experimental approach. *RSC Adv.* **2016**, *6*, 56249–56259. [CrossRef]

35. Sokolova, A.S.; Yarovaya, O.I.; Shernyukov, A.V.; Gatilov, Y.V.; Razumova, Y.V.; Zarubaev, V.V.; Tretiak, T.S.; Pokrovsky, A.G.; Kiselev, O.I.; Salakhutdinov, N.F. Discovery of a new class of antiviral compounds: Camphor imine derivatives. *Eur. J. Med. Chem.* **2015**, *105*, 263–273. [CrossRef] [PubMed]

36. Itani, W.S.; El-Banna, S.H.; Hassan, S.B.; Larsson, R.L.; Bazarbachi, A.; Gali-Muhtasib, H.U. Anti colon cancer components from Lebanese sage (*Salvia libanotica*) essential oil: Mechanistic basis. *Cancer Biol. Ther.* **2008**, *7*, 1765–1773. [CrossRef] [PubMed]

37. Kamoun, E.A.; Winkel, A.; Eisenburger, M.; Menzel, H. Carboxylated camphorquinone as visible-light photoinitiator for biomedical application: Synthesis, characterization, and application. *Arab. J. Chem.* **2016**, *9*, 745–754. [CrossRef]

38. Sokolova, A.S.; Morozova, E.A.; Vasilev, V.G.; Yarovaya, O.I.; Tolstikova, T.G.; Salakhutdinov, N.F. Curare-like camphor derivatives and their biological activity. *Russ. J. Bioorgan. Chem.* **2015**, *41*, 178–185. [CrossRef]

39. Wang, Y.; Busch-Petersen, J.; Wang, F.; Kiesow, T.J.; Graybill, T.L.; Jin, J.; Yang, Z.; Foley, J.J.; Hunsberger, G.E.; Schmidt, D.B.; et al. Camphor sulfonamide derivatives as novel, potent and selective CXCR3 antagonists. *Bioorg. Med. Chem. Lett.* **2009**, *19*, 114–118. [CrossRef] [PubMed]

40. Sokolova, A.S.; Yarovaya, O.I.; Baev, D.S.; Shernyukov, A.V.; Shtro, A.A.; Zarubaev, V.V.; Salakhutdinov, N.F. Aliphatic and alicyclic camphor imines as effective inhibitors of influenza virus H1N1. *Eur. J. Med. Chem.* **2017**, *127*, 661–670. [CrossRef] [PubMed]

41. Lo, V.K.-Y.; Chan, A.O.-Y.; Che, C.-M. Gold and silver catalysis: From organic transformation to bioconjugation. *Org. Biomol. Chem.* **2015**, *13*, 6667–6680. [CrossRef] [PubMed]

42. Cardoso, J.M.S.; Guerreiro, S.I.; Lourenço, A.; Alves, M.M.; Montemor, M.F.; Mira, N.P.; Leitão, J.H.; Carvalho, M.F.N.N. Ag(I) camphorimine complexes with antimicrobial activity towards clinically important bacteria and species of the *Candida* genus. *PLoS ONE* **2017**, *12*, e0177355. [CrossRef] [PubMed]

43. Cardoso, J.M.S.; Correia, I.; Galvão, A.M.; Marques, F.; Carvalho, M.F.N.N. Synthesis of Ag(I) camphor sulphonylimine complexes and assessment of their cytotoxic properties against cisplatin -resistant A2780cisR and A2780 cell lines. *J. Inorg. Biochem.* **2017**, *166*, 55–63. [CrossRef] [PubMed]

44. Cardoso, J.M.S.; Galvão, A.M.; Guerreiro, S.I.; Leitão, J.H.; Suarez, A.C.; Carvalho, M.F.N.N. Antibacterial activity of silver camphorimine coordination polymers. *Dalton Trans.* **2016**, *45*, 7114–7123. [CrossRef] [PubMed]

45. National Committee for Clinical Laboratory Standards. *Methods for Determining Bactericidal Activity of Antimicrobial Agents; Approved Guideline*; NCCLS Document M26-A; NCCLS: Wayne, PA, USA, 1999.

46. Sousa, S.A.; Feliciano, J.R.; Pita, T.; Guerreiro, S.I.; Leitão, J.H. *Burkholderia cepacia* complex regulation of virulence gene expression: A review. *Genes* **2017**, *8*, 43. [CrossRef] [PubMed]

47. Coutinho, C.P.; Barreto, C.; Pereira, L.; Lito, L.; Cristino, J.M.; Sá-Correia, I. Incidence of *Burkholderia contaminans* at a cystic fibrosis centre with an unusually high representation of *Burkholderia cepacia* during 15 years of epidemiological surveillance. *J. Med. Microbiol.* **2015**, *64*, 927–935. [CrossRef] [PubMed]

48. White, J.D.; Wardrop, D.J.; Sundermann, K.F. Camphorquinone and camphorquinone monoxime. *Org. Synth.* **2002**, *79*, 125.

49. Forster, M.O.; Zimmerli, A. CCXXV.—Studies in the camphane series. Part XXVIII. Stereoisomeric hydrazones and semicarbazones of camphorquinone. *J. Chem. Soc., Trans.* **1910**, *97*, 2156–2177. [CrossRef]

50. Carvalho, M.F.N.N.; Costa, L.M.G.; Pombeiro, A.J.L.; Schier, A.; Scherer, W.; Harbi, S.K.; Verfuerth, U.; Herrmann, R. Synthesis, structure, and electrochemistry of palladium complexes with camphor-derived chiral ligands. *Inorg. Chem.* **1994**, *33*, 6270–6277. [CrossRef]
51. Fernandes, T.A.; Mendes, F.; Roseiro, A.P.S.; Santos, I.; Carvalho, M.F.N.N. Insight into the cytotoxicity of polynuclear Cu(I) camphor complexes. *Polyhedron* **2015**, *87*, 215–219. [CrossRef]
52. Fitchett, C.M.; Steel, P.J. Chiral heterocyclic ligands. XII. Metal complexes of a pyrazine ligand derived from camphor. *Arkivoc* **2005**, *2006*, 218.
53. Sousa, S.A.; Ramos, C.G.; Leitão, J.H. *Burkholderia cepacia* complex: Emerging multihost pathogens equipped with a wide range of virulence factors and determinants. *Int. J. Microbiol.* **2011**, *2011*, 607575. [CrossRef] [PubMed]
54. Leitão, J.H.; Sousa, S.A.; Ferreira, A.S.; Ramos, C.G.; Silva, I.N.; Moreira, L.M. Pathogenicity, virulence factors, and strategies to fight against *Burkholderia cepacia* complex pathogens and related species. *Appl. Microbiol. Biotechnol.* **2010**, *87*, 31–40. [CrossRef] [PubMed]
55. Leitão, J.H.; Sousa, S.A.; Cunha, M.V.; Salgado, M.J.; Melo-Cristino, J.; Barreto, M.C.; Sá-Correia, I. Variation of the antimicrobial susceptibility profiles of *Burkholderia cepacia* complex clonal isolates obtained from chronically infected cystic fibrosis patients: A five-year survey in the major Portuguese treatment center. *Eur. J. Clin. Microbiol. Infect. Dis.* **2008**, *27*, 1101–1111. [CrossRef] [PubMed]

 *antibiotics*

Article

Silver Nanoparticle Conjugation-Enhanced Antibacterial Efficacy of Clinically Approved Drugs Cephradine and Vildagliptin

Abdulkader Masri [1], Ayaz Anwar [1], Dania Ahmed [2], Ruqaiyyah Bano Siddiqui [1], Muhammad Raza Shah [2] and Naveed Ahmed Khan [1,*]

[1] Department of Biological Sciences, School of Science and Technology, Sunway University, Bandar Sunway 47500, Malaysia; abdul@gmail.com or 17025941@imail.sunway.edu.my (A.M.); ayazanwarkk@yahoo.com (A.A.); ruqaiyyahs@sunway.edu.my (R.B.S.)

[2] HEJ Research Institute of Chemistry, International Center for Chemical and Biological Sciences, University of Karachi, Karachi 74600, Pakistan; dania.ahmed48@yahoo.com (D.A.); raza_shahm@yahoo.com (M.R.S.)

* Correspondence: naveed5438@gmail.com; Tel.: +60-(0)3-7491-8622 (ext. 7169); Fax: +60-(0)3-5635-8630

Received: 17 August 2018; Accepted: 19 October 2018; Published: 15 November 2018

Abstract: This paper sets out to determine whether silver nanoparticles conjugation enhance the antibacterial efficacy of clinically approved drugs. Silver conjugated Cephradine and Vildagliptin were synthesized and thoroughly characterized by ultraviolet visible spectrophotometry (UV-vis), Fourier transform infrared (FT-IR) spectroscopic methods, atomic force microscopy (AFM), and dynamic light scattering (DLS) analysis. Using antibacterial assays, the effects of drugs alone and drugs-conjugated with silver nanoparticles were tested against a variety of Gram-negative and Gram-positive bacteria including neuropathogenic *Escherichia coli* K1, *Pseudomonas aeruginosa*, *Klebsiella pneumoniae*, methicillin-resistant *Staphylococcus aureus* (MRSA), *Bacillus cereus* and *Streptococcus pyogenes*. Cytopathogenicity assays were performed to determine whether pretreatment of bacteria with drugs inhibit bacterial-mediated host cell cytotoxicity. The UV-vis spectra of both silver-drug nanoconjugates showed a characteristic surface plasmon resonance band in the range of 400–450 nm. AFM further confirmed the morphology of nanoparticles and revealed the formation of spherical nanoparticles with size distribution of 30–80 nm. FT-IR analysis demonstrated the involvement of Hydroxyl groups in both drugs in the stabilization of silver nanoparticles. Antibacterial assays showed that silver nanoparticle conjugation enhanced antibacterial potential of both Cephradine and Vildagliptin compared to the drugs alone. Pretreatment of bacteria with drugs inhibited *E. coli* K1-mediated host cell cytotoxicity. In summary, conjugation with silver nanoparticle enhanced antibacterial effects of clinically approved Cephradine. These findings suggest that modifying and/or repurposing clinically approved drugs using nanotechnology is a feasible approach in our search for effective antibacterial molecules.

Keywords: Cephradine; Vildagliptin; nanoparticles; antibacterial; nanotechnology

---

1. Introduction

Infectious diseases are a significant burden on public health, driven largely by socio-economic, environmental and ecological factors [1]. About 15 million of 57 million annual deaths worldwide are estimated to be caused by infectious diseases, principally due to bacterial pathogens [1]. The burden of morbidity and mortality falls most heavily on people in developing countries [2]. Drug-resistant microbes are a major factor causing microbial re-emergence [3]. Among numerous bacteria, *Escherichia coli*, *Staphylococcus aureus*, *Pseudomonas aeruginosa*, *Bacillus* species, *Klebsiella pneumoniae* are important human pathogens contributing to urinary tract infections, neonatal meningitis, gastroenteritis, wound and skin diseases, food poisoning and nosocomial infections [4–10]. The overuse of antibiotics has contributed to bacterial acquisition of drug resistance resulting in reduced efficacy of available drugs.

During the past decade, nanomedicine has shown great potential due to effectiveness of various nanoconjugates against pathogenic microbes [11]. Nanomaterials have been frequently used as effective coatings to prevent bacterial adhesion to surfaces as well as bactericidal agents [12]. He et al., showed the development of self-defensive and antibacterial adhesion surface coating based on bilayer hydrogel which can promote cell adhesion and proliferation [13,14]. Polymers-based antibacterial agents are also an important class of nanomaterials. Yuan et al. reported various types of hydroxyl-rich cationic derivatives of star-like poly (glycidyl methacrylate) as broad-spectrum antibacterial and antifouling surface coating agents [15]. Antibacterial activity of low molecular weight cationic polymers is shown to affect the membrane permeability and disruption against a broad range of bacteria [16]. In another report, a salivary statherin protein inspired poly(amidoamine) dendrimer is shown to exhibit antibacterial effects as effective coating on hydroxyapatite [17,18]. Similarly, dendrons have been shown as a clicking tool for generating nonleaching antibacterial materials [19]. Metal nanoparticles have been studied extensively because of their unique physicochemical characteristics including catalytic activity, optical properties, electronic properties, antimicrobial activity, and magnetic properties [20]. Among these, silver nanoparticles (AgNPs) have shown growth inhibitory as well as bactericidal effects [21]. The high surface area of AgNPs leads to high antimicrobial activity as compared with the silver metal [22]. With the limited discovery of novel antibacterial agents, a feasible approach is to modify clinically approved drugs to enhance their efficacy and/or drug repurposing to expedite discovery of effective formulation of antibacterial agents.

Cephradine (relative molecular mass 349.406 g mol^{-1}) is a first generation cephalosporins antibiotic drug that is widely used in the treatment of bacterial infections of the urinary and the respiratory tract, as well as ear, skin and soft tissues. It is used against both Gram-positive and Gram-negative bacteria. Its mode of action is inhibition of bacterial cell wall synthesis [23,24]. Vildagliptin (relative molecular mass 303.399 g mol^{-1}) is an antidiabetic drug, which is a small molecule and inhibits dipeptidyl peptidase-4 (DPP4). Vildagliptin has been shown to stimulate insulin secretion and inhibit glucagon secretion in a glucose-dependent manner [25–27]. Here we tested whether conjugation of AgNPs can enhance efficacy of the clinically approved drug, Cephradine.

2. Materials and Methods

2.1. Bacterial Cultures

The cultures of six bacterial isolates including neuropathogenic *Escherichia coli* K1 (a cerebrospinal fluid isolate from a meningitis patient; 018:K1:H7), strain E44, was used in the present study (Malaysian Type Culture Collection 710859), and methicillin-resistant *Staphylococcus aureus* (MRSA) was used as described previously (Malaysian Type Culture Collection 381123). The MRSA strain was originally derived from the blood cultures, obtained from the Luton & Dunstable Hospital NHS Foundation Trust, Luton, England, UK. *Pseudomonas aeruginosa*, *Klebsiella pneumoniae*, *Bacillus cereus* and *Streptococcus pyogenes* were obtained from the Microbiology Research Laboratory at Sunway University. Stock cultures were refreshed by subculturing every 15 days on nutrient agar plates and were maintained at 4 °C.

2.2. Synthesis of AgNPs Coated with Drugs

Cephradine conjugated silver nanoparticles (Ceph-AgNPs) were synthesized. Briefly, 5 mL (0.1 mM) Cephradine aqueous solution was reacted with 5 mL (0.1 mM) silver nitrate aqueous solution, and the reaction mixture was magnetically stirred for 10 min. Twenty µL of 5 mM freshly prepared Sodium borohydride aqueous solution (NaBH$_4$) was added in the above stirring reaction mixture. The color of solution turned yellow-brown from transparent upon addition of a reducing agent indicating the reduction of silver ions and the formation of Ceph-AgNPs [28,29]. For Vildagliptin-conjugated silver nanoparticles (Vgt-AgNPs), a similar procedure was repeated by optimizing different volume ratio (*v*/*v*) of silver solution and drugs. Stable Vgt-AgNPs were obtained at respective *v*/*v* of silver to drug at 1:1. The amount of drug loaded on the nanoparticles was also measured. Nanoparticles were centrifuged at 12,000× *g* for 1 h, supernatant was collected, freeze-dried, and the unloaded drugs was determined by weighing. The results are expressed as the percentage of the drug amount contained in 100 mg of the dried nanoparticle. The percentage of drug loading on nanoparticles was found to be 52% and 68% for Ceph-AgNPs and Vgt-AgNPs, respectively.

2.3. Characterization of AgNPs-Coated Drugs

After successful synthesis of nanoconjugates, Ceph-AgNPs and Vgt-AgNPs were subjected to complete analysis via ultraviolet-visible spectrophotometry (UV-vis), Fourier transformation infrared (FT-IR), atomic force microscopy (AFM), and dynamic light scattering (DLS) as described previously [28,29].

2.4. Bactericidal Assay

Antibacterial potential of AgNPs, Ceph and Ceph-AgNPs was determined by using bactericidal assay [30]. Briefly, bacterial cultures were adjusted to optical density (OD) of 0.22 at 595 nm using a spectrophotometer (OD_{595} = 0.22) which corresponds to 10^8 colony-forming units per mL (C.F.U. mL^{-1}). An inoculum of 10 µL of above bacteria culture (equivalent to approximately 10^6 C.F.U.) was incubated with various concentrations of either Ceph-AgNPs, and Vgt-AgNPs in 1.5 mL centrifuge tubes at 37 °C for 2 h. For negative controls untreated bacterial culture were incubated with phosphate buffer saline (PBS). Vildagliptin and Cephradine alone were used as additional controls, while bacteria incubated with 100 µg mL^{-1} of gentamicin were used as positive control. Next, bacteria were serially diluted and 10 µL of each dilution was plated on nutrient agar plates. These plates were incubated at 37 °C overnight, followed by viable bacterial C.F.U. count.

2.5. Cytopathogenicity Assay

Cytopathogenicity assays were performed as described previously [31]. Briefly, *E. coli* K1 were incubated with various concentrations of Cephradine, Vildagliptin, and their nanoconjugates for 2 h at 37 °C. Next, all test samples were incubated with confluent HeLa monolayers in supplemented medium. Plates were incubated at 37 °C for 24 h in a 5% CO_2 incubator and observed for cytotoxic effects. At the end of this incubation period, the supernatants were collected and cytopathogenicity was detected by measuring lactate dehydrogenase (LDH) release (Cytotoxicity Detection kit) as follows: % cytotoxicity = (sample value − control value)/total LDH release − control value) × 100. Control values were obtained from host cells incubated in RPMI-1640 medium alone. Total LDH release was determined from HeLa cells treated with 1% Triton X-100 for 30 min at 37 °C. The basis of this assay is that cell supernatant containing LDH catalyzes the conversion of lactate to pyruvate, generating reduced form of nicotinamide adenine dinucleotide (NADH) and H^+. In the second step, the catalyst (diaphorase, solution from kit) transfers H and H^+ from NADH and H^+ to the tetrazolium salt p-iodo-nitrotetrazolium violet (INT), which is reduced to formazan (dye), and absorbance is read at 490 nm.

3. Results

3.1. Characterization of Cephradine and Vildagliptin Coated Silver Nanoparticles

The UV-vis spectra of both silver-drug nanoconjugates showed characteristic surface plasmon resonance band in the range of 400–450 nm. AFM images were recorded to ascertain the morphology of these nanoparticles. The representative topographical images of Ceph-AgNPs and Vgt-AgNPs revealed the formation of spherical nanoparticles (Figure 1a,b). The size distribution of drugs nanoconjugates was determined by DLS (Figure 1c,d). Vgt-AgNPs was found to be a mixture of small and large particle size ranging from 2, 7, and 90 nm with an average size of 33 nm, as compared to Ceph-AgNPs (average size 85 nm). FT-IR analysis demonstrated the involvement of hydroxyl groups in both drugs in the stabilization of silver nanoparticles (Figure 2).

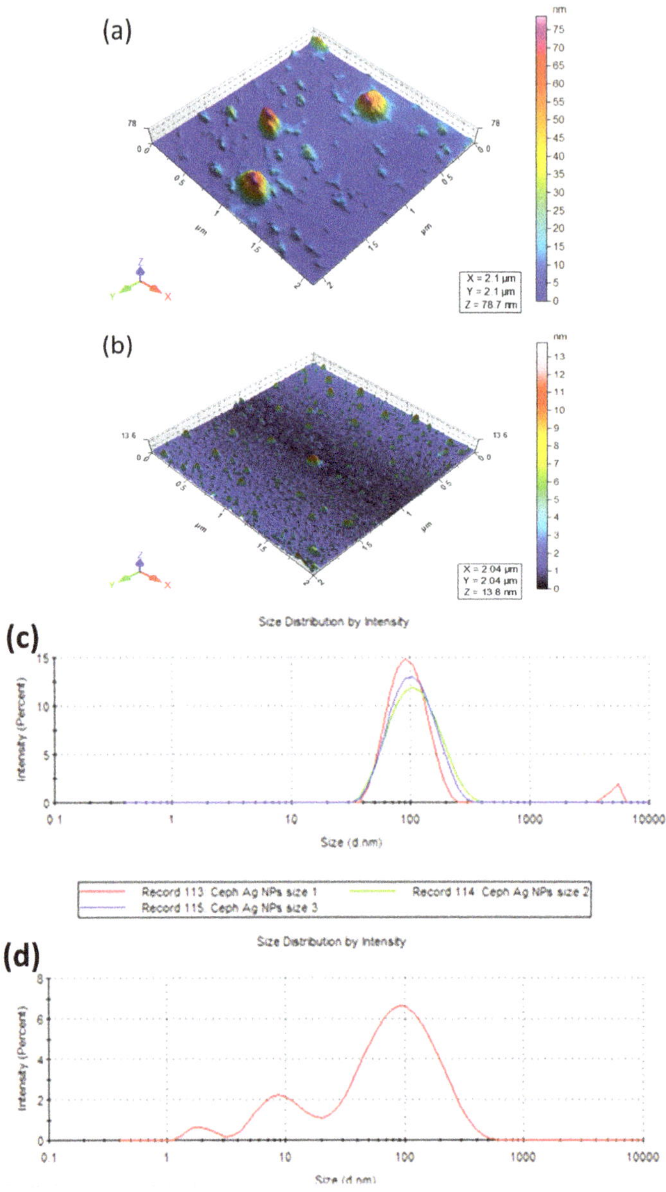

Figure 1. Representative images of atomic force microscopy (AFM) of Ceph-AgNPs (**a**); Vgt-AgNPs (**b**); AFM topographs were recorded on Agilent 5500 instrument used in tapping mode with silicon nitride cantilever. Ceph-AgNPs were found to be 85 nm (**c**); whereas, Vgt-AgNPs with the size distribution of 33 nm as measured by DLS (**d**). Both nanoparticles were spherical and polydisperesed.

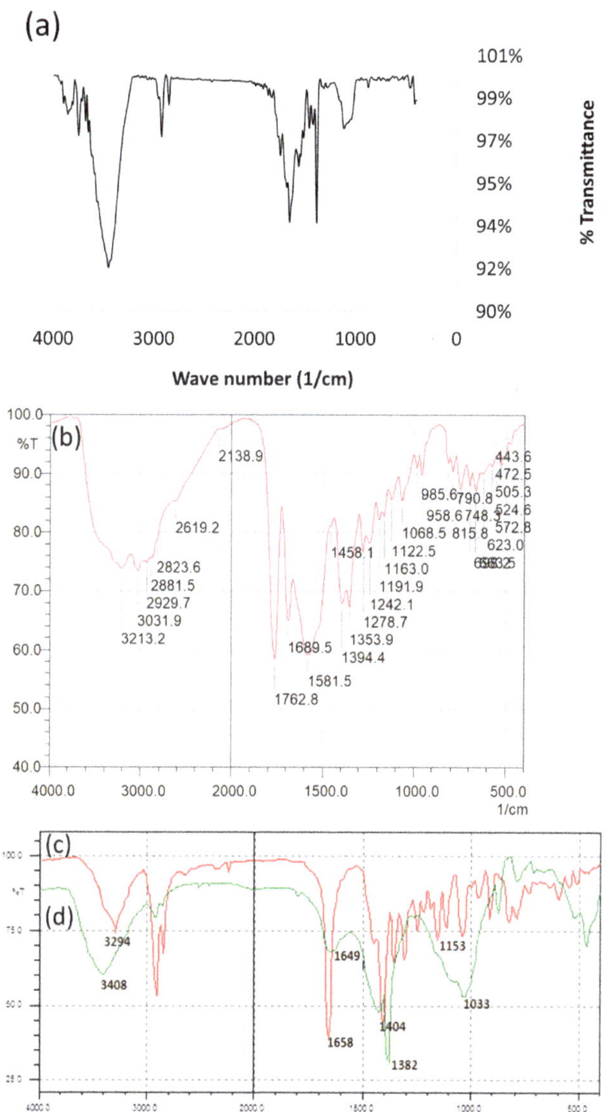

Figure 2. Representative spectra of FT-IR of Ceph-AgNPs (**a**); Vgt-AgNPs (**b**); shows the presence of hydroxyl, lactam, carboxylic acid and amino groups. However, while compared with drugs alone (**c**,**d**), hydroxyl groups peaks showed a shift in wave number which indicates the interaction with silver nanoparticles. Spectra were recorded at Bruker Vector 22 instrument using Potassium bromide (KBr) disc method.

3.2. Cephradine and Vildagliptine Conjugated with AgNPs Exhibited Increased Bactericidal Effects against E. coli K1 and MRSA Compared with AgNPs and Drugs Alone

Bactericidal assay was performed to determine the effects of drugs conjugated with AgNPs and drugs alone on *E. coli* K1 and MRSA. The concentrations of samples were adjusted to achieve MIC50 (minimum inhibitory concentration to kill 50% bacteria), which can be observed in MRSA and *S. pyogense*, and in the rest of bacteria MIC90 (minimum inhibitory concentration to kill 90% bacteria) is observed at tested concentrations. The results revealed that both drugs conjugated with AgNPs exhibited significant bactericidal effects ($p < 0.05$ using *t*-test, two-tailed distribution). In addition, treatment with bare AgNPs alone had limited effects on *E. coli* K1 and MRSA (Figure 3). Notably, Ceph-AgNPs showed significant bactericidal effects at 5 and 10 µM compared with drugs alone ($p < 0.05$). Vildagliptin alone did not show significant bactericidal effects at 5 or 10 µM, while, at 5 and 10 µM, Vgt-AgNPs, bacteria were significantly reduced (from 7.3×10^7 to 5.4×10^5) and (from 8.6×10^7 to 4.8×10^5) respectively for Gram-negative *E. coli* K1, and (from 1.0×10^7 to 4.3×10^6) (from 9.3×10^6 to 2.3×10^6) at 5 and 10 µM for Gram-positive MRSA (Figure 3).

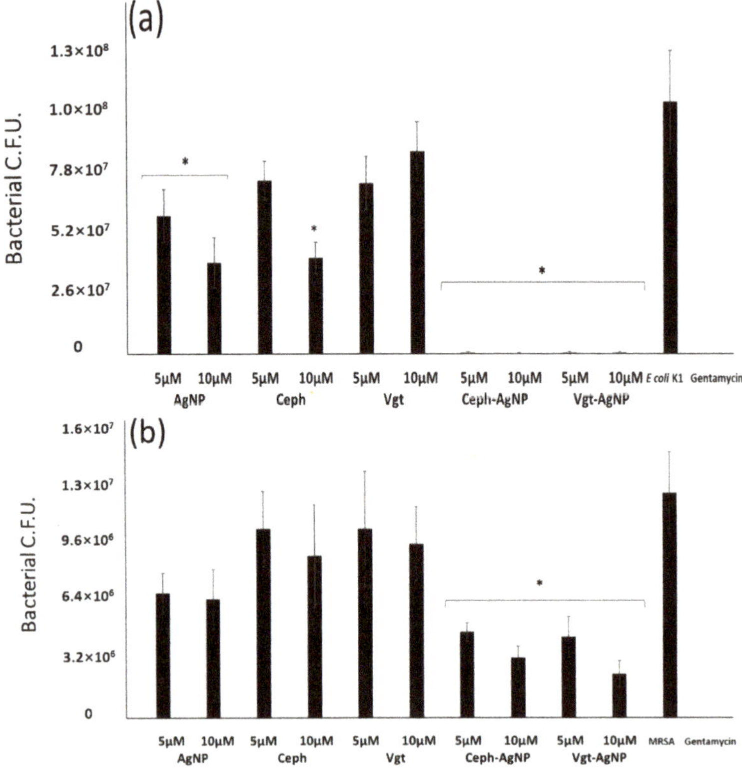

Figure 3. *E. coli* K1 (**a**); MRSA (**b**); colonies were determined following incubation with drugs alone, Ag alone, AgNPs conjugated with drugs. Briefly, 1×10^6 *E. coli* K1, MRSA colonies were incubated with drugs and controls at 37 °C for 24 h. Next, colonies were accounted. Both drug-conjugated AgNPs exhibited significant bactericidal effects ($p < 0.05$ using *t*-test, two-tailed distribution, as indicated by asterisk). The results are the mean ± standard error of three independent experiments performed in duplicate.

3.3. Cephradine and Vildagliptine Conjugated with AgNPs Exhibited Increased Bactericidal Effects against *P. aeruginosa, K. pneumoniae, B. cereus, S. pyogenes Compared with the Drugs Alone*

The results revealed that drugs conjugated with AgNPs exhibited significant bactericidal effects at 1- and 2-µM concentration against *P. aeruginosa, K. pneumoniae, B. cereus, S. pyogenes* ($p < 0.05$ using T test and two-tailed distribution). However, at 1 µM and 2 µM concentration, the treatment with AgNPs alone had similar effects on the number of bacteria. Notably, Ceph-AgNPs and Vgt-AgNPs showed significant effects at 1 µM and 2 µM compared with the drugs alone (Figure 4).

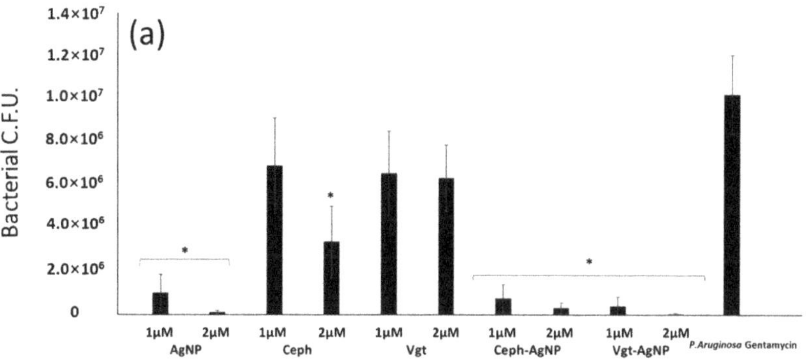

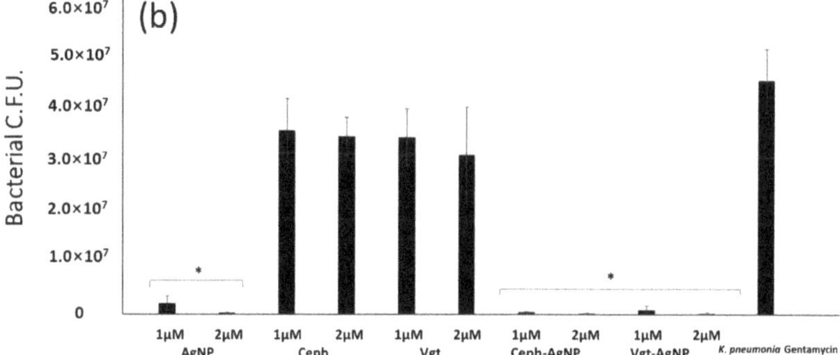

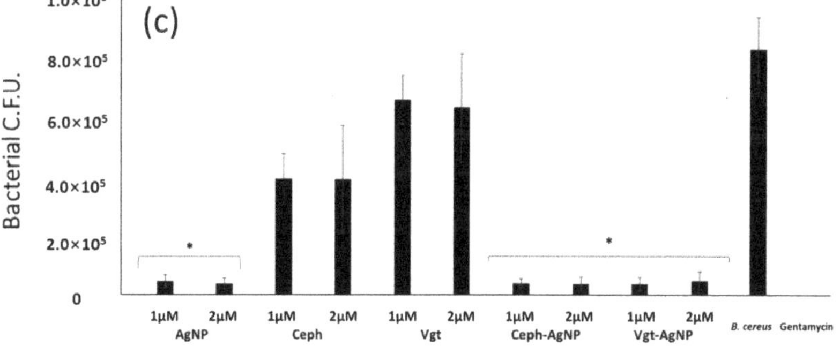

Figure 4. *Cont.*

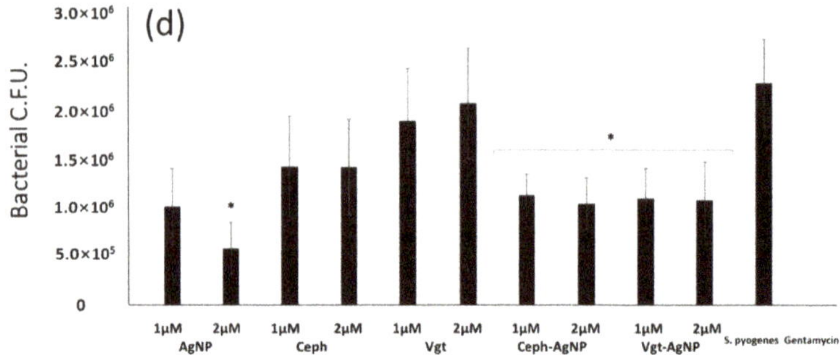

Figure 4. *P. aeruginosa* (**a**); *K. pneumoniae* (**b**); *B. cereus* (**c**); *S. pyogenes* (**d**). Bacterial colonies were determined following incubation with Cephradine and Vildagliptine. Briefly, 1×10^6 bacteria were incubated with drugs and controls at 37 °C for 24 h. Both drugs—conjugated AgNPs and Ag-NPS—alone exhibited significant bactericidal effects ($p < 0.05$ using *t*-test, two-tailed distribution, as indicated by asterisk), while as there was no significant effects of drugs alone. The results are the mean ± standard error of three independent experiments performed in duplicate.

3.4. Silver Nanoparticle-Conjugated Drugs Inhibited E. coli K1-Mediated Host Cell Cytotoxicity

To determine whether AgNP-conjugated drugs inhibited bacteria-mediated host cell cytotoxicity, assays were performed by incubating 10^6 *E. coli* K1 with HeLa cells for 24 h. *E. coli* K1 alone produced 60% host cell death. On the other hand, bacteria pretreated with AgNPs as well as Ceph-AgNPs and Vgt-AgNPs caused minimal host cell damage and host cell cytotoxicity was reduced to less than 15% (Figure 5). These findings also showed that Ceph-AgNPs and Vgt-AgNPs alone had minimal host cell cytotoxicity.

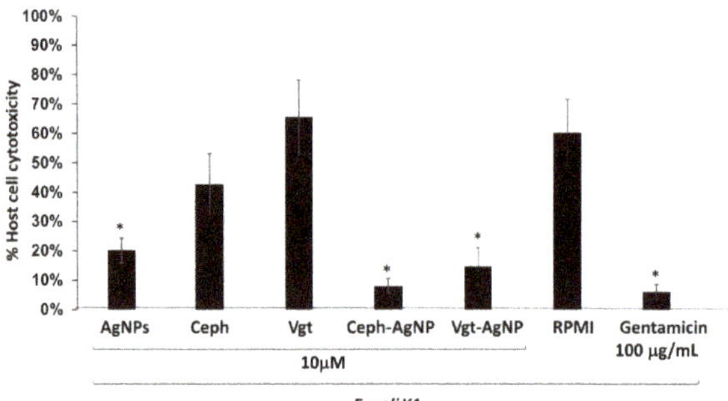

Figure 5. Cytopathogenicity assays were performed using *E. coli* K1 as described in materials and methods. Briefly, bacteria were incubated with test samples for 2 h before putting on HeLa cells monolayer with and without different drugs (10 µM), Cephradine (Ceph) and Vildagliptine (Vgt). Cells were then incubated for 24 h on standard conditions. Untreated bacteria, and bacteria treated with gentamicin were also treated with cells to evaluate their effects on cells cytotoxicity. Triton-X was used to lyse cells before determination of lactate dehydrogenase (LDH) release as marker of cell damage. LDH was determined by using Roche applied sciences LDH kit supplemented with enzyme and buffer. The % cytotoxicity was calculated by formula: % cytotoxicity = (sample value − control value)/total LDH release − control value) × 100. The results are representative of at least three independent experiments performed in duplicates. Asterisk indicates $p < 0.05$ using *t*-test, two-tailed distribution.

4. Discussion

The lack of development and approval of new and effective antibacterials as well as growing MDR microbes presents a major challenge in our ability to counter bacterial infections [31]. Nanotechnology offers great promise in the field of biomedicines, especially diagnosis and drug delivery. It offers opportunities for therapeutic agent delivery to specific cells and receptors. Nanomaterial-based drug delivery systems have the potential to improve pharmacokinetics and pharmacodynamics of the drugs [32]. The small size of nanoparticles provides them a greater surface area for maximum drug loading as well as high accessibility for specific targets. Recently, various drug-conjugated nanoparticles are being developed against infections caused by resistant microbes [33]. The most common metal carriers for nanoparticle-based drug delivery systems include gold, silver, and iron oxide due to their inertness and biocompatibility [33].

Though the mode of action of silver nanoparticles on the bacteria has been suggested to affect morphological and structural changes in the bacterial cells, the large surface area, provides better uptake by microorganisms [34]. Hence, silver nanoparticles have the ability to anchor to the bacterial cell wall and subsequently penetrate it, thereby causing structural changes leading to increased permeability of the cell membrane and cell death. In addition, the formation of free radicals by the silver nanoparticles have the ability to damage the cell membrane and make it porous resulting in bacterial cell death [35]. The bacterial membrane contains sulfur-containing proteins and the AgNPs interact with these proteins in the cell as well as with the phosphorus containing compounds. When AgNPs enter the bacterial cell, it forms a low molecular weight region in the center of the bacteria to which the bacteria conglomerates thus protecting the DNA from silver ions. Also, it generates reactive oxygen species, which are produced to attack the respiratory chain, cell division, and finally leading to cell death [36].

Silver conjugated Cephradine (mode of action involves binding and inactivation penicillin binding proteins leading to inhibition of peptidoglycan layer and causing cell lysis) and Vildagliptin (DPP4 inhibitor) were synthesized by reducing silver nitrate with sodium borohydride in the presence of drugs. These nanodrug conjugates were characterized by UV-visible spectrophotometry, FT-IR spectroscopy, and AFM. Cephradine and Vildagliptin successfully stabilized AgNPs and displayed surface plasmon resonance band in the range of 400–450 nm. FT-IR analysis showed the interaction of hydroxyl groups of drugs with silver nanoparticles for stabilization. Hence, the mode of stabilization is anticipated to be noncovalent interactions. Ceph-AgNPs and Vgt-AgNPs both were found to be spherical in shape and lie in a range of size distribution from 30–80 nm. Ceph-AgNPs were larger as compared to Vgt-AgNPs.

After characterization, these nanoparticles were subjected to antibacterial assays. Ceph-AgNPs and Vgt-AgNPs showed significant bactericidal effects against Gram-negative *E. coli* K1, *P. aeruginosa*, *K. pneumonia* and Gram-positive MRSA, *B. cereus*, and *S. pyogenes*. Nevertheless, these findings show tremendous potential in the development of new antibacterial formulations. As discussed previously, Cephradine is a cephalosporin first-generation antibiotic, which is not very effective against a number of bacteria because the different resistance mechanisms varied from the variation of the penicillin-binding protein, production of β-lactamase, existence of β-lactamase genes on plasmids or on bacterial chromosomes, and efflux pump mechanisms. For instance, MRSA produces abnormal penicillin binding protein 2A, which has low methicillin affinity (mediated through the *mecA* gene), which is carried on the staphylococcal cassette chromosome mec (SCCmec) by horizontal gene transfer and this results in resistance to Cephradine as an antibiotic [37]. Ceph-AgNPs provide a change in susceptibility of the drug along with, presumably, enhanced bioavailability. On the other hand, Vildagliptin (dipeptidyl peptidase-4 inhibitor) and its silver nanoparticles showed some promising effects as an antibacterial. Furthermore, these nanoparticles significantly and selectively reduced the pathogen-mediated host cell cytotoxicity caused by Gram-negative bacteria.

Drug-loaded nanoparticles enter in the bacteria by endocytosis showing either specific or non-specific type of interactions with cell membrane [38]. The positive charge of AgNPs interacts with lipopolysaccharides of Gram-negative bacteria with more affinity than the cellular wall of Gram-positive bacteria, that possess few sites for interactions, then releases the drugs intracellularly [38]. In previous reports, cephalosporin conjugated with AgNPs also showed enhancement in the antibacterial potency of Ceftriaxone and Cefixime against *E. coli* and *S. pyogenes*, respectively [39,40]. In conclusion, Cephradine and Vildagliptin showed bactericidal effects against the six tested bacteria in this study, but their conjugation with AgNPs enhanced their antibacterial efficacy. Moreover, Ceph-AgNPs and Vgt-AgNPs also significantly reduced the host cells cytotoxicity. The exact mechanism of action of these nanoparticles is not precisely understood and it is the subject of future studies along with testing their potential in vivo.

Author Contributions: Conceptualization, N.A.K., R.B.S., M.R.S.; Methodology, A.M., A.A. and D.A.; Software, A.M., D.A.; Validation, A.A., R.B.S. and N.A.K.; Formal Analysis, A.M. and A.A.; Investigation, A.M.; Resources, N.A.K.; Data Curation A.M., D.A.; Writing—Original Draft Preparation, A.M. and A.A.; Writing—Review & Editing, N.A.K. and R.B.S.; Visualization, M.R.S.; Supervision, R.B.S.; Project Administration, R.B.S.; Funding Acquisition, R.B.S.

Funding: This work is supported by University Research Award INT-2017-03 by Sunway University, Malaysia.

Acknowledgments: We are thankful to Sunway University for support.

Conflicts of Interest: The authors declare that they have no competing interests.

References

1. Garchitorena, A.; Sokolow, S.H.; Roche, B.; Ngonghala, C.N.; Jocque, M.; Lund, A.; Barry, M.; Mordecai, E.A.; Daily, G.C.; Jones, J.H.; et al. Disease ecology, health and the environment: A framework to account for ecological and socio-economic drivers in the control of neglected tropical diseases. *Philos. Trans. R. Soc. B* **2017**, *372*, 20160128. [CrossRef] [PubMed]

2. Wang, Y.; Yu, L.; Kong, X.; Sun, L. Application of nanodiagnostics in point-of-care tests for infectious diseases. *Int. J. Nanomed.* **2017**, *12*, 4789–4803. [CrossRef] [PubMed]

3. Anuj, S.A.; Gajera, H.P.; Hirpara, D.G.; Golakiya, B.A. Bactericidal assessment of nano-silver on emerging and re-emerging human pathogens. *J. Trace Elem. Med. Biol.* **2018**. [CrossRef] [PubMed]

4. Alanis, A.J. Resistance to antibiotics: Are we in the post-antibiotic era? *Arch. Med. Res.* **2005**, *36*, 697–705. [CrossRef] [PubMed]

5. Demain, A.L.; Sanchez, S. Microbial drug discovery: 80 years of progress. *J. Antibiot.* **2009**, *62*, 5–16. [CrossRef] [PubMed]

6. Nikaido, H. Multidrug resistance in bacteria. *Annu. Rev. Biochem.* **2009**, *78*, 119–146. [CrossRef] [PubMed]

7. Lee, K.; Silva, E.A.; Mooney, D.J. Growth factor delivery-based tissue engineering: General approaches and a review of recent developments. *J. R. Soc. Interface* **2011**, *8*, 153–170. [CrossRef] [PubMed]

8. Pantosti, A.; Venditti, M. What is, M.R.SA? *Eur. Respir. J.* **2009**, *34*, 1190–1196. [CrossRef] [PubMed]

9. Granum, P.E.; Lund, T. Bacillus cereus and its food poisoning toxins. *FEMS Microbiol. Lett.* **1997**, *157*, 223–228. [CrossRef] [PubMed]

10. Podschun, R.; Ullmann, U. Klebsiella spp. as nosocomial pathogens: Epidemiology, taxonomy, typing methods, and pathogenicity factors. *Clin. Microbiol. Rev.* **1998**, *11*, 589–603. [CrossRef] [PubMed]

11. Naqvi, S.Z.; Kiran, U.; Ali, M.I.; Jamal, A.; Hameed, A.; Ahmed, S.; Ali, N. Combined efficacy of biologically synthesized silver nanoparticles and different antibiotics against multidrug-resistant bacteria. *Int. J. Nanomed.* **2013**, *8*, 3187–3195. [CrossRef] [PubMed]

12. Grace, J.L.; Elliott, A.G.; Huang, J.X.; Schneider, E.K.; Truong, N.P.; Cooper, M.A.; Li, J.; Davis, T.P.; Quinn, J.F.; Velkov, T.; et al. Cationic acrylate oligomers comprising amino acid mimic moieties demonstrate improved antibacterial killing efficiency. *J. Mater. Chem. B* **2017**, *5*, 531–536. [CrossRef] [PubMed]

13. Dai, T.; Wang, C.; Wang, Y.; Xu, W.; Hu, J.; Cheng, Y. A Nanocomposite Hydrogel with Potent and Broad-Spectrum Antibacterial Activity. *ACS Appl. Mater. Interfaces* **2018**. [CrossRef] [PubMed]

14. He, M.; Wang, Q.; Zhao, W.; Li, J.; Zhao, C. A self-defensive bilayer hydrogel coating with bacteria triggered switching from cell adhesion to antibacterial adhesion. *Polym. Chem.* **2017**, *8*, 5344–5353. [CrossRef]

15. Yuan, H.; Yu, B.; Fan, L.H.; Wang, M.; Zhu, Y.; Ding, X.; Xu, F.J. Multiple types of hydroxyl-rich cationic derivatives of, P.G.MA for broad-spectrum antibacterial and antifouling coatings. *Polym. Chem.* **2016**, *7*, 5709–5718. [CrossRef]

16. Grace, J.L.; Huang, J.X.; Cheah, S.E.; Truong, N.P.; Cooper, M.A.; Li, J.; Davis, T.P.; Quinn, J.F.; Velkov, T.; Whittaker, M.R. Antibacterial low molecular weight cationic polymers: Dissecting the contribution of hydrophobicity, chain length and charge to activity. *RSC Adv.* **2016**, *6*, 15469–15477. [CrossRef] [PubMed]

17. Gou, Y.; Yang, X.; He, L.; Xu, X.; Liu, Y.; Liu, Y.; Gao, Y.; Huang, Q.; Liang, K.; Ding, C.; et al. Bio-inspired peptide decorated dendrimers for a robust antibacterial coating on hydroxyapatite. *Polym. Chem.* **2017**, *8*, 4264–4279. [CrossRef]

18. Rabanal, F.; Grau-Campistany, A.; Vila-Farrés, X.; Gonzalez-Linares, J.; Borràs, M.; Vila, J.; Manresa, A.; Cajal, Y. A bioinspired peptide scaffold with high antibiotic activity and low in vivo toxicity. *Sci. Rep.* **2015**, *5*, 10558. [CrossRef] [PubMed]

19. Bakhshi, H.; Agarwal, S. Dendrons as active clicking tool for generating non-leaching antibacterial materials. *Polym. Chem.* **2016**, *7*, 5322–5330. [CrossRef]

20. Daniel, M.C.; Astruc, D. Gold nanoparticles: Assembly, supramolecular chemistry, quantum-size-related properties, and applications toward biology, catalysis, and nanotechnology. *Chem. Rev.* **2004**, *104*, 293–346. [CrossRef] [PubMed]

21. Morones, J.R.; Elechiguerra, J.L.; Camacho, A.; Holt, K.; Kouri, J.B.; Ramírez, J.T.; Yacaman, M.J. The bactericidal effect of silver nanoparticles. *Nanotechnology* **2005**, *16*, 2346–2353. [CrossRef] [PubMed]

22. Shahverdi, A.R.; Fakhimi, A.; Shahverdi, H.R.; Minaian, S. Synthesis and effect of silver nanoparticles on the antibacterial activity of different antibiotics against *Staphylococcus aureus* and *Escherichia coli*. *Nanomed. Nanotechnol. Biol. Med.* **2007**, *3*, 168–171. [CrossRef] [PubMed]

23. Fernandes, R.; Amador, P.; Prudêncio, C. β-Lactams: Chemical structure, mode of action and mechanisms of resistance. *Rev. Med. Microbiol.* **2013**, *24*, 7–17. [CrossRef]

24. Barza, M.; Miao, P.V. Antimicrobial spectrum, pharmacology and therapeutic use of antibiotics. Part 3: Cephalosporins. *Am. J. Health-Syst. Pharm.* **1977**, *34*, 621–629.

25. Ahrén, B.; Schweizer, A.; Dejager, S.; Villhauer, E.B.; Dunning, B.E. Mechanisms of action of the dipeptidyl peptidase-4 inhibitor vildagliptin in humans. *Diabetes Obes. Metab.* **2011**, *13*, 775–783. [CrossRef] [PubMed]

26. Waghulde, M.; Naik, J. Comparative study of encapsulated vildagliptin microparticles produced by spray drying and solvent evaporation technique. *Dry. Technol.* **2017**, *35*, 1644–1654. [CrossRef]

27. Nauck, M. Incretin therapies: Highlighting common features and differences in the modes of action of glucagon-like peptide-1 receptor agonists and dipeptidyl peptidase-4 inhibitors. *Diabetes Obes. Metab.* **2016**, *18*, 203–216. [CrossRef] [PubMed]

28. Ahmed, D.; Shah, M.R.; Perveen, S.; Ahmed, S. Cephradine Coated Silver Nanoparticle their Drug Release Mechanism, and Antimicrobial Potential against Gram-Positive and Gram-Negative Bacterial Strains through AFM. *J. Chem. Soc. Pakistan* **2018**, *40*, 388–398.

29. Anwar, A.; Khalid, S.; Perveen, S.; Ahmed, S.; Siddiqui, R. Synthesis of 4-(dimethylamino) pyridine propylthioacetate coated gold nanoparticles and their antibacterial and photophysical activity. *J. Nanobiotechnol.* **2018**, *16*, 8. [CrossRef] [PubMed]

30. Khan, N.A.; Osman, K.; Goldsworthy, G.J. Lysates of Locusta migratoria brain exhibit potent broad-spectrum antibacterial activity. *J. Antimicrob. Chemother.* **2008**, *62*, 634–635. [CrossRef] [PubMed]

31. Ali, S.M.; Siddiqui, R.; Ong, S.K.; Shah, M.R.; Anwar, A.; Heard, P.J.; Khan, N.A. Identification and characterization of antibacterial compound (s) of cockroaches (*Periplaneta americana*). *Appl. Microbiol. Biotechnol.* **2017**, *101*, 253–286. [CrossRef] [PubMed]

32. Zazo, H.; Colino, C.I.; Lanao, J.M. Current applications of nanoparticles in infectious diseases. *J. Control. Release* **2016**, *224*, 86–102. [CrossRef] [PubMed]

33. Wilczewska, A.Z.; Niemirowicz, K.; Markiewicz, K.H.; Car, H. Nanoparticles as drug delivery systems. *Pharmacol. Rep.* **2012**, *64*, 1020–1037. [CrossRef]

34. Li, W.R.; Xie, X.B.; Shi, Q.S.; Zeng, H.Y.; Ou-Yang, Y.S.; Chen, Y.B. Antibacterial activity and mechanism of silver nanoparticles on *Escherichia coli*. *Appl. Microbiol. Biotechnol.* **2010**, *85*, 1115–1122. [CrossRef] [PubMed]

35. Prabhu, S.; Poulose, E.K. Silver nanoparticles: Mechanism of antimicrobial action, synthesis, medical applications, and toxicity effects. *Int. Nano Lett.* **2012**, *2*, 32. [CrossRef]

36. Rai, M.; Yadav, A.; Gade, A. Silver nanoparticles as a new generation of antimicrobials. *Biotechnol. Adv.* **2009**, *27*, 76–83. [CrossRef] [PubMed]

37. De Lencastre, H.; Oliveira, D.; Tomasz, A. Antibiotic resistant Staphylococcus aureus: A paradigm of adaptive power. *Curr. Opin. Microbiol.* **2007**, *10*, 428–435. [CrossRef] [PubMed]

38. Sondi, I.; Salopek-Sondi, B. Silver nanoparticles as antimicrobial agent: A case study on, *E. coli* as a model for Gram-negative bacteria. *J. Colloid Interface Sci.* **2004**, *275*, 177–182. [PubMed]
39. Shah, M.R.; Ali, S.; Ateeq, M.; Perveen, S.; Ahmed, S.; Bertino, M.F.; Ali, M. Morphological analysis of the antimicrobial action of silver and gold nanoparticles stabilized with ceftriaxone on *Escherichia coli* using atomic force microscopy. *New J. Chem.* **2014**, *38*, 5633–5640. [CrossRef]
40. Rasheed, W.; Shah, M.R.; Perveen, S.; Ahmed, S.; Uzzaman, S. Revelation of susceptibility differences due to Hg (II) accumulation in *Streptococcus pyogenes* against, C.X.-AgNPs and Cefixime by atomic force microscopy. *Ecotoxicol. Environ. Saf.* **2018**, *147*, 9–16. [CrossRef] [PubMed]

Article

In Vitro Synergism of Silver Nanoparticles with Antibiotics as an Alternative Treatment in Multiresistant Uropathogens

Montserrat Lopez-Carrizales [1], Karla Itzel Velasco [1], Claudia Castillo [2], Andrés Flores [3], Martín Magaña [3], Gabriel Alejandro Martinez-Castanon [4] and Fidel Martinez-Gutierrez [1,*]

[1] Laboratorio de Microbiología, Universidad Autónoma de San Luis Potosí, San Luis Potosí, CP 78210, Mexico; montsecarrizales@icloud.com (M.L.-C.); kivg56@hotmail.com (K.I.V.)
[2] Laboratorio de Células Neurales Troncales, CIACYT-Facultad de Medicina, Universidad Autónoma de San Luis Potosí, San Luis Potosí, CP 78210, Mexico; claudiacastillo@gmail.com
[3] Hospital Central Dr. Ignacio Morones Prieto, San Luis Potosí, CP 78290, Mexico; santosf2000@yahoo.com (A.F.); mmaganaa@hotmail.com (M.M.)
[4] Facultad de Estomatología, Universidad Autónoma de San Luis Potosí, San Luis Potosí, CP 78290, Mexico; mtzcastanon@fciencias.uaslp.mx
* Correspondence: fidel@uaslp.mx; Tel.: +52-(444)-826-2440 (ext. 6591)

Received: 16 March 2018; Accepted: 14 June 2018; Published: 19 June 2018

Abstract: The increase in the prevalence of bacterial resistance to antibiotics has become one of the main health problems worldwide, thus threatening the era of antibiotics most frequently used in the treatment of infections. The need to develop new therapeutic strategies against multidrug resistant microorganisms, such as the combination of selected antimicrobials, can be considered as a suitable alternative. The in vitro activities of two groups of conventional antimicrobial agents alone and in combination with silver nanoparticles (AgNPs) were investigated against a set of ten multidrug resistant clinical isolate and two references strains by MIC assays and checkerboard testing, as well as their cytotoxicity, which was evaluated on human fibroblasts by MTT assay at the same concentration of the antimicrobial agents alone and in combination. Interesting results were achieved when the AgNPs and their combinations were characterized by Dynamic Light Scattering (DLS), Zeta Potential, Transmission Electron Microscopy (TEM), UV–visible spectroscopy and Fourier Transforms Infrared (FTIR) spectroscopy. The in vitro activities of ampicillin, in combination with AgNPs, against the 12 microorganisms showed one Synergy, seven Partial Synergy and four Additive effects, while the results with amikacin and AgNPs showed three Synergy, eight Partial Synergy and one Additive effects. The cytotoxic effect at these concentrations presented a statistically significant decrease of their cytotoxicity ($p < 0.05$). These results indicate that infections caused by multidrug resistant microorganisms could be treated using a synergistic combination of antimicrobial drugs and AgNPs. Further studies are necessary to evaluate the specific mechanisms of action, which could help predict undesirable off-target interactions, suggest ways of regulating a drug's activity, and identify novel therapeutic agents in this health problem.

Keywords: antimicrobial activity; biofilm; urinary infection; silver nanoparticles; bacterial resistance

---

1. Introduction

Urinary tract infections (UTIs) are defined as inflammatory processes related to the invasion and multiplication of microorganisms that occur at any level of the urinary tract, including urethral (urethritis), bladder (cystitis), ureters (ureteritis) and kidney infections (pyelonephritis) [1].

Approximately between 150 and 250 million cases of UTIs occur each year worldwide [2]. In 2011, more than 8 million cases were reported in the U.S. [1], of which 93,300 were acquired in intensive care units [3]. Recently, the epidemiological bulletin of the Ministry of Health reported in 2016 a total of 4,023,432 cases of UTIs in Mexico, of which 76.68% were in women and only 23.31% in men.

Gram-negative intestinal bacteria are the most common etiology agents of UTIs, where uropathogenic *Escherichia coli* (UPEC) is the major microorganism isolate, which is a member of the Enterobacteriacea family [4]. Other commonly associated pathogens include *Klebsiella* sp. and *Proteus mirabilis*, both of which are characterized by their urease enzyme and Gram-positive bacteria such as *Staphylococcus saprophyticus* and *Enterococcus faecalis* [5].

Urinary Tract Infections Associated with Catheters (CAUTI) are one of the most frequent explanations of nosocomial infections [6]. Patients with urinary catheters show an increment of bacteriuria in relation to duration of catheterization [7], however, the most important factor is the biofilm formation along the catheter surface [7,8]. A biofilm is a resistance mechanism that consists in a self-organized community of microorganisms embedded in a matrix of extracellular polymeric substances synthesized by themselves [9]. Many bacterial species show growth in the form of biofilms, which gives them various advantages [10]. Some of the benefits are metabolic cooperation (nutrients) [11], horizontal gene transfer [12], protection against environmental stresses, lower susceptibility to antimicrobial agents [13,14] and prevention of host defense mechanisms (immune system) [15]. The most common organisms that contaminate the urinary catheter and develop biofilms are strains of *Escherichia coli, Pseudomonas aeruginosa, Enterococcus, Proteus mirabilis, Klebsiella pneumoniae* and coagulase-negative staphylococci [10,16].

The antibiotic resistance of bacteria is a global health problem that is continually expanding, and is recognized as a medical problem that increases morbidity and mortality rates, which implies length of hospital stays as well as cost and bad prognosis [17,18]. In fact, the speed at which bacteria are establishing resistance to current antibiotics is faster than the development of new molecules with antimicrobial features. Unfortunately, it is very difficult to identify new bacterial targets that can be used to develop new classes of antimicrobial agents that are safe and effective.

In this context, nanotechnology opens new possibilities, allowing new solutions with old resources. Nanoscale materials such as silver nanoparticles (AgNPs) have emerged as novel agents due to their unique physicochemical properties and remarkable antimicrobial activities that confer a great advantage for the development of alternative products against, for example, multi-drug resistant microorganisms [19,20]. Due to the above, it has been proposed to implement the use of AgNPs on different devices for medical use. One of the strategies is the modification of surfaces of the devices to inhibit the formation of bacterial biofilms [21].

Recently, several studies have indicated that AgNPs can enhance the effect of antibiotics against susceptible and resistant bacteria, [22] as a well as decrease bacterial adhesion in the early stages of biofilm formation. In 2016, Rajendran and et al., impregnated urinary catheters with antibiotics (amikacin and nitrofurantoin) and a synergistic combination of antibiotics and AgNPs (synthesized by a biological method) in order to evaluate antibiofilm activity. The authors reported that the synergistic combination showed a 90% inhibition of bacterial adhesion, whereas functionalization with antibiotics showed only 25% inhibition [23].

Some authors have reported that AgNPs have toxic effects on mammalian cells; for example, impairment of normal mitochondrial function, increased membrane permeability and generation of reactive oxygen species [24,25].

In this study, the synergistic activity of AgNPs was evaluated with conventional antibiotics against Gram-positive and Gram-negative multi-drug resistant isolates from clinical samples. Results presented here show that AgNPs, in combination with antibiotics, increase the antimicrobial effect in an additive or synergistic manner. Furthermore, MTT assays suggest that at low concentrations the AgNPs and their combinations do not present cytotoxic effects in eukaryotic cells.

2. Results and Discussion

2.1. Synthesis and Characterization of AgNPs

TEM micrograph revealed that the AgNPs were of spherical and pseudospherical shapes (Figure 1). Based on the particle size distribution histogram evaluated from the corresponding TEM micrograph (n = 100), the mean ($\pm$SD) size of AgNPs 8.57 $\pm$ 1.17 nm was calculated. The mean size for the combination of AgNPs with ampicillin (AgNPs + AMP) was 4.01 $\pm$ 0.80 nm and for the combination of AgNPs with amikacin (AgNPs + AMK) was 6.03 $\pm$ 0.87 nm (Table 1).

The particle size distributions of AgNPs + AMP and AgNPs + AMK were also evaluated in Dulbecco's Modified Eagle Medium (DMEM).

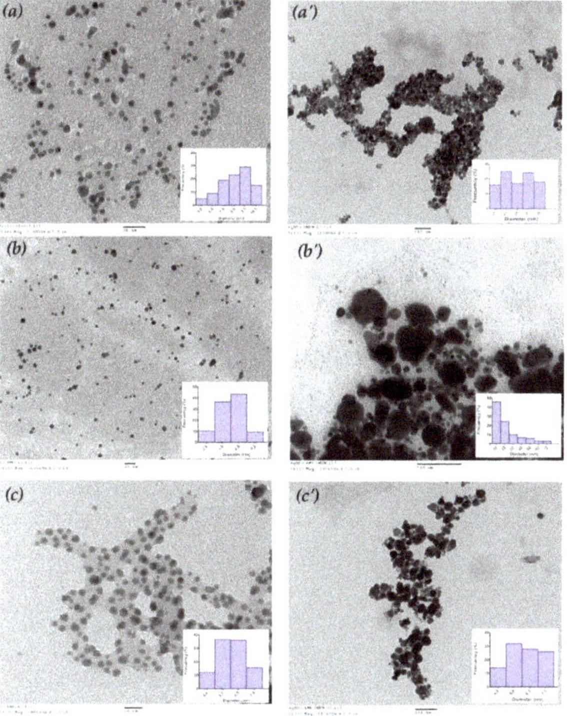

Figure 1. Morphological characterization of silver nanoparticles and their combinations with antibiotics. Transmission electron micrographs showing the formation of spherical and pseudospherical nanoparticles. (**a**) AgNPs; (**a′**) AgNPs in DMEM; (**b**) AgNPs + AMP; (**b′**) AgNPs + AMP in DMEM; (**c**) AgNPs + AMK; (**c′**) AgNPs + AMK in DMEM. Insets: Particle size distribution histogram. DMEM: Dulbecco's Modified Eagle Medium. AMP: Ampicillin. AMK: Amikacin.

AgNPs synthesized in aqueous solution and their combinations with antibiotics were characterized by DLS (Table 1). The results of the dialyzed AgNPs showed a narrow size distribution with a hydrodynamic diameter of ($\pm$SD) 8.23 nm $\pm$ 0.91 nm and a zeta potential value of -40.80 mV $\pm$ 9.54 mV. The combination of AgNPs + AMP also showed a narrow size distribution with a hydrodynamic diameter of 4.69 nm $\pm$ 0.51 nm. This decrease in size was attributed to the fact that ampicillin favors the homogeneous dispersion of the nanoparticles and gives it a greater stability when it obtains a zeta potential value of -51.00 mV $\pm$ 20.20 mV. The results of combination of AgNPs + AMK showed a narrow distribution of sizes with a hydrodynamic diameter of 947.90 nm $\pm$ 65.30 nm and a value of -21.10 mV $\pm$ 4.63 mV for zeta potential. The increase of size of the nanoparticles are attributed to the addition of amikacin, which favors agglomeration of the AgNPs and causes an increase in the zeta potential, which translates in less stable nanoparticles.

The hydrodynamic diameters and zeta potentials of AgNPs and their combinations were also evaluated in DMEM.

Table 1. Characterization of silver nanoparticles and their combinations with antibiotics by DLS and TEM.

	DLS				TEM	
	Hydrodynamic Diameter (nm)	Hydrodynamic Diameter in DMEM (nm)	Zeta Potential (mV)	Zeta Potential in DMEM (mV)	Diameter (nm)	Diameter in DMEM (nm)
AgNPs	8.23 ± 0.91	39.25 ± 6.42	−40.80 ± 9.54	−16.20 ± 0.0	8.57 ± 1.17	25.08 ± 2.74
AgNPs + AMP	4.69 ± 0.51	26.25 ± 5.55	−51.00 ± 20.20	−15.60 ± 0.0	4.01 ± 0.80	24.17 ± 16.29
AgNPs + AMK	947.90 ± 65.30	222.50 ± 47.10	−21.10 ± 4.63	−18.10 ± 7.56	6.03 ± 0.87	6.14 ± 0.99

Data are expressed as mean and standard deviation. DLS: Dynamic Light Scattering. TEM: Transmission Electron Microscopy. DMEM: Dulbecco's Modified Eagle Medium. AMP: Ampicillin. AMK: Amikacin.

The UV–visible spectrum revealed a peak at a wavelength of 412 nm for AgNPs whereas in the combination AgNPs + AMP, the peak was observed at 410 nm and in the combination AgNPs + AMK, the peak appeared at a wavelength of 450 nm (Figure 2). These peaks correspond to the excitation of the surface plasmon of AgNPs. The plasmonic resonance depends on several parameters, like the nature, size and geometry of the nanoparticles and the physical properties of the medium in which the nanoparticles are dispersed. In the case of AgNPs, the plasmon peak appears at a wavelength around 400 nm [26].

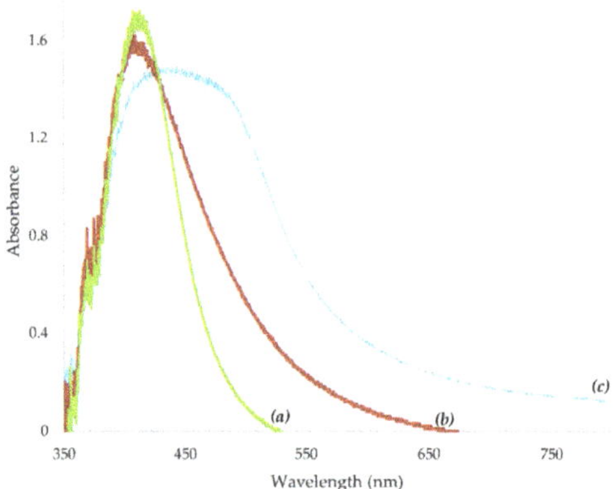

Figure 2. UV–visible absorption spectra of silver nanoparticles and their combinations with antibiotics. UV–visible spectrum showed the maximum absorbance at (**a**) 412 nm for AgNPs, (**b**) 410 nm for AgNPs + AMP and (**c**) 450 nm for AgNPs + AMK. AMP: Ampicillin. AMK: Amikacin.

In this study, gallic acid was used as a reducing and stabilizing agent, a molecule with a carboxylic group and rich in hydroxyl functional groups. Yoosaf et al. proposed that AgNPs are stabilized by gallic acid through electrostatic interactions through their oxidized carboxylic group and the afore-cited hydroxyl groups were capable of forming hydrogen bonds [27].

To study the possible interactions between antibiotics and the surface of the AgNPs, FTIR was performed (Figure 3). FTIR is useful for determining the chemical composition of antibiotics involved in the coating of AgNPs. The observed intense bands were compared with standard values to identify the functional groups. The FTIR spectra of the antibiotics (Figure 3a,b) changed greatly upon the combination with AgNPs (AgNPs + AMP and AgNPs + AMK), as displayed in Figure 3a',b'. Ampicillin is a molecule that has carbonyl and amine functional groups, while amikacin is a molecule rich in hydroxyl and amine groups. In the case of ampicillin, the band at 1759 cm^{-1} disappeared completely (Figure 3a′), which suggests that the antibiotic interacts with the AgNPs through its carbonyl group (C=O). Furthermore, the peak of the primary amine at 3380 cm^{-1} (Figure 3a) shifted to 3130 cm^{-1} (Figure 3a′) after the combination with nanoparticles, indicating that the amine functional group was involved in the interaction with the surface of the AgNPs. Therefore, the spectrum of the amikacin showed that the bands at 1626 cm^{-1} corresponding to a carbonyl group and 3400 cm^{-1} for a primary amine disappeared completely (Figure 3b′) after being combined with the nanoparticles. Those results of the FTIR suggest that the functional groups of the antibiotics could be involved in the interaction by hydrogen bonds with gallic acid [28,29]. The results showed in Figure 3 are in concordance with the previous results of the Hua Deng et al., 2016, who carried out UV–Vis and Raman spectroscopy studies reveal that amikacin can form complexes with AgNPs, while ampicillin do not [30]. The authors reported that no Raman enhancement is observed when AgNPs are combined with ampicillin at any test concentrations. This implies that the antibiotics do not strongly interact with AgNPs to replace the stabilizer molecules on the surface of AgNPs. Moreover, they infer that the combinations of antibiotics and AgNPs have different ways to develop antimicrobial activities.

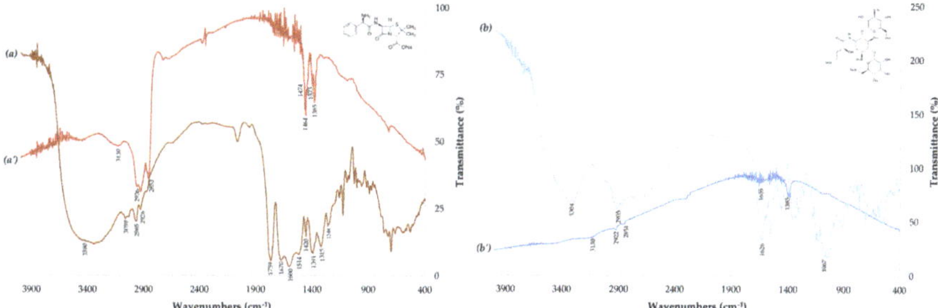

Figure 3. FTIR spectra of the antibiotics and their combinations with silver nanoparticles. (**a**) AMP: Ampicillin; (**a′**) AgNPs + AMP; (**b**) AMK: Amikacin; (**b′**) AgNPs + AMK. Insets: AMP and AMK structure.

2.2. Samples and Bacterial Strains

The multidrug resistance clinical strains used for this experiment were isolated from the urine of patients with CAUTI; microbiological analysis showed that the clinical pathogenic strains isolated were in accordance with the main etiologic agents causing CAUTI; these results are in accordance with previously results reported when the UTI were evaluated on hospitalized patients in Kolkata, an eastern region of India, as well as in studies where complicate and non-complicate UTI were studied [4,31].

2.3. Antimicrobial Test

A set of ten clinical pathogenic strains resistant to antibiotics associated with CAUTI were evaluated, of which two corresponded to Gram-positive strains and the rest to Gram-negative strains. The results showed that all clinical isolates (*E. faecium*, *S. aureus*, *A. baumannii*, *E. cloacae*, three different isolates of *E. coli*, *K. pneumoniae*, *M. morgannii* and *P. aeruginosa*) showed a MIC to AgNPs between 4 and 16 μg/mL. The bacterial strains showed MIC values of 4–128 μg/mL for amikacin and all Gram-negative strains were resistant to ampicillin.

2.4. Checkerboard Synergy

Table 2 shows the MIC archived with the ten multidrug resistance clinical strains grown in Mueller Hinton Broth with amikacin and ampicillin, both in the absence of AgNPs and when present; when the bacteria were incubated with the combination of AgNPs and antibiotic (AgNPs + AMK and AgNPs + AMP), the ampicillin and amikacin MICs decreased drastically for all strains. The combination of AgNPs + AMK reduce the MIC by 2 to 32-fold. By contrast, with the combination of AgNPs + AMP reduce the MIC just with *S. aureus* and *E. cloacae* by 1- and 4-fold respectively; with the other microorganisms, the MICs were reduced by 16- and 32-fold. These results show a better antimicrobial activity of the combination of AgNPs + AMP, which could be explained by the interaction between the AgNPs and ampicillin and, therefore, arrangement of the molecules in a new compound, which could work in both ways, like independent chemical entities or a new compound; more experiments are needed to explain the role and proportion of each one of them.

Table 2. Efficacy of silver nanoparticles, antibiotics and their combinations against clinical strains.

Clinical Strains	MIC (µg/mL)			Fold Change	MIC (µg/mL)		Fold Change
	AgNPs	AMK	AgNPs + AMK *		AMP	AgNPs + AMP *	
Gram-positive							
Enterococcus faecium	8	128	4	32	128	8	16
Staphylococcus aureus	8	4	2	2	4	4	1
Gram-negative							
Acinetobacter baumannii	16	128	4	32	128	8	16
Enterobacter cloacae	16	16	8	2	128	32	4
Escherichia coli (501) **	8	64	4	16	128	8	16
Escherichia coli (508) **	8	4	1	4	128	8	16
Escherichia coli (515) **	8	32	8	4	128	8	16
Klebsiella pneumoniae	4	4	2	2	128	4	32
Morganella morganii	8	8	4	2	128	8	16
Pseudomonas aeruginosa	4	32	4	8	128	4	32

* Minimum Inhibitory Concentration (MIC) represents the concentration of antibiotic (amikacin or ampicillin) present in the combination. AMK: Amikacin. AMP: Ampicillin. ** The numbers in parentheses indicate that *E. coli* corresponds to a different clinical sample.

The synergistic effects of AgNPs and conventional antibiotics were evaluated by determination of the Fractional Inhibitory Concentration (FIC) index (Figure 4). Synergistic interactions of AgNPs and amikacin were observed against *Acinetobacter baumannii*, *Escherichia coli* (508) and *E. coli* (ATCC 25922). Synergistic interactions of AgNPs and ampicillin were found only against *Acinetobacter baumannii*. Other combinational activities of AgNPs and antibiotics were considered as partially synergistic interactions. These synergistic activities of AgNPs in combination with conventional antibiotics suggest that it may be possible to reduce the viability of bacterial strains at lower antibiotic concentrations (Table 3).

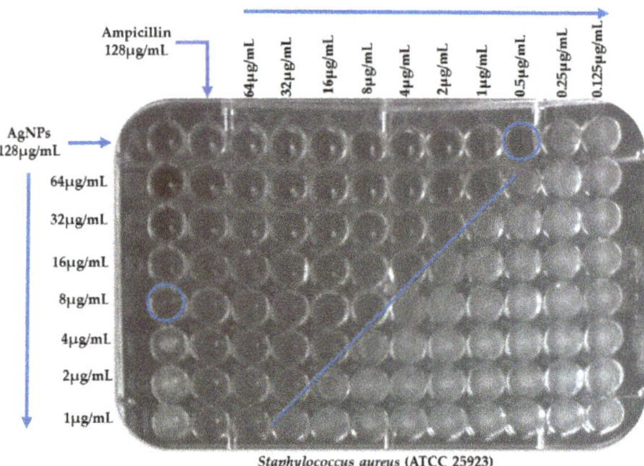

Figure 4. Example of checkerboard testing. Blue circles denote the MIC of antimicrobial agents (alone) and blue line denote the FIC (combination of both). MIC: Minimum Inhibitory Concentration. FIC: Fractional Inhibitory Concentration.

Table 3. FIC index of combinations among silver nanoparticles and antibiotics against clinical and reference strains.

Bacterial Strains	FIC Index			
	AgNPs + AMK		AgNPs + AMP	
Gram-positive				
Staphylococcus aureus (ATCC 25923)	1.06	(AD)	1.03	(AD)
Enterococcus faecium	0.53	(PS)	0.56	(PS)
Staphylococcus aureus	0.63	(PS)	1.25	(AD)
Gram-negative				
Escherichia coli (ATCC 25922)	0.31	(S)	1.50	(AD)
Acinetobacter baumannii	0.28	(S)	0.31	(S)
Enterobacter cloacae	0.75	(PS)	1.25	(AD)
Escherichia coli (501) **	0.56	(PS)	0.56	(PS)
Escherichia coli (508) **	0.31	(S)	0.56	(PS)
Escherichia coli (515) **	0.75	(PS)	0.56	(PS)
Klebsiella pneumoniae	0.75	(PS)	0.53	(PS)
Morganella morganii	0.75	(PS)	0.56	(PS)
Pseudomonas aeruginosa	0.63	(PS)	0.53	(PS)

FIC: Fractional Inhibitory Concentration. AMK: Amikacin. AMP: Ampicillin. ATCC: American Type Culture Collection. The FIC index was interpreted as follows: FIC $\leq$ 0.5, Synergy (S); 0.5 $\leq$ FIC < 1, Partial Synergy (PS); FIC = 1, Additive (AD); 2 $\leq$ FIC < 4, Indifferent (I); FIC > 4, Antagonism (AN) [32,33]. ** The numbers in parentheses indicate that *E. coli* corresponds to a different clinical sample.

2.5. Cytotoxicity of AgNPs

The cytotoxicity of AgNPs and antibiotics was evaluated separately and in combination by the MTT assay. AgNPs were tested at concentrations of 0.25, 1, 4, 16, 64 and 128 µg/mL in human fibroblasts. The percentages of living and dead cells were determined after 24 h of being exposed in contact with the AgNPs. AgNPs concentrations less than 4 µg/mL showed a cytotoxic effect that resulted in a death rate of 13.8% or less. However, concentrations greater than 64 µg/mL caused significant cell death of approximately 67%. In addition, antibiotics (ampicillin and amikacin) were tested at concentrations of 64, 32, 8, 2, 0.5 and 0.125 µg/mL in human fibroblasts. It was found that the viability percentage for each of the concentrations of ampicillin was greater than 80% and for amikacin greater than 76%. To evaluate the cytotoxic effects of the combination of AgNPs and conventional antibiotics, ten combinations of different concentrations (AgNPs µg/mL + antibiotic µg/mL: 128 + 64; 64 + 32; 32 + 16; etc.) were tested. It was found that there is no statistically significant difference between the two treatments (AgNPs + AMP and AgNPs + AMK) with respect to cell viability when the two-way ANOVA was performed ($p < 0.05$). However, combinations of 128 µgAgNPs/mL + 64 µg of antibiotic/mL and 64 µgAgNPs/mL + 32 µg of antibiotic/mL caused a statistically significant decrease in cell viability when compared with the rest of the combinations tested ($p < 0.05$) evidenced by the reduction of the mitochondrial activity. On the other hand, it is important to highlight that when AgNPs were combined with antibiotics, at concentrations equal to or less than 8 µg AgNPs/mL showed a viability percentage between 90–95% (Figure 5).

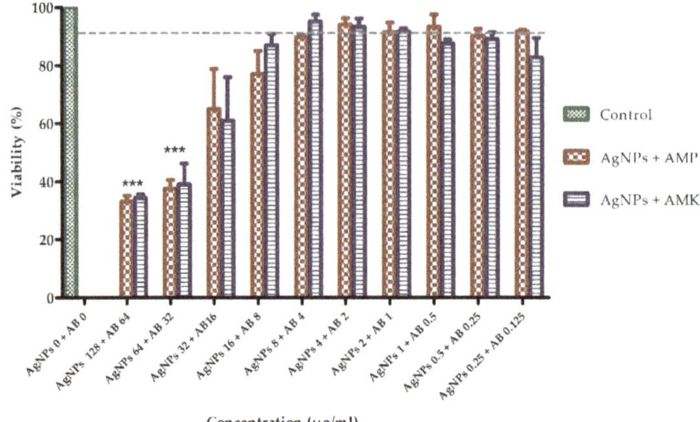

Figure 5. Viability of cells treated with combinations of silver nanoparticles and antibiotics. To measure cytotoxicity, fibroblasts were treated with increasing concentrations of AgNPs + AMP (red) or AgNPs + AMK (blue) (n = 3). Twenty-four hours after of addition of treatment cell viability was determined using MTT. Results are expressed as mean and standard deviation. * $p < 0.05$, ** $p < 0.01$, *** $p < 0.001$ by two-way ANOVA. Control: DMEM. AB: Antibiotic = Amikacin (AMK) or Ampicillin (AMP).

Interesting results were archived when two specific multi-resistant strains, *E. faecium* with resistance to vancomycin and *A. baumannii* with resistance to meropenem, were tested and both showed, with the combination of AgNPs + AMK, a reduction of the MIC by 32-fold, as well as with the combination of AgNPs + AMP of the MIC by 16-fold. In both cases, the cytotoxicity to fibroblasts, of the concentrations of AgNPs + Antibiotic which showed a reduction of MIC, showed a reduction with statistical significance (Table 4).

Table 4. Viability of fibroblasts treated with silver nanoparticles, antibiotics and their combinations using concentrations corresponding to the MIC value.

Clinical Strains	Viability of Fibroblasts (%)				
	AgNPs	**AMK**	**AgNPs + AMK**	**AMP**	**AgNPs + AMP**
E. faecium	>80	≈55	>90	>90	85–95
A. baumannii	72	≈55	>90	>90	85–95

MIC: Minimum Inhibitory Concentration. AMK: Amikacin. AMP: Ampicillin.

3. Materials and Methods

3.1. Synthesis of AgNPs

For the synthesis of nanoparticles, $AgNO_3$ (0.01 M) was used as a metallic precursor and gallic acid (0.1 g) was used as a reducer and stabilizer agent. NaOH (3 M) was used for pH regulation. AgNPs were synthesized by dissolving 0.0169 g of $AgNO_3$ in 90 mL of deionized water and this solution was placed in a 250 mL reaction vessel. A total of 0.01 g gallic acid was dissolved in 10 mL of deionized water and was added to the $AgNO_3$ solution with magnetic stirring. After the addition of gallic acid, the pH value of the solution was adjusted using a solution of NaOH 3 M. At the end of the synthesis, approximately 100 mL of nanoparticles were obtained with a pH of 12.66, of which 50 mL were dialyzed for 48 h on a 12 kDa nitrocellulose membrane.

3.2. Characterization of AgNPs

AgNPs were characterized by Dynamic Light Scattering (DLS), the hydrodynamic diameter and zeta potential were determined using a Malvern Zetasizer Nano ZS (Malvern Panalytical, Malvern, United Kingdom) operating with a He-Ne laser at a wavelength of 633 nm and a detection angle of 90°. Samples were analyzed for 60 s at 25 °C. To confirm the shape, each sample was diluted with deionized water and 50 µL of each suspension was placed on a copper wire for Transmission Electron Microscopy (TEM). All samples were analyzed by Transmission Electron Microscopy (JEOL JEM-1230, Tokyo, Japan) at an acceleration voltage of 100 kV. Afterwards, AgNPs were characterized by UV-visible spectroscopy using an S2000 UV-Vis spectrophotometer from OceanOptics Inc. (Dunedin, FL, USA). The functional groups present in the antibiotics were identified by Fourier Transform Infrared Spectroscopy (Shimadzu, IRaffinity-1, Osaka, Japan). A certain amount of nanopowder was collocated in the equipment and the spectrum was taken in the range of 400–4000 cm^{-1} with a resolution of 2 cm^{-1} and 200 times scanning using the attenuated total reflection (ATR) method.

3.3. Preparation and Characterization of Combinations of AgNPs with Antibiotics

To study the effect of ampicillin and amikacin on the size, shape and stability of the AgNPs, an aqueous solution containing a 1:1 ratio of antibiotic (128 µg/mL) and nanoparticles (128 µg/mL) was prepared for each antibiotic. These solutions were characterized by TEM, DLS, zeta potential and UV-visible spectroscopy. On the other hand, the chemical interaction between the AgNPs and antibiotics was carried out by FTIR, we prepared an aqueous solution containing higher concentrations of antimicrobials (500 µg/mL), the combinations preserved the ratio of 1:1. Subsequently, these solutions were centrifuged, keeping only the precipitate, which was left to dry for 24 h at room temperature.

3.4. Patients

The study protocol and the letter of informed consent were approved by the local Hospital's Ethics and Science Committee with the number 102-16.

The study included urine samples from patients with urinary catheters treated at the local Hospital. Patients were selected in basis to NOM-045-SSA2-2005. For the purposes of this study, only samples of patients older than 15 years, of any genus, whose culture had a bacterial count greater than or equal to 50,000 CFU/mL were selected, and also with an antimicrobial resistance profile.

3.5. Microbiological Analysis

3.5.1. Sample Collection and Bacterial Culture

Urine samples were collected in 5 mL syringes using the probe puncture technique under aseptic conditions by trained personnel. Microbiological analysis was performed according to the established criteria by the American Society for Microbiology (ASM) for urine samples. A count greater than or equal to 50 colonies (equivalent to 50,000 CFU/mL) was considered as the cutoff point for diagnosing infection. Plaques without development at 24-h incubation were discarded and indicated absence of urinary tract infection.

3.5.2. Identification and Antimicrobial Susceptibility Profile of Clinical Strains

The identification and antimicrobial susceptibility of the microorganisms isolated were determined by VITEK2 equipment. From a pure culture grown for 24 h, an inoculum was transferred to a test tube with 3 mL of solution sterile saline (0.45% to 0.5% NaCl aqueous solution, pH 4.5 to 7.0). Subsequently, the turbidity was adjusted to 0.50–0.63 units of the McFarland scale with the densitometer. The bacterial suspension was placed inside the cassette. The identification and susceptibility cards were placed in the nearby slot, inserting the transfer tube into the test tube with the corresponding suspension. The cassette was placed with the samples in the VITEK2 system. The resistance profiles are shown in Table S1.

3.5.3. Conservation of Strains

The conservation of the strains was carried out by the freeze conservation method. Glycerol was used as the cryoprotective agent. The samples were stored at $-20\,^{\circ}$C.

3.6. Antimicrobial Test

The Minimum Inhibitory Concentration (MIC) was determined by broth microdilution assay in accordance with the procedures recommended by the Clinical and Laboratory Standards Institute (CLSI) [34]. MICs were determined by incubating the microorganisms in 96-well microplates for 24 h at 37 °C. Microorganisms were exposed to serial dilutions of the antimicrobial agent (AgNPs, ampicillin and amikacin), and the end points were determined when no turbidity in the well was observed. The standardization of the method was made based on the criteria established by the CLSI. The strains used were *Staphylococcus aureus* ATCC 25923 and *Escherichia coli* ATCC 25922. The antibiotics used were oxacillin (64 µg/mL) and ceftazidime (32 µg/mL). The assay was performed in duplicate for four days. The results of the standardization of the MIC are shown in Table S2.

3.7. Checkerboard Synergy

The MIC of each antimicrobial substance alone or in combination was determined by a broth microdilution method in accordance with CLSI standards. The assay was performed in 96-well microtiter plates, a two-fold dilution of the antibiotic was distributed into each well to obtain a varying concentration of 128, 64, 32, 16, 8, 4, 2, 1, 05, 0.25 and 0.125 µg/mL in the wells of the first row, while those of the AgNPs were similarly distributed among the first column (128 to 1 µg/mL). The AgNPs dilutions were started from the columns to the right and the antibiotic dilutions were started from the first row downwards. Thus, each of the wells held a unique combination of concentrations of AgNPs and the antibiotic. The broth microdilution plates were inoculated with each test microorganism to yield the appropriate density (10^5 CFU/mL) in 100 µL Mueller–Hinton broth and incubated at the optimum temperature and time of growth conditions (37 °C/24 h). The MIC was determined as the least dilution without any turbidity. The MICs of single antimicrobial A and B ($\mathrm{MIC_A}$ and $\mathrm{MIC_B}$) and in combination were determined. The ten multidrug resistant clinical strains were exposed to serial dilutions of the antimicrobial agents. Subsequently, we calculated the proportion: $\mathrm{MIC_{Antibiotic\ alone}}/\mathrm{MIC_{Antibiotic\ in\ combination}}$ (fold-change) to describe the number of times that MIC decreased from an initial to final value.

3.8. Cytotoxicity of AgNPs

To evaluate the toxicity of AgNPs in combination with antibiotics, human fibroblasts (baby foreskin) were used. Cells were dispensed in 96-well microplates at a density of 5000 cells per well in DMEM supplemented with 1% fetal bovine serum (FBS) for 24 h at 37 °C. After 24 h the cells were incubated with the established concentrations of AgNPs and antibiotics for 24 h at 37 °C. Treatment was withdrawn and 100 µL MTT (3-(4,5-cimethylthiazol-2-yl)-2,5-diphenyl tetrazolium bromide) (0.5 mg/mL) was added in DMEM without FBS for 4 h at 37 °C for the formation of formazan crystals. MTT was removed from the wells and 100 µL of DMSO was added to read absorbance in a Synergy H1 microplate reader with Gen 5 software (Biotek Instruments, Winooski, VT, USA) at a wavelength of 595 nm. As a positive control, cells were treated with hydrogen peroxide and as a negative control, they were only treated with medium. The assay was performed in triplicate.

3.9. Statistical Analysis

Each assay was repeated three times. Data are presented as mean $\pm$ standard deviation (SD). The comparison between the effects of the two sources of variation was made using the two-way analysis of variance (ANOVA). The analysis was performed with the statistical software SPSS 23.0 (IBM, New York, NY, USA). A value of $p < 0.05$ was considered significant.

4. Conclusions

These results indicate that infections due to multidrug resistant microorganisms could be treated by the use of a synergistic combination of antimicrobial drugs and AgNPs. Further studies are necessary to evaluate the specific mechanisms of action, which could help predict undesirable off-target interactions, suggest ways of regulating a drug's activity, and identify novel therapeutics to this health problem.

Supplementary Materials: The following are available online at http://www.mdpi.com/2079-6382/7/2/50/s1, Table S1: Resistance profile of the set of clinical strains, Table S2: Standardization of the method MIC.

Author Contributions: F.M.-G. is the intellectual author of the project who designed the microbiological experiments. G.A.M.-C. designed the experiments related to the synthesis and characterization of silver nanoparticles and contributed to the analysis and interpretation of the data. A.F. performed the isolation of the pathogens and the sensitivity profile. M.L.-C. performed the microbiological experiments and the statistical analysis of the data. C.C. designed the experiments related to the study of the cytotoxic activity of silver nanoparticles and K.I.V. performed these experiments. M.M. oversaw the follow-up of the patients. F.M.-G. and M.L.-C. wrote this paper.

Acknowledgments: This work was supported by the Fondo de Apoyo a la Investigación (FAI) UASLP, C16-FAI-09-23.23 and Doctorado en Ciencias Odontológicas, Facultad de Estomatología, UASLP.

Conflicts of Interest: The authors declare no conflict of interest.

Appendix FIC Calculation

One strategy used to overcome the resistance mechanism of the microorganisms is the use of the combination of drugs. The way to measure their effectiveness is through the calculation of the Fractional Inhibitory Concentration (FIC) index.

The FIC index was calculated as follows:

FIC of AgNPs = $\text{MIC}_{\text{AgNPs}}$ in combination/$\text{MIC}_{\text{AgNPs}}$ alone

FIC of Antibiotic = $\text{MIC}_{\text{Antibiotic}}$ in combination/$\text{MIC}_{\text{Antibiotic}}$ alone

FIC index = FIC of AgNPs + FIC of Antibiotic

The FIC index was interpreted as follows: FIC < 0.5, Synergy (S); $0.5 \leq$ FIC <1, Partial Synergy (PS); FIC = 1, Additive (AD); $2 \leq$ FIC < 4, Indifferent (I); FIC > 4, Antagonism (AN).

References

1. Dielubanza, E.J.; Schaeffer, A.J. Urinary tract infections in women. *Med. Clin. N. Am.* **2011**, *95*, 27–41. [CrossRef] [PubMed]

2. Stamm, W.E.; Norrby, S.R. Urinary tract infections: Disease panorama and challenges. *J. Infect. Dis.* **2001**, *183* (Suppl. 1), S1–S4. [CrossRef] [PubMed]

3. Magill, S.S.; Edwards, J.R.; Bamberg, W.; Beldavs, Z.G.; Dumyati, G.; Kainer, M.A.; Lynfield, R.; Maloney, M.; McAllister-Hollod, L.; Nadle, J.; et al. Multistate point-prevalence survey of health care-associated infections. *N. Engl. J. Med.* **2014**, *370*, 1198–1208. [CrossRef] [PubMed]

4. Mukherjee, M.; Basu, S.; Mukherjee, S.K.; Majumder, M. Multidrug-Resistance and Extended Spectrum Beta-Lactamase Production in Uropathogenic *E. coli* which were Isolated from Hospitalized Patients in Kolkata, India. *J. Clin. Diagn. Res.* **2013**, *7*, 449–453. [CrossRef] [PubMed]

5. Ronald, A. The etiology of urinary tract infection: Traditional and emerging pathogens. *Am. J. Med.* **2002**, *113* (Suppl. 1A), 14S–19S. [CrossRef]

6. Iacovelli, V.; Gaziev, G.; Topazio, L.; Bove, P.; Vespasiani, G.; Finazzi Agrò, E. Nosocomial urinary tract infections: A review. *Urologia* **2014**, *81*, 222–227. [CrossRef] [PubMed]

7. Saint, S.; Lipsky, B.A.; Goold, S.D. Indwelling urinary catheters: A one-point restraint? *Ann. Intern. Med.* **2002**, *137*, 125–127. [CrossRef] [PubMed]

8. Stickler, D.J. Bacterial biofilms in patients with indwelling urinary catheters. *Nat. Clin. Pract. Urol.* **2008**, *5*, 598–608. [CrossRef] [PubMed]

9. Costerton, J.W.; Stewart, P.S.; Greenberg, E.P. Bacterial biofilms: A common cause of persistent infections. *Science* **1999**, *284*, 1318–1322. [CrossRef] [PubMed]

10. Subramanian, P.; Shanmugam, N.; Sivaraman, U.; Kumar, S.; Selvaraj, S. Antiobiotic resistance pattern of biofilm-forming uropathogens isolated from catheterised patients in Pondicherry, India. *Australas. Med. J.* **2012**, *5*, 344–348. [PubMed]

11. Goller, C.C.; Romeo, T. Environmental influences on biofilm development. *Curr. Top. Microbiol. Immunol.* **2008**, *322*, 37–66. [PubMed]

12. Mah, T.F.; O'Toole, G.A. Mechanisms of biofilm resistance to antimicrobial agents. *Trends Microbiol.* **2001**, *9*, 34–39. [CrossRef]

13. Mittal, S.; Sharma, M.; Chaudhary, U. Biofilm and multidrug resistance in uropathogenic Escherichia coli. *Pathog. Glob. Health* **2015**, *109*, 26–29. [CrossRef] [PubMed]

14. Stewart, P.S.; Costerton, J.W. Antibiotic resistance of bacteria in biofilms. *Lancet* **2001**, *358*, 135–138. [CrossRef]

15. Trautner, B.W.; Darouiche, R.O. Role of biofilm in catheter-associated urinary tract infection. *Am. J. Infect. Control* **2004**, *32*, 177–183. [CrossRef] [PubMed]

16. Stickler, D.J. Clinical complications of urinary catheters caused by crystalline biofilms: Something needs to be done. *J. Intern. Med.* **2014**, *276*, 120–129. [CrossRef] [PubMed]

17. Theuretzbacher, U. Antibiotic innovation for future public health needs. *Clin. Microbiol. Infect.* **2017**, *23*, 713–717. [CrossRef] [PubMed]

18. Walsh, T.R.; Toleman, M.A. The emergence of pan-resistant Gram-negative pathogens merits a rapid global political response. *J. Antimicrob. Chemother.* **2012**, *67*, 1–3. [CrossRef] [PubMed]

19. Martinez-Gutierrez, F.; Olive, P.L.; Banuelos, A.; Orrantia, E.; Nino, N.; Sanchez, E.M.; Ruiz, F.; Bach, H.; Av-Gay, Y. Synthesis, characterization, and evaluation of antimicrobial and cytotoxic effect of silver and titanium nanoparticles. *Nanomedicine* **2010**, *6*, 681–688. [CrossRef] [PubMed]

20. Morones, J.R.; Elechiguerra, J.L.; Camacho, A.; Holt, K.; Kouri, J.B.; Ramírez, J.T.; Yacaman, M.J. The bactericidal effect of silver nanoparticles. *Nanotechnology* **2005**, *16*, 2346–2353. [CrossRef] [PubMed]

21. Hajipour, M.J.; Fromm, K.M.; Ashkarran, A.A.; Jimenez de Aberasturi, D.; de Larramendi, I.R.; Rojo, T. Antibacterial properties of nanoparticles. *Trends Biotechnol.* **2012**, *30*, 499–511. [CrossRef] [PubMed]

22. Birla, S.S.; Tiwari, V.V.; Gade, A.K.; Ingle, A.P.; Yadav, A.P.; Rai, M.K. Fabrication of silver nanoparticles by *Phoma glomerata* and its combined effect against *Escherichia coli, Pseudomonas aeruginosa* and *Staphylococcus aureus*. *Lett. Appl. Microbiol.* **2009**, *48*, 173–179. [CrossRef] [PubMed]

23. Mala, R.; Annie Aglin, A.; Ruby Celsia, A.S.; Geerthika, S.; Kiruthika, N.; VazagaPriya, C.; Srinivasa Kumar, K. Foley catheters functionalised with a synergistic combination of antibiotics and silver nanoparticles resist biofilm formation. *IET Nanobiotechnol.* **2017**, *11*, 612–620. [CrossRef] [PubMed]

24. Braydich-Stolle, L.; Hussain, S.; Schlager, J.J.; Hofmann, M.C. In vitro cytotoxicity of nanoparticles in mammalian germline stem cells. *Toxicol. Sci.* **2005**, *88*, 412–419. [CrossRef] [PubMed]

25. Hussain, S.M.; Hess, K.L.; Gearhart, J.M.; Geiss, K.T.; Schlager, J.J. In vitro toxicity of nanoparticles in BRL 3A rat liver cells. *Toxicol. In Vitro* **2005**, *19*, 975–983. [CrossRef] [PubMed]

26. Tauran, Y.; Brioude, A.; Coleman, A.W.; Rhimi, M.; Kim, B. Molecular recognition by gold, silver and copper nanoparticles. *World J. Biol. Chem.* **2013**, *4*, 35–63. [CrossRef] [PubMed]

27. Yoosaf, K.; Ipe, B.I.; Suresh, C.H.; Thomas, K.G. In situ synthesis of metal nanoparticles and selective naked-eye detection of lead ions from aqueous media. *J. Phys. Chem. C* **2007**, *111*, 12839–12847. [CrossRef]

28. Hur, Y.E.; Kim, S.; Kim, J.-H.; Cha, S.-H.; Choi, M.-J.; Cho, S.; Park, Y. One-step functionalization of gold and silver nanoparticles by ampicillin. *Mater. Lett.* **2014**, *129*, 185–190. [CrossRef]

29. Rogowska, A.; Rafińska, K.; Pomastowski, P.; Walczak, J.; Railean-Plugaru, V.; Buszewska-Forajta, M.; Buszewski, B. Silver nanoparticles functionalized with ampicillin. *Electrophoresis* **2017**, *38*, 2757–2764. [CrossRef] [PubMed]

30. Deng, H.; McShan, D.; Zhang, Y.; Sinha, S.S.; Arslan, Z.; Ray, P.C.; Yu, H. Mechanistic Study of the Synergistic Antibacterial Activity of Combined Silver Nanoparticles and Common Antibiotics. *Environ. Sci. Technol.* **2016**, *50*, 8840–8848. [CrossRef] [PubMed]
31. Flores-Mireles, A.L.; Walker, J.N.; Caparon, M.; Hultgren, S.J. Urinary tract infections: Epidemiology, mechanisms of infection and treatment options. *Nat. Rev. Microbiol.* **2015**, *13*, 269–284. [CrossRef] [PubMed]
32. Odds, F.C. Synergy, antagonism, and what the chequerboard puts between them. *J. Antimicrob. Chemother.* **2003**, *52*. [CrossRef] [PubMed]
33. Hwang, I.S.; Hwang, J.H.; Choi, H.; Kim, K.J.; Lee, D.G. Synergistic effects between silver nanoparticles and antibiotics and the mechanisms involved. *J. Med. Microbiol.* **2012**, *61*, 1719–1726. [CrossRef] [PubMed]
34. Clinical and Laboratory Standards Institute. *Performance Standards for Antimicrobial Susceptibility Testing*; CLSI Supplement M100; CLSI: Wayne, PA, USA, 2017.

Article

Toxicological Assessment of a Lignin Core Nanoparticle Doped with Silver as an Alternative to Conventional Silver Core Nanoparticles

Cassandra E. Nix [1], Bryan J. Harper [1], Cathryn G. Conner [2], Alexander P. Richter [2], Orlin D. Velev [2] and Stacey L. Harper [1,3,4,*]

[1] Department of Environmental & Molecular Toxicology, Oregon State University, Corvallis, OR 97331, USA; nixc@oregonstate.edu (C.E.N.); Bryan.Harper@oregonstate.edu (B.J.H.)
[2] Department of Chemical and Biomolecular Engineering, North Carolina State University, Raleigh, NC 27606, USA; cgconner@ncsu.edu (C.G.C.); aprichte@ncsu.edu (A.P.R.); odvelev@ncsu.edu (O.D.V.)
[3] Oregon Nanoscience and Microtechnologies Institute, Corvallis, OR 97330, USA
[4] School of Chemical, Biological and Environmental Engineering, Oregon State University, Corvallis, OR 97331, USA
* Correspondence: stacey.harper@oregonstate.edu; Tel.: +1-(541)-737-2791

Received: 29 March 2018; Accepted: 2 May 2018; Published: 4 May 2018

Abstract: Elevated levels of silver in the environment are anticipated with an increase in silver nanoparticle (AgNP) production and use in consumer products. To potentially reduce the burden of silver ion release from conventional solid core AgNPs, a lignin-core particle doped with silver ions and surface-stabilized with a polycationic electrolyte layer was engineered. Our objective was to determine whether any of the formulation components elicit toxicological responses using embryonic zebrafish. Ionic silver and free surface stabilizer were the most toxic constituents, although when associated separately or together with the lignin core particles, the toxicity of the formulations decreased significantly. The overall toxicity of lignin formulations containing silver was similar to other studies on a silver mass basis, and led to a significantly higher prevalence of uninflated swim bladder and yolk sac edema. Comparative analysis of dialyzed samples which had leached their loosely bound Ag^+, showed a significant increase in mortality immediately after dialysis, in addition to eliciting significant increases in types of sublethal responses relative to the freshly prepared non-dialyzed samples. ICP-OES/MS analysis indicated that silver ion release from the particle into solution was continuous, and the rate of release differed when the surface stabilizer was not present. Overall, our study indicates that the lignin core is an effective alternative to conventional solid core AgNPs for potentially reducing the burden of silver released into the environment from a variety of consumer products.

Keywords: nanotechnology; environmentally-friendly; pesticide; antimicrobial; zebrafish

---

1. Introduction

Silver nanoparticles (AgNPs) are an effective antimicrobial agent and the most widely commercialized engineered nanomaterial, incorporated into half of all reported consumer and medical products in the Nanotechnology Consumer Products Inventory [1]. Prominent examples include cosmetics, clothing, shoes, detergents, water filters, phones, laptops, and toys [2–4]. AgNP use has risen steadily in the past decade (~52 new consumer products per year), and global production is estimated based on surveys of European producers to be between 12–1216 tons per year by 2020, assuming the number of products on the market continues to increase at the current rate [5,6]. With the increasing production and use of AgNPs, the fate and the subsequent release of silver in nanomaterial and ionic form into the environment are of concern.

Research indicates that AgNPs can enter aqueous environments from discharges at the point of production, by erosion from household products, and from disposal of silver-containing products [7–11]. These studies have prompted the investigation of AgNP interactions in the environment [12], particularly aquatic systems, to determine which general intrinsic and extrinsic properties are important in determining fate [9,13–15]. Extrinsic properties include environmental factors and processes that can impact the fate of the particles in aquatic systems, such as pH, temperature, ionic strength of the water, and natural organic matter, as well as processes like sedimentation, deposition, dissolution, agglomeration, and/or particle sulfidation [16–19]. Intrinsic factors address inherent particle characteristics, such as size, shape, chemical composition, surface structure, and surface charge [12,20–23]. Extrinsic factors can interact with intrinsic features of nanoparticles to alter particle behavior with concomitant effects on properties, such as the bioavailability of AgNPs to living organisms; thus, a more comprehensive understanding is needed [13,24].

AgNPs are known to be toxic to many aquatic organisms, including algae, bacteria, invertebrates, and fish [2]. Several mechanisms of action have been proposed, mainly attributing the toxicity of AgNPs to silver ions released from the nanoparticle. However, nanoparticle-specific mechanisms are also being investigated, with data suggesting that mechanistic differences exist compared to dissolved silver [5,25]. Silver ion specific mechanisms include interactions with thiols and electron donor groups, which can impact enzymes and DNA, which makes them unavailable for cellular processes [26–28], denaturing of DNA and RNA, which ultimately affects protein synthesis [29,30], and production of superoxide radicals and other reactive oxygen species via reactions with oxygen [29]. Particle-specific mechanisms have been suggested that focus on the ability of AgNPs to cause cell membrane damage, leading to disruption in the ion efflux system in cells [31,32], as well as by intracellular ion release elicited by the acidic conditions of the lysosomal cellular compartment where particles are internalized (Trojan horse effect) [33]. Since multiple aquatic organisms may be at risk due to an increased prevalence of silver in the environment, it is important to consider ways to reduce the environmental silver burden related to AgNP production and use.

By applying the principles of green chemistry during nanomaterial design and synthesis, harmful effects to the environment can be limited while maintaining the desired antimicrobial activity [34]. In order to reduce silver ion release into the environment, a silver-doped lignin nanoparticle was engineered, which is anticipated to have lower environmental impacts upon release into the environment [35]. During the synthesis of these particles, we replaced the silver core with lignin, which was chosen as it is a natural biodegradable biopolymer [36]. Similar synthesized lignin nanoparticles have been shown to have no impact on algae and yeast survival, suggesting they have a high level of biocompatibility [37]. The lignin is easily precipitated into nanosized particles using environmentally-friendly solvents, and the resulting nanoparticles can be infused with up to ten times lower (Ag^+) than silver core nanoparticles, and still maintain the same antimicrobial efficacy [35]. The particles are then surface-functionalized with a polycationic electrolyte layer to stabilize the particle, as well as to provide additional antimicrobial impact. The lignin nanoparticles exhibit both high and low affinity binding regions for silver ions, and these differing affinities, as well as the electrostatic barrier provided by the surface stabilizer, impact the rate of silver ion release to the surrounding solution [35,36]. It is expected that the low affinity binding sites will primarily release the majority of the weakly bound silver in the first 24 h [35]; however, we also wanted to investigate the long-term release from the high affinity binding sites, so two of the formulated samples were dialyzed to remove the weakly bound silver. When compared to their non-dialyzed counterparts, this allowed us to determine whether there are any differences in toxicological responses after the release of the loosely bound silver ions to quantify the potential environmental risks of these particles.

Our aim was to elucidate which aspects of the formulation contribute most to the toxicity of the formulation, and to discover whether these nanoparticles exhibit any toxicity after dialysis, which intended to simulate post-consumer use. We hypothesized that (1) released silver ions from the lignin particle and the surface stabilizer are the main contributors to the aquatic toxicity of these nanoparticles; and (2) once the particles have been dialyzed to remove the ionic silver from the low affinity lignin-binding sites, there would be a reduction in toxicity of the formulated particles. To test these hypotheses, we utilized the embryonic zebrafish assay, which is a widely-used model for toxicity testing as it provides a suite of developmental endpoints that are critical to the survival of the organism [38,39]. Zebrafish also develop quickly and are optically transparent, which allows for easy observations of phenotypic responses [38]. Additionally, they share similar homology to humans, so observed effects of chemical stressors from this assay can potentially be extrapolated to human physiological responses [40].

2. Materials and Methods

2.1. Materials and Characterization

Reference component solutions of silver nitrate ($AgNO_3$) salt (CAS# 7761-88-8, Fisher Scientific, Hampton, NH, USA) at 50 mg/L of Ag^+ dissolved in ultrapure water and poly (diallyldimethylammonium chloride) (PDAC, MW 100,000–200,000, CAS# 26062-79-3, Sigma Aldrich, St. Louis, MO, USA) at 200 mg/L in ultrapure water were prepared and refrigerated at 4 °C until use. The lignin (Indulin AT) for the nanoparticle core was extracted from biomass as a by-product of kraft pulping processes [36,41]. The Indulin AT lignin powder (lot MB05) and supporting documentation were obtained from MeadWestVaco, SC. The size range of the particles after synthesis with the pH-drop flash precipitation method (Figure S1c) was 84 ± 5 nm in ultrapure water [35]. PDAC was chosen to provide a cationic surface charge to the particles, such that they would be attracted to the negatively charged bacterial cell membranes. Stock nanomaterial suspensions of the lignin nanoparticle (NP), the silver functionalized lignin nanoparticle (NP + Ag), the silver functionalized particle with the cationic PDAC surface (NP + Ag + PDAC), and the lignin nanoparticle with PDAC alone (NP + PDAC) were prepared as previously described, and tested for antimicrobial efficacy, as is reported in previous publications [35,36]. Stock concentrations of each component were as follows: 500 mg/L lignin nanoparticle, 5 mg/L Ag^+, and 200 mg/L PDAC. All stock materials were stored in distilled water at 4 °C until use. Seven-fold dilutions of stock nanomaterial suspensions were performed with fishwater to prepare the varied exposure solutions. Fishwater was prepared by dissolving 260 mg/L Instant Ocean salts (Aquatic Ecosystems, Apopka, FL, USA) in reverse osmosis water and adjusting pH to 7.2 ± 0.2 using ~0.1 g sodium bicarbonate (conductivity 480–600 μS/cm) [39]. Experimental materials were stored under the same conditions as the reference materials. The NP + Ag and NP + PDAC formulations were solely used for comparative purposes, whereas the NP + Ag + PDAC is the proposed complete product formulation.

The samples to be dialyzed (NP + Ag and NP + Ag + PDAC) were placed in deionized water for 24 h which included a Slide-A-Lyzer MINI Dialysis Device (Thermo Scientific, Waltham, MA, USA) with a 10 K molecular weight cutoff membrane to remove dissolved silver from solution prior to dilution and testing. A second sample of NP + Ag was also dialyzed and allowed to age for 5 months prior to testing. Thus, the dialyzed samples included NP + Ag Aged, NP + Ag Fresh, and NP + Ag + PDAC Fresh, with the "Fresh" and "Aged" designations referring to when the sample was tested relative to when it was dialyzed.

The hydrodynamic diameter (HDD) and the zeta potential of each formulation that contained particles were measured in triplicate using a Zetasizer Nano ZS (Malvern Instruments Ltd., Worcestershire, UK) at 26.8 °C after dilution with fishwater to 50 mg/L. Aliquots (1 mL) were stored in an incubator under the same conditions as the embryonic zebrafish, until ready for analysis. Measurements were made over a five-day period, which also included an initial measurement (Day 0) which correlates with the exposure time of the experiment. Metadata associated with the zeta potential measurements can be found in Table S1.

2.2. Embryonic Zebrafish Assay

Exposure solutions of reference and nanomaterial suspensions were dispensed into 96-well plates, with each row having 12 wells of a given concentration of test material. Each well was filled with 200 µL of test solution and one of the eight rows on the plate was reserved for fishwater alone as a control. Adult zebrafish (*Danio rerio*) were maintained at the Sinnhuber Aquatic Research Laboratory (SARL) at Oregon State University, Corvallis, OR, USA. Embryos received from SARL were approximately 6–8 h post-fertilization (hpf) and were inspected under a dissecting microscope to ensure viability and developmental stage, then placed individually into wells of a 96-well plate. The chorionic membrane surrounding the zebrafish was preserved. Two replicate exposures were conducted over two weeks for each material, which allowed us to have a total sample size of 24 fish per concentration, per material. After plating, the exposure wells were covered with Parafilm to reduce evaporation, and embryos were incubated at 26.8 °C under a 14:10 light/dark photoperiod.

2.3. Toxicological Evaluations of Embryonic Zebrafish

Fish were observed at 24 hpf and 120 hpf for mortality, as well as a suite of developmental, morphological, and physiological abnormalities. At 24 hpf, embryos were evaluated for mortality, presence of spontaneous movement, delayed developmental progression, and notochord malformations. At 120 hpf, mortality was evaluated in conjunction with malformations of the snout, brain, pectoral and caudal fin, eye, jaw, otic structures, axis, trunk, somites, swim bladder, and body pigmentation. In addition, physiological and behavioral endpoints evaluated at 120 hpf include the presence of pericardial or yolk sac edema, impaired circulation and active touch response [39]. Hatching success was measured between 48 and 120 hpf, with embryos that hatched between 48 and 72 hpf being considered normal, and any individuals hatching after 72 hpf were considered delayed [42]. All endpoints were reported as either absent or present. Representative images of control fish and any individuals that displayed developmental abnormalities at 24 and 120 hpf were taken with an Olympus SZX10 microscope (Tokyo, Japan) fitted with an Olympus SC100 high resolution digital color camera (Olympus Corporation, Center Valley, PA, USA), and representative images are included in the Supplementary Materials (Figure S1). All experiments were performed in compliance with national care and use guidelines, and approved by the Institutional Animal Care and Use Committee (IACUC) at Oregon State University (ACUP #4764).

2.4. Measurement of Dissolved Silver and Particle-Associated Silver

Both the concentration of dissolved silver released from the nanoparticles and the silver associated with the particle itself were quantified by inductively coupled plasma-optical emission spectroscopy or mass spectrometry (ICP-OES or ICP-MS). To quantify silver content in solution, acid digestion of particles was performed using established methods [18,43]. Triplicate 0.5 mL samples of stock suspensions were centrifuged at $13,000 \times g$ for 10 min in a 3 kDa centrifugal filter (VWR, Radnor, PA, USA) with a polyethersulfone (PES) membrane, to separate the lignin particles from the filtrate. A total of 0.45 mL of filtrate sample was collected, diluted 10-fold with ultrapure water, and adjusted with 70% trace-metal grade HNO_3 to a final concentration of 3% HNO_3. For the lignin particle digestion, 0.1 mL of stock solution was digested in the same manner as the filtrate samples, without the centrifugation step. All samples were digested in Teflon tubes at 200 °C with 3 mL 70% trace-metal grade HNO_3. The acid was allowed to completely evaporate, and the process was repeated three times. Final digested samples were dissolved in 5 mL of 3% HNO_3 prior to ICP-OES/MS analysis. The silver ICP standard was purchased from RICCA Chemical Company (Ricca Chemical Company, Arlington, TX, USA) and diluted to six concentrations spanning the expected concentrations. All samples, including standards, were analyzed in triplicate with ICP-OES

(Teledyne Leeman Labs, Hudson, NH, USA) for silver content, except the filtrate from the NP + Ag + PDAC sample, which was analyzed by ICP-MS (Thermo-Fisher, Waltham, MA, USA) to provide a lower level of detection ($\geq$5 µg/L).

2.5. Statistical Analyses

All statistical analyses were conducted with SigmaPlot version 13.0 (Systat Software, San Jose, CA, USA), unless otherwise noted, and all differences were considered significant at $p \leq 0.05$. For measurements of zeta potential and HDD, significant differences were determined with repeated measures analysis of variance (ANOVA) and Tukey's post hoc analysis. Two-way ANOVA was conducted to ensure that there was no significant difference in mortality between replicate exposure plates prior to pooling of the data. Concentration–response curves were generated with the Environmental Protection Agency's Toxicity Relationship Analysis Program (EPA TRAP v. 1.30, March 2015). EPA TRAP was also used to calculate the concentration at which fifty percent of exposed zebrafish perished (LC_{50}) for each material, and the Litchfield/Wilcoxon formula was utilized for LC_{50} comparisons between treatments [44]. Significant sublethal endpoints were determined by Fisher's Exact Test, by comparing the control (fishwater alone) response to each concentration response tested. To determine whether there were significant differences in the concentration of silver associated with the particle versus the filtrate in the ICP analysis, a paired t-test was utilized.

3. Results and Discussion

3.1. Particle Characterization

Average zeta potential and HDD for the formulated particles in fishwater were measured over a five-day period. As expected, the zeta potential of the formulated particles varied with the presence of the surface stabilizer with the lignin NP alone, and the NP + Ag, both having negative zeta potentials in fishwater (-24.6 and -29.0 mV, respectively), while the particles with the cationic surface stabilizer had positive zeta potentials with and without silver ions present (28.1 and 25.5 mV respectively. Over the five day incubation in fishwater, only relatively minor changes in particle zeta potential were identified (Figure S2a). Zeta potentials for the dialyzed particles in fishwater were similarly consistent over time, and correlated well with their non-dialyzed counterparts (Figure S2b).

The initial HDD of the various particle formulations in fishwater were similar to the 84 nm primary particle size ranging from 80 to 95 nm, depending on the particle type (Figure S2c). Most of the formulated particles had consistent HDDs over time, however, the NP + Ag sample had the largest increase in size, reaching 124 nm by the end of the five day incubation. The dialyzed NP + Ag samples (both freshly dialyzed and aged dialyzed samples) had much more consistent HDD than the non-dialyzed counterparts (Figure S2d). These data suggest that the rapid loss of silver ions from the non-dialyzed components to the solution can lead to some particle swelling, however, the surface stabilizer effectively stabilized the particles from agglomeration.

3.2. Analysis of Dissolved Silver and Particle-Associated Silver

The concentration of silver in solution and the silver associated with the particles was quantified for the five nanoparticle samples that included silver. Figure 1 shows the concentration of silver present in the particle and in the solution which, when combined, matches the nominal concentration provided for each material. In all cases, the silver associated with the particle was greater than the silver present in solution (1.62 to 132 times greater. The full formulation (NP + Ag + PDAC) contains approximately 11 times more silver associated with the particle than the dialyzed full formulation (NP + Ag + PDAC Fresh). Additionally, the age of the NP + Ag sample played a role in silver distribution, as the older formulation contained approximately 10 times more silver in the filtrate than the particle.

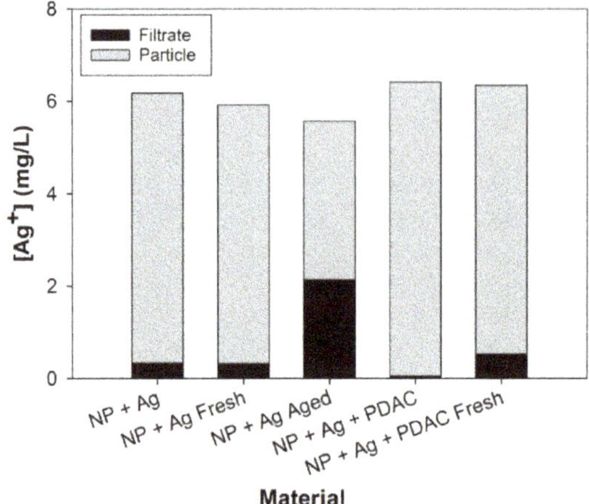

Figure 1. Concentration of silver associated with the filtrate and the particle as determined by ICP analysis. "Fresh" and "Aged" designations relate to the amount of time since the samples were dialyzed. All samples were analyzed via ICP-OES, except for the filtrate in the NP + Ag + PDAC sample, which was analyzed with ICP-MS.

Previous analyses of similar particles by Richter and colleagues [35] found that the concentration of silver associated with the particle after dialysis was approximately 18%; however, in this study, we found much higher concentrations associated with the particle (61.7–99.2%). This may have been due to variance between batches of the lignin nanoparticle stock solutions, as well as the differences in digestion techniques. Although the dialysis process is effective at releasing the loosely bound silver from the particles, the electrostatic barrier in the samples that contained PDAC may have impacted the rate at which silver was released. The full formulation had the lowest release of dissolved silver, although there was significantly more silver released from the freshly dialyzed full formulation sample, suggesting that PDAC may retard the release of ionic silver by repulsive electrostatic interactions. Additionally, through previous characterization of the lignin particle functional groups by Richter and colleagues [35,36], there is a higher proportion of organically-bound sulfur compared to other lignin types (nine times that of high-purity lignin), which would likely provide strong binding sites for dissolved silver [45].

3.3. Comparative Analysis of Formulation Toxicity

No significant differences were found between replicate exposure plates; therefore, replicates were pooled to increase the sample size to 24 fish per concentration, per material. To encompass all possible comparisons, but for clarity in interpreting the data, two groupings were made, which parallel our hypotheses. These two groupings are formulation comparisons and dialyzed sample comparisons. Concentration-response curves for the two groupings are illustrated in the SI (Figure S3). Additionally, a modeled concentration-response curve was generated for the reference material silver nitrate, which is included in the SI (Figure S4).

3.3.1. Formulation Components

As represented in Figure 2a, the lignin core nanoparticle (NP) itself was the least toxic component (LC_{50} = 323 mg/L), and when the NP was combined with the other aspects of the formulation, LC_{50} values decreased significantly in all cases (NP + Ag LC_{50} = 164 mg/L, NP + PDAC LC_{50} = 33 mg/L, NP + Ag + PDAC LC_{50} = 32 mg/L). The presence of silver in the full formulation (NP + Ag + PDAC) did not change the overall toxicity relative to NP + PDAC. Additionally, when PDAC was present in the formulation (NP + PDAC or NP + Ag + PDAC), a significant increase in mortality events occurred. PDAC and Ag^+ alone were the two most toxic constituents, with LC_{50}'s of 5.39 mg/L and 1.53 mg/L, respectively.

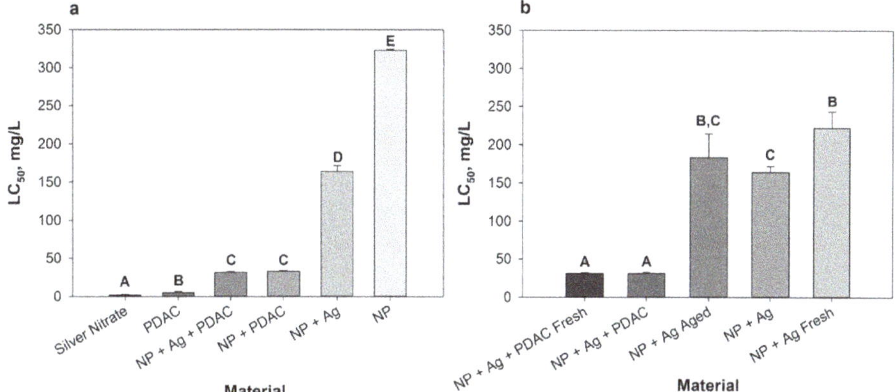

Figure 2. LC_{50}'s for nanoparticles and components (**a**) and dialyzed samples (**b**) with standard error of two experimental replicates with 12 embryos exposed at each concentration in each replicate test (24 embryos per concentration total). Significant differences between LC_{50} values are indicated with a change in letter above the bar.

The estimated LC_{50} for Ag^+ is greater than many published literature values for zebrafish [46–50], however, exposure time and conditions differ in these studies, which may explain the observed differences in toxicity. Our zebrafish embryos were exposed at 8 hpf with the chorion intact, whereas some of the referenced studies did not expose the fish until after hatching, or even as adults. Th chorionic membrane can modulate silver toxicity by sequestering ions to prevent them from entering the perivitelline fluid [51], and removing the chorion has been shown to increase toxic responses [40,51]. It is likely that the presence of the chorion may have played a large role in modulating silver toxicity, but the exposure media may have also played a role as well.

The hardness of our prepared fishwater may have altered the toxicity of silver nitrate, as well as nanoparticle-containing formulations to the embryonic zebrafish. Based on dissolved magnesium and calcium concentrations, our fishwater is categorized as soft water (<60 mg/L $CaCO_3$), whereas many of the above studies utilize moderately hard to hard water when exposing their zebrafish (up to 148 mg/L $CaCO_3$). It has been reported that LC_{50}'s tend to be higher in the presence of dissolved organic matter, which has the greatest effect on silver toxicity, followed by Cl^-, Na^+, and Ca^{2+} [52]. This is based on the coalescence effect, which leads to complexation and/or formation of nanoparticle agglomerates and/or aggregates, which can decrease apparent toxicity by minimizing particle uptake by the organism [53]. As our fishwater was categorized as soft water (low concentrations of Na^+ and Ca^{2+}), we then determined the concentration of chloride ions present, as silver ion bioavailability can be impacted due to complexation and subsequent precipitation of silver chloride [24,47].

In the Instant Ocean salt formulation used to make the fishwater, the majority of the cations are paired with chloride, and we determined the chloride concentration in our fishwater to be 142 mg/L, which is approximately 55% of the dissolved ion content. It is possible that when the fishwater was used to dilute the silver-containing treatments, the silver complexed with the chloride and precipitated out of solution, which may have led to a greater LC_{50} value. To determine whether precipitation was a significant factor, we utilized Visual MINTEQ (v.3.1) for each silver-containing formulation. We used the filtrate concentrations from the ICP data for each of the formulations that contained silver, and determined that nearly all of the dissolved silver (98.2–99.8%) would complex with the chloride present in the fishwater to form a precipitate of silver chloride, except the NP + Ag + PDAC sample, as the concentration of Ag^+ in solution was very low. However, as the exposures occurred over a five day period, and the movement of silver from the particle to the surrounding solution is a dynamic process, dissolved silver could have still been bioavailable to the zebrafish.

Considering our nanoparticle and silver combination formulations, comparisons to published LC_{50} values for conventional silver core AgNPs differ significantly. Reported AgNP LC_{50}'s for fish generally range between 0.05 and 20 mg/L [2,54]. Variations in reported LC_{50} values may relate to differences in the type of exposure, exposure time, age of the fish, the presence of the chorionic membrane, the use of bare or coated nanoparticles, and/or differences in exposure media. A study completed by Bar-Ilan and colleagues [55] matched our exposure conditions most consistently, and had reported LC_{50}'s within the range above, although the LC_{50}'s differed from our findings. They exposed embryonic zebrafish to different sizes of colloidal AgNPs (3–100 nm), and found a range of LC_{50}'s, from 10.1 to 14.7 mg/L. Although the sizes of the particles, length and timing of the exposure, and retention of the chorionic membrane is consistent with our experiment, our LC_{50}'s were an order of magnitude less toxic based on total particle mass (for example, the measured LC_{50} for the NP + Ag formulation was 164 mg/L). Thus, the silver doped lignin nanoparticles have a lower mass-based toxicity than solid core AgNPs, while maintaining a similar antimicrobial efficacy [35]. The difference may be due to the use of the lignin particle, which was shown via ICP-OES/MS analysis to retain bound silver, probably due to the higher binding affinity sites (Figure 1). It should be noted that there was a 100:1 ratio of lignin to Ag^+ in the nanoparticles, so the amount of silver in each particle was only a hundredth of the mass of the total particle, making our LC_{50} values, based on silver content alone, similar to those measured for solid core AgNPs reported in other studies. Therefore, potential exposure to silver ions would be reduced in the presence of the lignin particles, leading to an apparent increase in the LC_{50}. This suggests that by replacing the silver core typical of conventional AgNPs, the concentration of silver released to the environment may be reduced, which was one of the goals of formulating the nanoparticle with a lignin core.

PDAC alone was the second most toxic component tested, and formulations that contained PDAC were significantly more toxic than formulations that did not contain PDAC (Figure 2a). Although PDAC is a high charge density cationic polymer commonly used as a flocculant/coagulant in wastewater treatment, it is also cited as a cytotoxin that interacts with cell membranes to elicit cell damage, and eventually necrosis [56,57]. Our results correspond with this literature finding, as embryos that were exposed to PDAC alone progressively blackened and disintegrated, starting at 5.75 mg/L. Other formulations that contained PDAC did not elicit this response, perhaps because the PDAC is electrostatically associated with the lignin particle, or was "complexed". Free, or "non-complexed" PDAC, can interact with blood components, such as erythrocytes and plasma proteins, cell membranes, and extracellular matrix proteins, leading to undesired side effects not seen with complexed polycations [58]. Our experimental observations support this, as we see that the uncomplexed PDAC sample is indeed more toxic than any formulation that contains a nanoparticle–PDAC complex (Figure 2a). Additionally, research suggests polycationic polymers like PDAC can disrupt the lipid bilayer, with larger polymers leading to the formation of holes in the lipid membrane that increased membrane permeability [59,60]. Given that information, it is possible that PDAC made the fish more susceptible to both dissolved and particle-bound silver as a result of changes in membrane permeability. The positively charged PDAC-coated particles (SI, Figure S2) may have also been attracted to the negatively-charged membranes of the zebrafish, which could have increased the exposure to silver associated with the particle.

3.3.2. Dialyzed Formulations

The purpose of dialyzing the samples was to simulate post-consumer use of the nanoparticle by purging the surrounding solution of excess silver ions. In the LC_{50} comparisons of dialyzed materials, two results can be observed (Figure 2b). First, there was no difference in LC_{50} between the dialyzed complete formulation (NP + Ag + PDAC Fresh) and its non-dialyzed counterpart (NP + Ag + PDAC). This may be due to similar nominal concentrations of silver, as calculated by adding the silver associated with the particle and silver present in the filtrate (Figure 1). Second, the NP + Ag samples showed a significant decrease in toxicity post-dialysis, with LC_{50} values increasing immediately after dialysis (NP + Ag Fresh, LC_{50} = 222 mg/L), however, the dialyzed aged sample had a similar toxicity to the non-dialyzed sample (NP + Ag LC_{50} = 164 mg/L and NP + Ag Aged LC_{50} = 184 mg/L). Perhaps over time, the higher affinity binding sites on the lignin release more silver into solution compared to the freshly dialyzed sample, leading to the slight decrease in the aged LC_{50} (Figure 1).

3.4. Analysis of Sublethal Endpoints

There were several endpoints that were significant ($p \leq 0.05$), which included morphological abnormalities, including uninflated swim bladder and snout malformations, developmental endpoints, such as delay in hatching and delayed developmental progression, and physiological anomalies, such as impaired circulation, yolk sac edema, and pericardial edema (Figure 3a–f). Exposure to the silver nitrate at concentrations similar to the amount of silver in the NPs did not cause any significant malformations in the embryos. Overall, there was a significant increase in the types of sublethal responses observed in the dialyzed samples compared to the non-dialyzed samples. The dialyzed samples, particularly the aged sample, had proportionally less silver associated with the particle than the non-dialyzed samples (Figure 1), which could explain the increase in sublethal impacts, as well as the lack of sublethal endpoints for NP + Ag + PDAC (Figure 3).

Swim bladder malformations occurred when silver or PDAC were included in the particle formulation; however, this did not occur when silver and PDAC were both present in the particle formulation (NP + Ag + PDAC), except when freshly dialyzed (Figure 3e). Exposure to silver ions during embryonic zebrafish development has been described as impacting cholinergic signaling, which is important in swim bladder formation [61]. Swim bladder malformations were not significant in the NP + Ag + PDAC sample, likely due to high ratio of silver associated with the particle as compared to the filtrate (Figure 1).

Malformations of the snout were only significant in the freshly dialyzed full formulation (Figure 3e). Although others have reported silver nanoparticles causing snout malformations [54], we do not see this malformation in silver nitrate (Figure S5) or any other formulation containing silver, suggesting some other mechanisms of malformation may be involved. Perhaps the presence of all three formulation components may have contributed to the prevalence of snout malformations, in addition to the increased concentration of silver in the filtrate, as it exceeded the other treatments by a factor of six (Figure 1), except for the aged dialyzed particle (NP + Ag Aged).

Yolk sac edema was present in all fish exposed to formulations containing silver, and impaired circulation was significant in both freshly dialyzed formulations (Figure 3d,e). Significant pericardial edema responses were only noted in the NP + Ag and NP + Ag + PDAC Fresh formulations. Similarly, several other studies have reported that as a result of silver nanoparticle exposure, pericardial edema, yolk sac edema, and impaired circulation are prevalent in the early developmental stages of zebrafish [62–65]. As the development of the circulatory system and the formation of the heart typically occur between 21 and 24 hpf in zebrafish embryos [42,64], it is likely that the release of dissolved silver from the nanoparticles resulted in these endpoints being prevalent, however, particle-specific responses cannot be ruled out. The chorion has been reported to modulate metal toxicity [51], but there has been recent evidence that nanoparticles (30–72 nm) can move through the chorionic membrane pores and distribute to numerous parts of the fish, including the brain, heart, yolk, and blood [66]. Should the distribution of silver include the yolk and heart, edemas would be expected to occur due to disturbances in osmoregulation [55,62,63,67]. Once these developmental pathways are disturbed during early development, normal embryogenesis can be impacted, resulting in numerous defects [68]. In addition, silver nanoparticles may agglomerate in exposure media, which can alter oxygen exchange through chorionic pores, affecting oxygen tension and osmotic balance, which could then result in edemas similar to those observed in our study [66].

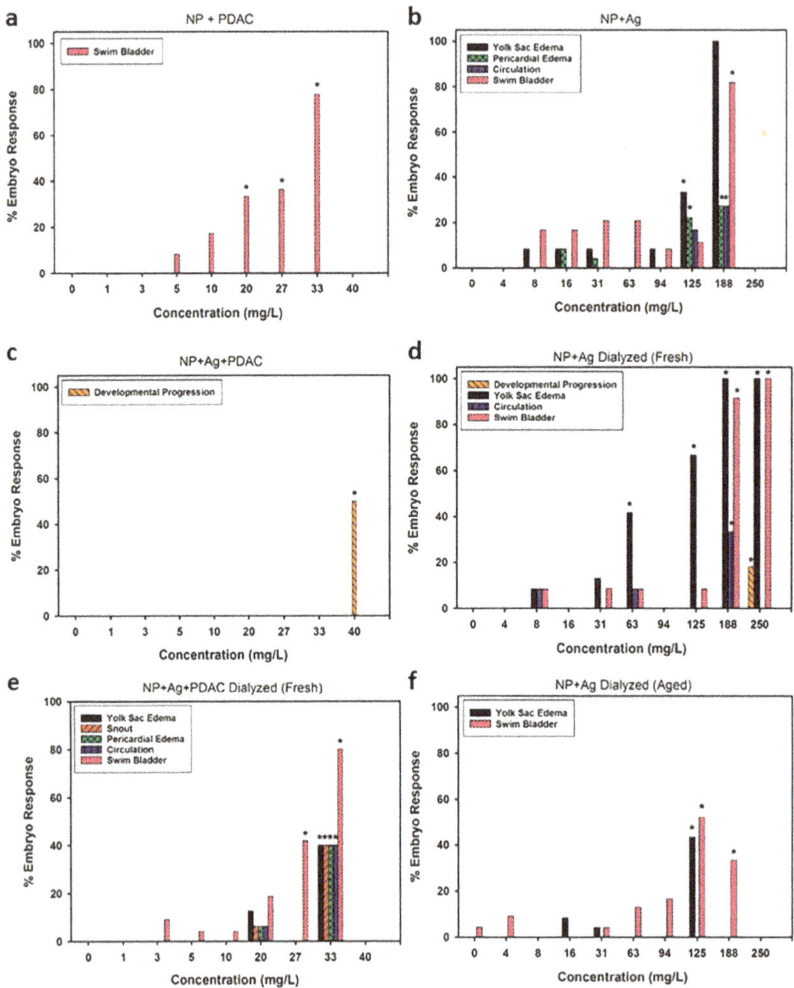

Figure 3. Types of developmental and morphological abnormalities observed in 120 hpf zebrafish embryos exposed to various formulations of the nanoparticles including (**a**) NP + PDAC; (**b**) NP + Ag; (**c**) NP + Ag + PDAC; (**d**) NP + Ag Dialyzed (Fresh); (**e**) NP + Ag + PDAC Dialyzed (Fresh) and (**f**) NP + Ag Dialyzed (Aged). Asterisk represents significant increase relative to unexposed (control) fish embryos at $p \leq 0.05$.

Pericardial edema coupled with impaired circulation following AgNP exposure has been shown to be concentration dependent [63], with an increase in prevalence from 10–100 mg/L. This was the trend we observed in our study, although our study saw increases up to 125 mg/L with a maximum of 17% responding. The most reasonable explanation for the slight difference in observations may be due to differences between solid silver nanoparticles with surface stabilizers, and the lignin-coated particles with specific silver binding sites on the particle core. Polyvinyl alcohol was used as the surface coating in Asharani and colleagues [63], which may have altered the dynamics of silver ion release relative to our samples. Our samples contained PDAC as a surface stabilizer, which provides an electrostatic barrier that could impede silver ion release, which we did see in our samples (Figure 1); however, when freshly dialyzed, PDAC had limited impact on silver ion release, suggesting solution equilibrium may be controlled by PDAC. Additionally, although there was silver present in the filtrate of the freshly dialyzed samples, the lignin core bound the majority of the silver, probably due to the higher affinity silver ion binding sites being utilized.

Significant delay in hatching and delayed developmental progression were the two developmental abnormalities observed in our study following exposure to silver salts (Figure S5) and high concentrations of the bare lignin nanoparticles (Figure S6), respectively. As delayed developmental progression was found following exposure to relatively high concentrations of the bare lignin nanoparticle and no other samples, the bare particle may interact with necessary ions in the solution, making them limited for supporting embryo development at concentrations of 350 mg/L and higher (Figure S6). This is further supported by the finding that this process does not occur when the particle is functionalized with silver, PDAC, or both, as the binding sites on the lignin particle are already occupied. Silver nitrate was the only material tested that led to a delay in hatching. A delay in hatching is primarily caused by deactivation of the ZHE1 enzyme, which prevents chorionic degradation [69]. Lin and colleagues have shown that dissolved metals can interfere with ZHE1, and although silver was included in their assay, it did not lead to a significant decrease in ZHE1 activity [69]. Asharani and colleagues exposed zebrafish to Ag^+ at 2.14 mg/L, and observed a delay in hatching at 4% compared to controls, but this was not significant [62]. Approximately 50% of our surviving fish exhibited a delay in hatching compared to controls following exposure to silver salts, however, this was not seen in lignin nanoparticle formulations containing silver.

4. Conclusions

The results of this study provide several insights into a nanoparticle engineered to be an environmentally friendly alternative to solid silver core nanoparticles. Our data shows that the use of lignin as the nanoparticle core could be a viable alternative, as it did not pose a significant toxicological hazard to our test organism. Since our reported toxicity was similar to other findings when compared on the basis of silver content, the toxicity of silver-enabled nanoparticles may be predictable, based on the silver concentration of the particle. Ionic silver and PDAC alone were the most toxic components of the formulation, which may be attributed to their higher diffusivity and propensity to interact with cell membranes relative to silver and/or PDAC associated with the particle. The inclusion of PDAC not only adds antimicrobial activity to the particle, but also seems to delay the release of silver ions, so in situations where time release of antimicrobial agents is desired, stabilizing the particles with PDAC may be warranted. This data also encourages further development of similar nanomaterials to minimize their impact on the environment, as well as testing the current particle under environmentally relevant conditions to evaluate toxicity. One way of reducing the environmental impact of these engineered nanomaterials is to design them in a way to minimize the release of soluble components, or to replace these components with less toxic ingredients. We are presently investigating the use of an alternative nanoparticle coating which is biologically derived, that may have the potential to be less toxic in comparison to PDAC.

5. Associated Content

Supplemental Information Is Available for This Publication

Representative images of zebrafish with and without significant developmental impacts; Average zeta potential and hydrodynamic diameter measurements for particle-containing formulations over a five day period; Metadata associated with zeta potential measurements; Concentration-response curves for formulation components and dialyzed materials based on zebrafish mortality at 120 hpf;

Modeled concentration-response curve for the reference material silver nitrate based on zebrafish mortality at 120 hpf; Visual MINTEQ output for all silver-containing formulations.

Supplementary Materials: The following are available online at http://www.mdpi.com/2079-6382/7/2/40/s1, Figure S1: Representative images of zebrafish with and without significant developmental impacts, Figure S2: Average zeta potential and hydrodynamic diameter (HDD) of particle formulations over a five day period in fishwater. Figure S1a,b are average zeta potential measurements for the formulation components (a) and the dialyzed formulations (b), and Figure S1c,d are the average HDD measurements for the formulation components (c) and the dialyzed formulations (d), Figure S3: Concentration-response comparisons for formulation components (a) and dialyzed materials (b) based on zebrafish mortality at 120 hpf (a) Significant differences ($p \leq 0.05$) existed between materials in the formulation component treatments. The lignin particle exhibited the lowest toxicity, followed by silver, and PDAC. (b) No significant differences ($p > 0.05$) existed in the dialyzed sample treatments. Comparisons included the two full formulations (NP + Ag + PDAC) and the three NP + Ag formulations, Figure S4: Concentration–response curve for silver nitrate based on zebrafish mortality at 120 hpf, Figure S5: Frequency of delayed hatching in zebrafish embryos exposed to Ag^+ as silver nitrate in fishwater. Asterisk represents significant increase ($p \leq 0.05$) relative to unexposed (control) fish embryos, Figure S6: Frequency of delayed developmental progression in zebrafish embryos exposed to lignin nanoparticles in fishwater. Asterisk represents significant increase ($p \leq 0.05$) relative to unexposed (control), Table S1: Metadata associated with zeta potential measurements.

Author Contributions: Alexander P. Richter and Orlin D. Velev conceived the materials, Cathryn G. Conner synthesized and provided the materials, Bryan J. Harper and Stacey L. Harper designed the experiments and contributed to the analysis of the data and Cassandra E. Nix performed the toxicological experiments, contributed to the data analysis and wrote the paper.

Acknowledgments: We would like to thank Alicea Meredith, Lindsay Denluck and Teresa Peterson for their assistance on the project and the Sinnhuber Aquatic Research Laboratory (Grant #P30 ES000210) for providing the zebrafish embryos. This work was supported by the United States Department of Agriculture National Institute of Food and Agriculture (USDA-NIFA), Grant 2013-67021-21181, partially supported by an Oregon State University Agricultural Research Foundation Grant (ARF8301A), as well as a grant from the National Institute of Environmental and Health Sciences (Grant # R01ES017552) to Stacey Harper. Alexander P. Richter thanks the Lemelson Foundation and the Lemelson-MIT program for support. We also acknowledge the support provided to Cathryn G. Conner by the Molecular Biotechnology Training Program (MBTP) sponsored by the National Institutes of Health and the Graduate School at North Carolina State University (5 T32 GM008776-15).

Conflicts of Interest: The authors declare no conflict of interest.

References

1. Vance, M.E.; Kuiken, T.; Vejerano, E.P.; McGinnis, S.P.; Hochella, M.F.; Rejeski, D.; Hull, M.S. Nanotechnology in the real world: Redeveloping the nanomaterial consumer products inventory. *Beilstein J. Nanotechnol.* **2015**, *6*, 1769–1780. [CrossRef] [PubMed]

2. Bondarenko, O.; Juganson, K.; Ivask, A.; Kasemets, K.; Mortimer, M.; Kahru, A. Toxicity of ag, cuo and zno nanoparticles to selected environmentally relevant test organisms and mammalian cells in vitro: A critical review. *Arch. Toxicol.* **2013**, *87*, 1181–1200. [CrossRef] [PubMed]

3. Bystrzejewska-Piotrowska, G.; Golimowski, J.; Urban, P.L. Nanoparticles: Their potential toxicity, waste and environmental management. *Waste Manag.* **2009**, *29*, 2587–2595. [CrossRef] [PubMed]

4. Marambio-Jones, C.; Hoek, E.M.V. A review of the antibacterial effects of silver nanomaterials and potential implications for human health and the environment. *J. Nanopart. Res.* **2010**, *12*, 1531–1551. [CrossRef]

5. Massarsky, A.; Trudeau, V.L.; Moon, T.W. Predicting the environmental impact of nanosilver. *Environ. Toxicol. Pharmacol.* **2014**, *38*, 861–873. [CrossRef] [PubMed]

6. Piccinno, F.; Gottschalk, F.; Seeger, S.; Nowack, B. Industrial production quantities and uses of ten engineered nanomaterials in europe and the world. *J. Nanopart. Res.* **2012**, *14*, 1109. [CrossRef]

7. Benn, T.; Cavanagh, B.; Hristovski, K.; Posner, J.D.; Westerhoff, P. The release of nanosilver from consumer products used in the home. *J. Environ. Qual.* **2010**, *39*, 1875–1882. [CrossRef] [PubMed]

8. Benn, T.M.; Westerhoff, P. Nanoparticle silver released into water from commercially available sock fabrics. *Environ. Sci. Technol.* **2008**, *42*, 4133–4139. [CrossRef] [PubMed]

9. Gottschalk, F.; Sun, T.; Nowack, B. Environmental concentrations of engineered nanomaterials: Review of modeling and analytical studies. *Environ. Pollut.* **2013**, *181*, 287–300. [CrossRef] [PubMed]

10. Kaegi, R.; Sinnet, B.; Zuleeg, S.; Hagendorfer, H.; Mueller, E.; Vonbank, R.; Boller, M.; Burkhardt, M. Release of silver nanoparticles from outdoor facades. *Environ. Pollut.* **2010**, *158*, 2900–2905. [CrossRef] [PubMed]

11. Mackevica, A.; Olsson, M.E.; Hansen, S.F. The release of silver nanoparticles from commercial toothbrushes. *J. Hazard. Mater.* **2016**, *322*, 270–275. [CrossRef] [PubMed]

12. Dobias, J.; Bernier-Latmani, R. Silver release from silver nanoparticles in natural waters. *Environ. Sci. Technol.* **2013**, *47*, 4140–4146. [CrossRef] [PubMed]

13. Handy, R.D.; Owen, R.; Valsami-Jones, E. The ecotoxicology of nanoparticles and nanomaterials: Current status, knowledge gaps, challenges, and future needs. *Ecotoxicology* **2008**, *17*, 315–325. [CrossRef] [PubMed]

14. Maurer-Jones, M.A.; Gunsolus, I.L.; Murphy, C.J.; Haynes, C.L. Toxicity of engineered nanoparticles in the environment. *Anal. Chem.* **2013**, *85*, 3036–3049. [CrossRef] [PubMed]

15. Selck, H.; Handy, R.D.; Fernandes, T.F.; Klaine, S.J.; Petersen, E.J. Nanomaterials in the aquatic environment: A european union-united states perspective on the status of ecotoxicity testing, research priorities, and challenges ahead: Nanomaterials in the aquatic environment. *Environ. Toxicol. Chem.* **2016**, *35*, 1055–1067. [CrossRef] [PubMed]

16. Furtado, L.M.; Bundschuh, M.; Metcalfe, C.D. Monitoring the fate and transformation of silver nanoparticles in natural waters. *Bull. Environ. Contam. Toxicol.* **2016**, *97*, 445–449. [CrossRef] [PubMed]

17. Furtado, L.M.; Norman, B.C.; Xenopoulos, M.A.; Frost, P.C.; Metcalfe, C.D.; Hintelmann, H. Environmental fate of silver nanoparticles in boreal lake ecosystems. *Environ. Sci. Technol.* **2015**, *49*, 8441–8450. [CrossRef] [PubMed]

18. Kim, K.-T.; Truong, L.; Wehmas, L.; Tanguay, R.L. Silver nanoparticle toxicity in the embryonic zebrafish is governed by particle dispersion and ionic environment. *Nanotechnology* **2013**, *24*, 115101. [CrossRef] [PubMed]

19. Peijnenburg, W.J.G.M.; Baalousha, M.; Chen, J.; Chaudry, Q.; Von der kammer, F.; Kuhlbusch, T.A.J.; Lead, J.; Nickel, C.; Quik, J.T.K.; Renker, M.; et al. A review of the properties and processes determining the fate of engineered nanomaterials in the aquatic environment. *Crit. Rev. Environ. Sci. Technol.* **2015**, *45*, 2084–2134. [CrossRef]

20. Lacave, J.M.; Retuerto, A.; Vicario-Parés, U.; Gilliland, D.; Oron, M.; Cajaraville, M.P.; Orbea, A. Effects of metal-bearing nanoparticles (Ag, Au, CdS, ZnO, SiO_2) on developing zebrafish embryos. *Nanotechnology* **2016**, *27*, 325102. [CrossRef] [PubMed]

21. Nel, A.; Xia, T.; Mädler, L.; Li, N. Toxic potential of materials at the nanolevel. *Science* **2006**, *311*, 622–627. [CrossRef] [PubMed]

22. Sharma, V.K.; Siskova, K.M.; Zboril, R.; Gardea-Torresdey, J.L. Organic-coated silver nanoparticles in biological and environmental conditions: Fate, stability and toxicity. *Adv. Colloid Interface Sci.* **2014**, *204*, 15–34. [CrossRef] [PubMed]

23. Shin, S.; Song, I.; Um, S. Role of physicochemical properties in nanoparticle toxicity. *Nanomaterials* **2015**, *5*, 1351–1365. [CrossRef] [PubMed]

24. Groh, K.J.; Dalkvist, T.; Piccapietra, F.; Behra, R.; Suter, M.J.F.; Schirmer, K. Critical influence of chloride ions on silver ion-mediated acute toxicity of silver nanoparticles to zebrafish embryos. *Nanotoxicology* **2015**, *9*, 81–91. [CrossRef] [PubMed]

25. Ivask, A.; ElBadawy, A.; Kaweeteerawat, C.; Boren, D.; Fischer, H.; Ji, Z.; Chang, C.H.; Liu, R.; Tolaymat, T.; Telesca, D.; et al. Toxicity mechanisms in escherichia coli vary for silver nanoparticles and differ from ionic silver. *ACS Nano* **2014**, *8*, 374–386. [CrossRef] [PubMed]

26. Clement, J.; Jarrett, P. Antibacterial silver. *Met. Based Drugs* **1994**, *1*, 467–482. [CrossRef] [PubMed]

27. Gordon, O.; Vig Slenters, T.; Brunetto, P.S.; Villaruz, A.E.; Sturdevant, D.E.; Otto, M.; Landmann, R.; Fromm, K.M. Silver coordination polymers for prevention of implant infection: Thiol interaction, impact on respiratory chain enzymes, and hydroxyl radical induction. *Antimicrob. Agents Chemother.* **2010**, *54*, 4208–4218. [CrossRef] [PubMed]

28. Morones, J.R.; Elechiguerra, J.L.; Camacho, A.; Holt, K.; Kouri, J.B.; Ramírez, J.T.; Yacaman, M.J. The bactericidal effect of silver nanoparticles. *Nanotechnology* **2005**, *16*, 2346–2353. [CrossRef] [PubMed]

29. Feng, Q.L.; Wu, J.; Chen, G.Q.; Cui, F.Z.; Kim, T.N.; Kim, J.O. A mechanistic study of the antibacterial effect of silver ions on escherichia coli and staphylococcus aureus. *J. Biomed. Mater. Res.* **2000**, *52*, 662–668. [CrossRef]

30. Fong, J.; Wood, F. Nanocrystalline silver dressings in wound management: A review. *Int. J. Nanomed.* **2006**, *1*, 441–449. [CrossRef]

31. Hwang, E.T.; Lee, J.H.; Chae, Y.J.; Kim, Y.S.; Kim, B.C.; Sang, B.-I.; Gu, M.B. Analysis of the toxic mode of action of silver nanoparticles using stress-specific bioluminescent bacteria. *Small* **2008**, *4*, 746–750. [CrossRef] [PubMed]

32. Sharma, V.K.; Yngard, R.A.; Lin, Y. Silver nanoparticles: Green synthesis and their antimicrobial activities. *Adv. Colloid Interface Sci.* **2009**, *145*, 83–96. [CrossRef] [PubMed]

33. Sabella, S.; Carney, R.P.; Brunetti, V.; Malvindi, M.A.; Al-Juffali, N.; Vecchio, G.; Janes, S.M.; Bakr, O.M.; Cingolani, R.; Stellacci, F.; et al. A general mechanism for intracellular toxicity of metal-containing nanoparticles. *Nanoscale* **2014**, *6*, 7052–7061. [CrossRef] [PubMed]

34. Anastas, P.; Eghbali, N. Green chemistry: Principles and practice. *Chem. Soc. Rev.* **2010**, *39*, 301–312. [CrossRef] [PubMed]

35. Richter, A.P.; Brown, J.S.; Bharti, B.; Wang, A.; Gangwal, S.; Houck, K.; Cohen Hubal, E.A.; Paunov, V.N.; Stoyanov, S.D.; Velev, O.D. An environmentally benign antimicrobial nanoparticle based on a silver-infused lignin core. *Nat. Nanotechnol.* **2015**, *10*, 817–823. [CrossRef] [PubMed]

36. Richter, A.P.; Bharti, B.; Armstrong, H.B.; Brown, J.S.; Plemmons, D.; Paunov, V.N.; Stoyanov, S.D.; Velev, O.D. Synthesis and characterization of biodegradable lignin nanoparticles with tunable surface properties. *Langmuir* **2016**, *32*, 6468–6477. [CrossRef] [PubMed]

37. Frangville, C.; Rutkevičius, M.; Richter, A.P.; Velev, O.D.; Stoyanov, S.D.; Paunov, V.N. Fabrication of environmentally biodegradable lignin nanoparticles. *ChemPhysChem* **2012**, *13*, 4235–4243. [CrossRef] [PubMed]

38. Hill, A.J. Zebrafish as a model vertebrate for investigating chemical toxicity. *Toxicol. Sci.* **2005**, *86*, 6–19. [CrossRef] [PubMed]

39. Truong, L.; Harper, S.; Tanguay, R. Evaluation of embryotoxicity using the zebrafish model. In *Drug Safety Evaluation: Methods and Protocols*; Humana Press: New York, NY, USA, 2011; pp. 271–279.

40. Kim, K.-T.; Tanguay, R.L. The role of chorion on toxicity of silver nanoparticles in the embryonic zebrafish assay. *Environ. Health Toxicol.* **2014**, *29*, e2014021. [CrossRef] [PubMed]

41. Duval, A.; Lawoko, M. A review on lignin-based polymeric, micro- and nano-structured materials. *React. Funct. Polym.* **2014**, *85*, 78–96. [CrossRef]

42. Kimmel, C.B.; Ballard, W.W.; Kimmel, S.R.; Ullmann, B.; Schilling, T.F. Stages of embryonic development of the zebrafish. *Dev. Dyn.* **1995**, *203*, 253–310. [CrossRef] [PubMed]

43. Wu, F.; Harper, B.J.; Harper, S.L. Differential dissolution and toxicity of surface functionalized silver nanoparticles in small-scale microcosms: Impacts of community complexity. *Environ. Sci. Nano* **2017**, *4*, 359–372. [CrossRef]

44. Sprague, J.; Fogels, A. *Watch the y in Bioassay*; EPS-5-AR-77-1; Procedural 3rd Aquatic Toxicology Workshop: Halifax, NS, Canada, 1977; pp. 107–118.

45. Bielmyer, G.K.; Grosell, M.; Paquin, P.R.; Mathews, R.; Wu, K.B.; Santore, R.C.; Brix, K.V. Validation study of the acute biotic ligand model for silver. *Environ. Toxicol. Chem.* **2007**, *26*, 2241–2246. [CrossRef] [PubMed]

46. Alsop, D.; Wood, C.M. Metal uptake and acute toxicity in zebrafish: Common mechanisms across multiple metals. *Aquat. Toxicol.* **2011**, *105*, 385–393. [CrossRef] [PubMed]

47. Bielmyer, G.; Brix, K.; Grosell, M. Is Cl⁻ protection against silver toxicity due to chemical speciation? *Aquat. Toxicol.* **2008**, *87*, 81–87. [CrossRef] [PubMed]

48. Bilberg, K.; Hovgaard, M.B.; Besenbacher, F.; Baatrup, E. In vivo toxicity of silver nanoparticles and silver ions in zebrafish (*Danio rerio*). *J. Toxicol.* **2012**, *2012*, 1–9. [CrossRef] [PubMed]

49. Powers, C.M.; Slotkin, T.A.; Seidler, F.J.; Badireddy, A.R.; Padilla, S. Silver nanoparticles alter zebrafish development and larval behavior: Distinct roles for particle size, coating and composition. *Neurotox. Teratol.* **2011**, *33*, 708–714. [CrossRef] [PubMed]

50. Powers, C.M.; Yen, J.; Linney, E.A.; Seidler, F.J.; Slotkin, T.A. Silver exposure in developing zebrafish (*danio rerio*): Persistent effects on larval behavior and survival. *Neurotoxicol. Teratol.* **2010**, *32*, 391–397. [CrossRef] [PubMed]

51. Rombough, P. The influence of zona radiata on the toxicities of zinc, lead, mercury, copper and silver ions to embryos of steelhead trout salmo gairdneri. *Comp. Biochem. Physiol.* **1985**, *82*, 115–117. [CrossRef]

52. McGeer, J.C.; Playle, R.C.; Wood, C.M.; Galvez, F. A physiologically based biotic ligand model for predicting the acute toxicity of waterborne silver to rainbow trout in freshwaters. *Environ. Sci. Technol.* **2000**, *34*, 4199–4207. [CrossRef]

53. Lapresta-Fernández, A.; Fernández, A.; Blasco, J. Nanoecotoxicity effects of engineered silver and gold nanoparticles in aquatic organisms. *TrAC Trends Anal. Chem.* **2012**, *32*, 40–59. [CrossRef]

54. Reidy, B.; Haase, A.; Luch, A.; Dawson, K.; Lynch, I. Mechanisms of silver nanoparticle release, transformation and toxicity: A critical review of current knowledge and recommendations for future studies and applications. *Materials* **2013**, *6*, 2295–2350. [CrossRef] [PubMed]

55. Bar-Ilan, O.; Albrecht, R.M.; Fako, V.E.; Furgeson, D.Y. Toxicity assessments of multisized gold and silver nanoparticles in zebrafish embryos. *Small* **2009**, *5*, 1897–1910. [CrossRef] [PubMed]

56. Fischer, D.; Li, Y.; Ahlemeyer, B.; Krieglstein, J.; Kissel, T. In vitro cytotoxicity testing of polycations: Influence of polymer structure on cell viability and hemolysis. *Biomaterials* **2003**, *24*, 1121–1131. [CrossRef]

57. Wandrey, C.; Hernandez-Barajas, J.; Hunkeler, D. Diallyldimethylammonium chloride and its polymers. *Adv. Polym. Sci.* **1999**, *145*, 123–177.

58. Kircheis, R.; Wightman, L.; Wagner, E. Design and gene delivery activity of modified polyethylenimines. *Adv. Drug Del. Rev.* **2001**, *53*, 341–358. [CrossRef]

59. Hong, S.; Leroueil, P.R.; Janus, E.K.; Peters, J.L.; Kober, M.-M.; Islam, M.T.; Orr, B.G.; Baker, J.R.; Banaszak Holl, M.M. Interaction of polycationic polymers with supported lipid bilayers and cells: Nanoscale hole formation and enhanced membrane permeability. *Bioconjug. Chem.* **2006**, *17*, 728–734. [CrossRef] [PubMed]

60. Mecke, A.; Majoros, I.J.; Patri, A.K.; Baker, J.R.; Banaszak Holl, M.M.; Orr, B.G. Lipid bilayer disruption by polycationic polymers: The roles of size and chemical functional group. *Langmuir* **2005**, *21*, 10348–10354. [CrossRef] [PubMed]

61. Robertson, G.N.; McGee, C.A.S.; Dumbarton, T.C.; Croll, R.P.; Smith, F.M. Development of the swimbladder and its innervation in the zebrafish,danio rerio. *J. Morphol.* **2007**, *268*, 967–985. [CrossRef] [PubMed]

62. Asharani, P.V.; Lian Wu, Y.; Gong, Z.; Valiyaveettil, S. Toxicity of silver nanoparticles in zebrafish models. *Nanotechnology* **2008**, *19*, 255102. [CrossRef] [PubMed]

63. Asharani, P.V.; Lian Wu, Y.; Gong, Z.; Valiyaveettil, S. Comparison of the toxicity of silver, gold and platinum nanoparticles in developing zebrafish embryos. *Nanotoxicology* **2011**, *5*, 43–54. [CrossRef] [PubMed]

64. Lee, K.J.; Browning, L.M.; Nallathamby, P.D.; Osgood, C.J.; Xu, X.-H.N. Silver nanoparticles induce developmental stage-specific embryonic phenotypes in zebrafish. *Nanoscale* **2013**, *5*, 11625–11636. [CrossRef] [PubMed]

65. Osborne, O.J.; Johnston, B.D.; Moger, J.; Balousha, M.; Lead, J.R.; Kudoh, T.; Tyler, C.R. Effects of particle size and coating on nanoscale ag and tio2 exposure in zebrafish (*danio rerio*) embryos. *Nanotoxicology* **2013**, *7*, 1315–1324. [CrossRef] [PubMed]

66. Liu, W.; Long, Y.; Yin, N.; Zhao, X.; Sun, C.; Zhou, Q.; Jiang, G. Toxicity of engineered nanoparticles to fish. In *Engineered Nanoparticles and the Environment: Biophysicochemical Processes and Toxicity*; John Wiley & Sons: Hoboken, NJ, USA, 2016; Volume 4.

67. Lee, K.J.; Nallathamby, P.D.; Browning, L.M.; Osgood, C.J.; Xu, X.-H.N. In vivo imaging of transport and biocompatibility of single silver nanoparticles in early development of zebrafish embryos. *ACS Nano* **2007**, *1*, 133–143. [CrossRef] [PubMed]

68. Kiener, T.K.; Selptsova-Friedrich, I.; Hunziker, W. TJP3/ZO-3 is critical for epidermal barrier function in zebrafish embryos. *Dev. Biol.* **2008**, *316*, 36–49. [CrossRef] [PubMed]

69. Lin, S.; Zhao, Y.; Ji, Z.; Ear, J.; Chang, C.H.; Zhang, H.; Low-Kam, C.; Yamada, K.; Meng, H.; Wang, X.; et al. Zebrafish high-throughput screening to study the impact of dissolvable metal oxide nanoparticles on the hatching enzyme, ZHE1. *Small* **2013**, *9*, 1776–1785. [CrossRef] [PubMed]

Article

Silver Nanoparticles and Polyphenol Inclusion Compounds Composites for *Phytophthora cinnamomi* Mycelial Growth Inhibition

Petruta Mihaela Matei [1,2], Jesús Martín-Gil [2], Beatrice Michaela Iacomi [1], Eduardo Pérez-Lebeña [2], María Teresa Barrio-Arredondo [3] and Pablo Martín-Ramos [4,*]

[1] Department of Bioengineering of Horticultural and Viticultural Systems, University of Agricultural Sciences and Veterinary Medicine of Bucharest, Bulevardul Mărăşti 59, Bucureşti 011464, Romania; petruta.matei@horticultura-bucuresti.ro (P.M.M.); b.iacomi@yahoo.fr (B.M.I.)

[2] Agriculture and Forestry Engineering Department, ETSIIAA, Universidad de Valladolid, Avenida de Madrid 44, 34004 Palencia, Spain; mgil@iaf.uva.es (J.M.-G.); eplebena@gmail.com (E.P.-L.)

[3] Centro de Salud Barrio España, Sanidad de Castilla y León (SACYL), Calle de la Costa Brava, 4, 47010 Valladolid, Spain; jesusmartingil@gmail.com

[4] Department of Agricultural and Environmental Sciences, EPS, Instituto de Investigación en Ciencias Ambientales (IUCA), University of Zaragoza, Carretera de Cuarte, s/n, 22071 Huesca, Spain

* Correspondence: pmr@unizar.es; Tel.: +34-(974)-292668

Received: 29 June 2018; Accepted: 15 August 2018; Published: 16 August 2018

Abstract: *Phytophthora cinnamomi*, responsible for "root rot" or "dieback" plant disease, causes a significant amount of economic and environmental impact. In this work, the fungicide action of nanocomposites based on silver nanoparticles and polyphenol inclusion compounds, which feature enhanced bioavailability and water solubility, was assayed for the control of this soil-borne water mold. Inclusion compounds were prepared by an aqueous two-phase system separation method through extraction, either in an hydroalcoholic solution with chitosan oligomers (COS) or in a choline chloride:urea:glycerol deep eutectic solvent (DES). The new inclusion compounds were synthesized from stevioside and various polyphenols (gallic acid, silymarin, ferulic acid and curcumin), in a [6:1] ratio in the COS medium and in a [3:1] ratio in the DES medium, respectively. Their in vitro response against *Phytophthora cinnamomi* isolate MYC43 (at concentrations of 125, 250 and 500 µg·mL^{-1}) was tested, which found a significant mycelial growth inhibition, particularly high for the composites prepared using DES. Therefore, these nanocomposites hold promise as an alternative to fosetyl-Al and metalaxyl conventional systemic fungicides.

Keywords: antifungal; chitosan oligomers; composites; deep eutectic solvents; phenolic compounds; *Phytophthora cinnamomi*; root rot; silver nanoparticles

1. Introduction

Nanotechnology has shown remarkable applications in biomedicine, diagnosis and antibacterial treatments, and is now transforming the agricultural sector, particularly with the development of novel nanopesticides and nanofertilizers [1]. The increase in the frequency of resistant or tolerant pathogenic agents, which has in turn led to an excessive application of pesticides, has resulted in an increase in the presence of residues in food products, which may pose a major risk to health. The design and testing of safe, effective and environmentally sustainable formulations based on nanoemulsions, nanocomposites and nanoparticles to control agricultural pests and pathogens has become a burgeoning field of research in the past few years.

Silver, which has long been used as a disinfectant for pathogenic microorganisms [2], has become one of the best exponents of this transition. Silver nanoparticles (AgNPs), which feature antibacterial, antifungal and antitumor activities [3–5], are one of the most popular active ingredients employed to enhance the efficacy of plant protection products. Furthermore, they can be prepared through green synthesis procedures with the aid of plant extracts [6], which act as reducing and stabilizing agents. Polyol components, polysaccharides, and water-soluble heterocycles (such as those from *Stevia rebaudiana* [7], *Curcuma longa* [8], *Pongamia pinnata* [9], *Gliricidia sepium* [10], *Eucalyptus hybrida* [11], or *Quercus brantii* [12]) have been reported to lead to a synergistic effect in the resulting phytonanocomposites [13–15].

Nonetheless, bioactive compounds from plants (that include phenolic acids, flavonoids, curcuminoids, coumarins, quinones, tannins and lignans), in spite of having a wide range of activities, generally suffer from a number of drawbacks derived from their inherent physicochemical characteristics (poor water solubility, low bioavailability, chemical instability, photodegradation, rapid metabolism and short half-life) [16], which limit their applications. To stabilize them and improve their bioavailability, one well-known approach is to use biopolymers, such as chitin, chitosan, starch, and cellulose, or other macromolecular systems [17]. Binary composites based on chitosan with polyphenols (e.g., gallic acid or curcumin [18–20]) and ternary composites that also include AgNPs [21,22] with a broadened bio-activity have been reported in the literature.

Other approaches to improve solubility are based on forming inclusion compounds with terpene glycosides (such as rubusoside, stevioside, rebaudioside, or steviol monoside) or cyclodextrins, which result in an enhancement of the solubility of polyphenols [23,24], or on using deep eutectic solvents (DES). DES are an excellent extraction medium for phenolic compounds [25] and may be used in combination with inclusion compounds [26] or with chitosan [27,28].

Phytophthora cinnamomi is a pathogen with over 1000 host species, transmitted by the soil and which causes rotting of the roots of many horticultural and forestry crops [29]. *P. cinnamomi* can collapse, which cause sudden death of plants and a decrease in fruit yield and size. The infection by *P. cinnamomi* can also occur together with other species of *Phytophthora*, mainly *P. cambivora*, *P. cryptogea*, *P. citricola* and *P. cactorum*. Its eradication by means of fungicides is expensive and causes damage to the environment, and fumigation is not always effective for deeper roots [30,31]. Consequently, the European Union is promoting the development of new natural bioactive compounds to replace conventional systemic fungicides, such as the organophosphorus compound fosetyl-Al or acylalanines such as metalaxyl.

It has been shown, in vitro, that AgNPs synthesized using aqueous plant extracts have had antifungal effect on *Phytophthora* pathogens [32], and so do chitosan [33], the binary combinations of the two [34] and their ternary combinations with propolis [35]. Nonetheless, to the best of the authors' knowledge, no studies have explored the use of composites of AgNPs with polyphenol inclusion compounds, combined either with chitosan oligomers (COS) or with DES, for the control of *P. cinnamomi* or any other oomycetes.

In the present study, four polyphenols (gallic acid, silymarin, ferulic acid and curcumin) were assessed for the microwave-assisted formation of the new inclusion compounds with stevioside. AgNPs were subsequently incorporated into the composites [36]. Two types of host matrices were tested, namely COS in a hydroalcoholic solution, and a DES based on a choline chloride and urea solution (1:2 *v/v*) in glycerol, evaluating in vitro their response against *P. cinnamomi* at different concentrations.

2. Results

The antifungal activity of the different treatments in aforementioned two preparation media (COS in hydroalcoholic solution and DES) against *P. cinnamomi* (isolate MYC43) was studied in vitro by monitoring the radial growth of the mycelium (Figures 1 and 2).

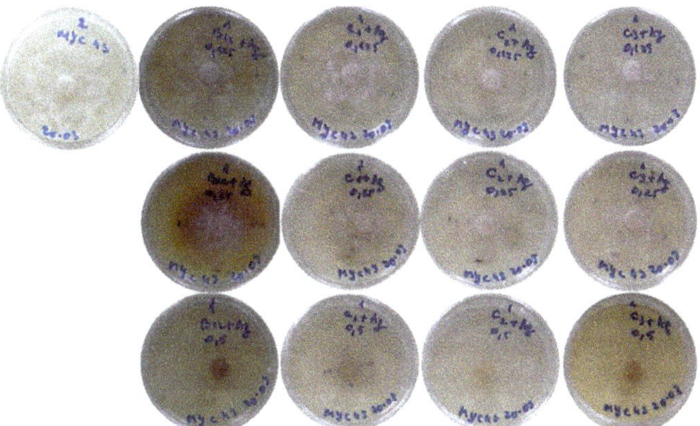

Figure 1. *Phytophthora cinnamomi* growth inhibition assays with the nanocomposites based on the chitosan oligomers in hydroalcoholic solution preparation medium. *From left to right*: Control (no treatment) and treatments with AgNPs combined with gallic acid, silymarin, ferulic acid and curcumin inclusion compounds. *From top to bottom*: 125 µg·mL^{-1}, 250 µg·mL^{-1}, and 500 µg·mL^{-1} concentrations. Only one repetition per treatment is shown.

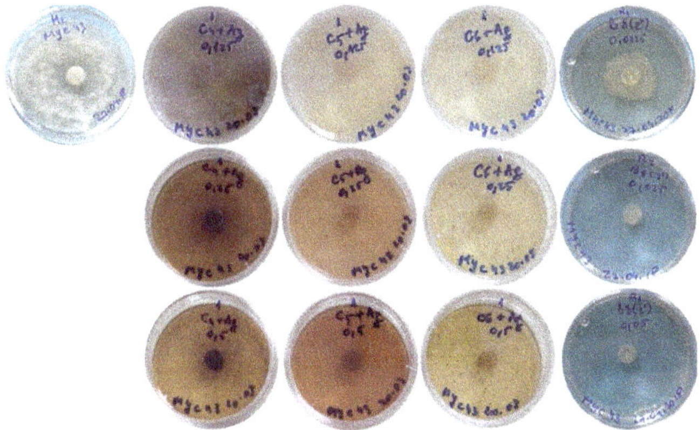

Figure 2. *P. cinnamomi* growth inhibition assays with the nanocomposites based on the deep eutectic solvent preparation medium. *From left to right*: Control (no treatment) and treatments with AgNPs combined with gallic acid, silymarin, ferulic acid and curcumin inclusion compounds. *From top to bottom*: 125 µg·mL^{-1}, 250 µg·mL^{-1} and 500 µg·mL^{-1} concentrations. Only one repetition per treatment is shown. The blue background in the control and curcumin-treated samples is due to the blue color of the paper on which the plates were on.

As shown in Figure 3, the increase in the concentration of the inclusion complexes from 125 µg·mL^{-1} to 500 µg·mL^{-1} resulted in a reduction in the radial growth of the mycelium in all cases. It may be observed that 100% mycelial growth inhibition occurred with the COS medium at the highest concentration of 500 µg·mL^{-1} for the composites with gallic acid, ferulic acid and curcumin (but not for silymarin). On the other hand, at lower concentrations (125 and 250 µg·mL^{-1}), silymarin and ferulic acid-based treatments were more effective than those based on gallic acid and curcumin.

As regards the nanocomposites with a DES preparation medium, total inhibitory activities were attained for the composites based on the four polyphenols under study concentrations of 250, and 500 µg·mL^{-1}. Further, at the lowest concentration of 125 µg·mL^{-1}, the antifungal performance of the composites in the DES medium was close to 90% for all polyphenols. Thus, the product efficacies were clearly higher than those of the composites based on the hydroalcoholic solution of COS.

Upon comparison with the treatments without phenolic inclusion compounds, it could be observed that the AgNPs-only treatment attained a lower inhibition than the composites for the COS medium, with mycelium growth even at the highest dose. On the other hand, the DES-based AgNPs-only treatment performed noticeably better than its COS counterpart, albeit it did not completely inhibit growth at a concentration of 250 mg·mL^{-1} (whereas all the composites did). Therefore, an enhanced fungal growth control activity of the ternary mixtures was evidenced.

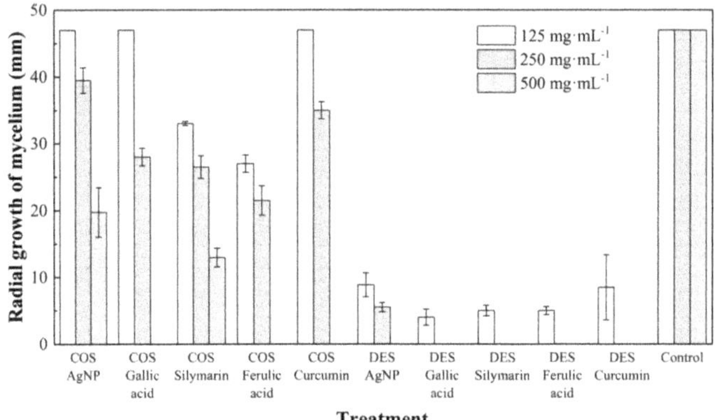

Figure 3. Radial growth values of *P. cinnamomi* in the presence of the composites, which consisted of different polyphenol inclusion compounds and silver nanoparticles at different concentrations, either in a chitosan hydroalcoholic solution (COS) or in a deep eutectic solvent (DES). A control (no treatment) and two treatments with AgNPs (in COS and DES) without the inclusion compounds are shown for comparison purposes. Error bars represent the standard deviation across three replicates.

The results from the sensitivity tests for *P. cinnamomi* may also be expressed with the help of EC_{50} and EC_{90} indicators (Table 1). The sensitivity of the isolate mainly varied as a function of the preparation media, but also according to the phenolic compound used. In line with the discussion presented above, *P. cinnamomi* (MYC43) was found to be remarkably more sensitive to the treatments prepared in DES, with EC_{50} values ranging from 0.1 to 8.9 µg·mL^{-1} and EC_{90} values between 77.9 and 184.3 µg·mL^{-1}. For comparison purposes, for the COS in hydroalcoholic solution medium treatments the EC_{50} values ranged from 171.6 to 279.9 µg·mL^{-1} and those of EC_{90} from 450.4 to 963.7 µg·mL^{-1}.

Table 1. Effective concentrations that inhibited mycelial growth by 50% and 90% (EC_{50} and EC_{90}, respectively).

Treatment	EC_{50} (µg·mL^{-1})	EC_{90} (µg·mL^{-1})
COS AgNPs	458.4	1192.8
COS Gallic acid	261.3	455.6
COS Silymarin	261.8	963.7
COS Ferulic acid	171.6	450.4
COS Curcumin	279.9	487.4
DES AgNPs	13.3	253.3
DES Gallic acid	0.1	77.9
DES Silymarin	0.6	107.8
DES Ferulic acid	0.6	107.8
DES Curcumin	8.9	184.3

The highest sensitivity of the *P. cinnamomi* isolate (MYC43) corresponded to the inclusion compound with gallic acid in DES (EC_{50} = 0.1 µg·mL^{-1}), followed by the inclusion compounds with silymarin and ferulic acid, and finally by the one with curcumin, with EC_{50} values of 0.6, 0.6 and 8.9 µg·mL^{-1}, respectively.

3. Discussion

With a view to comparing the efficacy of the proposed nanocomposites verses other phenolic-based products against *Phytophthora* spp. discussed in the literature, it should be noted that Kim et al. [37] reported strong fungicidal activities of *Curcuma longa* L. rhizome-derived curcumin in ethyl acetate and hexane fractions against *P. infestans* with 100% and 84% control values at a concentration of 1000 µg·mL^{-1}. Apart from the higher concentrations used in those experiments as compared to the ones presented herein, it should be noted that a contribution of the pronounced cytotoxic activities of the solvents could not be excluded.

Pompimon et al. [38] assessed the anti-*P. capsici* activity of *C. longa* in acetone fraction, finding an inhibition of mycelial growth of ca. 90% at a concentration of 300 µg·mL^{-1}, higher than the EC_{90} of 184 µg·mL^{-1} of the curcumin-based treatment in DES media reported in this study.

The nanocomposites of the four polyphenols in DES medium would also be more active than, for instance, cuminic acid, which featured an EC_{50} value against mycelial growth of *P. capsici* of 19.7 µg· mL^{-1} (which in turn was lower than the EC_{50} value of other benzoic acid derivatives in previous reports) [39]. Other natural compounds, such as furanocoumarins (e.g., psoralen or isopsoralen) would require concentrations of 500 µg·mL^{-1} to attain 82–84% disease control against *P. infestans* [40].

An EC_{50} value of amphopolycarboxyglycinate-stabilized AgNPs against *P. infestans* of 3.1 µg·mL^{-1} was reported by Krutyakov et al. [41], i.e., a 30 and 6 times higher concentration than those obtained for the AgNPs combined with gallic acid, silymarin and ferulic acid inclusion compounds in DES, respectively.

Banik and Pérez-de-Luque [42] found that the integration of copper nanoparticles (CuNPs) with non-nano copper like copper oxychloride, both at a 50 µg·mL^{-1} concentration, resulted in a 76% growth inhibition in vitro of the oomycete *P. cinnamomi* as compared to the control. Since in comparative assays between NPs the concentrations of AgNPs are usually 10 times higher than those CuNPs [43,44], the equivalent concentration of AgNPs to attain aforementioned effects should be 500 ca. µg·mL^{-1}, four times higher than that required by the DES treatments to attain a comparable mycelial growth inhibition (ca. 90%).

Chitosan has also been assayed against *P. infestans* [33], finding that concentrations of 500 µg· mL^{-1} would be required to fully inhibit mycelial growth, similar to those of the COS-based nanocomposites with gallic acid, ferulic acid and curcumin in this study. On the other hand, *N*-(6-carboxyl cyclohex-3-ene carbonyl) chitosan with different degrees of substitution achieved an EC_{50} of 298 µg·mL^{-1} for *P. infestans* [45], better than the ones for the COS composites based on the three aforementioned polyphenols (in the 450–490 µg·mL^{-1} range).

The overall efficacy of the reported nanocomposites should be referred to the combination of the properties afforded by each of its constituents and their synergies.

According to Kim et al. [46], nanosilver may exert an antifungal activity by disrupting the structure of the cell membrane and inhibiting the normal budding process due to the destruction of the membrane integrity. Silver nanoparticles antifungal action may also result from the release of silver ions into the intracellular matrix of the pathogen [47]. Reports on the mechanism of inhibitory action of silver ions on microorganisms have shown that upon treatment with Ag$^+$, DNA loses its ability to replicate, resulting in inactivated expression of ribosomal subunit proteins, as well as certain other cellular proteins and enzymes essential to ATP production. It has also been hypothesized that Ag$^+$ would affect the function of membrane-bound enzymes, such as those in the respiratory chain [4].

Apropos of the role of the stevioside, the improvement in the solubility and bioavailability of the polyphenolic compounds should be ascribed to the formation of a nanocomposite structure comprising a transglycosylated compound, which includes the insoluble compounds. Transglycosylated materials have been reported to self-associate into particular micelle-like structures with a core-shell-like architecture, in which the hydrophobic skeleton is segregated from the aqueous exterior to form a novel drug-loading core, surrounded by a hydrophilic shell of sugar groups [48]. For instance, Kadota, Okamoto, Sato, Onoue, Otsu and Tozuka [24] found a 13000× increase in curcumin solubility when the tri-component system curcumin/α-glucosyl stevia/polyvinylpyrrolidone was used.

In relation to the phenolic compounds, they have been reported to have toxic activities against fungi involved in the deterioration of agricultural products by interfering with the development of mycelia [49]. They affect membrane functions such as electron transport, nutrition, enzyme activity, protein and nucleic acid synthesis, and they interact with membrane proteins, causing disruption of the structures and functionality. For instance, curcumin's efficacy would be influenced by its lipophilic nature, which leads to an adequate transmembrane permeability [16]. Its antifungal mechanism has been ascribed to the disruption of plasma membrane integrity, causing leakage of potassium ion from the cytosol and change in membrane potential [50]. On the other hand, gallic acid would exhibit both antioxidant as well as pro-oxidant characteristics, displaying a dual-edge sword behavior, which turns it into an efficient apoptosis inducing agent [51].

Regarding the inhibition mode of chitosan, three mechanisms have been proposed [52]: (1) Its positive charge can interact with negatively charged phospholipid components of fungi membrane, increasing its permeability and causing the leakage of cellular contents, which subsequently leads to cell death; (2) it can act as a chelating agent by binding to trace elements, causing the essential nutrients unavailable for normal growth of fungi; and (3) it may be able to penetrate the cell wall of fungi and bind to its DNA, inhibiting the synthesis of mRNA and, thus, affecting the production of essential proteins and enzymes.

As far as DES are concerned, they would act as a plasticizer, affecting the apparent viscosity of the solutions and enhancing water vapor permeability, water solubility and water sorption capability, as reported by Almeida, Magalhães, Souza and Gonçalves [27]. Nonetheless, it worth noting that, while choline chloride and urea show no inhibition as individual materials, their final product as ChCl:U DES has been reported to show inhibition towards *Candida cylindracea* [53]. This behavior could be due to the synergistic effect of forming DES [54] and can be used to prove that occasionally DES have a higher toxicological behavior than its original components.

The antimicrobial activity of DES is still not fully understood [55]. Some reports have noted that DES would increase the permeability of the lipid membrane of eukaryotic cells [56,57], as chitosan does. Since the mechanism for COS and DES would tentatively be similar, the differences in the performance of the composites presented herein in terms of fungal growth control should then be ascribed to differences in their ability to solubilize a wide range of solutes (e.g., components in the fungal cell membrane), pH, osmolality or chelation of membrane-bound divalent cations [58].

Since one of the possible mechanisms of action of silver requires that the silver ions enter the fungal cell for efficient killing, the enhancement of permeability driven by COS, DES and polyphenols would support that their interaction should be synergistic rather than simply additive.

From our work, the best results of mycelial growth inhibition at the lowest concentration (125 mg·mL^{-1}) in DES (GI 91.5%), attained for the composite based on gallic acid, may be ascribed to the fact that gallic acid is extremely well absorbed, and very soluble in water as compared with other polyphenols [51]. Moreover, the introduction of the hydroxyl group on the cation in the chloride of choline salt has also been reported to significantly improve the extraction capacity of ionic liquids for gallic acid [59].

4. Materials and Methods

4.1. Reagents

Stevioside standard was purchased from Wako (Osaka, Japan). Gallic acid, silymarin, ferulic acid, curcumin, choline chloride, urea, glycerol, and silver nanoparticles (40 nm particle size

(TEM), 0.02 mg·mL^{-1} in aqueous buffer, with sodium citrate as a stabilizer) were purchased from Sigma-Aldrich/Merck KGaA (Darmstadt, Germany).

Chitosan oligomers were obtained from medium molecular weight chitosan (supplied by Hangzhou Simit Chemical Technology Co., Ltd., Hangzhou, China), dissolving 10 g in 500 mL of acetic acid (1%) under constant stirring at 60 °C. Once dissolved, hydrogen peroxide (0.3 mol·L^{-1}) was added for the degradation of the polymer chains, obtaining oligomers of less than 2000 Da [60].

Liquefaction of the choline chloride and urea (1:2 *v/v*) eutectic mixture occurred at 80 °C under stirring in a hot-plate magnetic stirrer for 10 min, in good agreement with Biswas et al. [61].

4.2. Microwave-Assisted Preparation of the Polyphenol Inclusion Compounds

The aqueous biphasic system separation technique was used for the formation of the inclusion compounds. This technology is a liquid-liquid extraction system for bioseparation and is frequently used to process all types of biotechnological materials, such as proteins, enzymes, phytochemicals, nucleic acids and pigments [62]. In the case under study, the inclusion complexes formed with stevioside were recovered from the upper part of the reactor.

4.3. Chitosan Oligomers (COS)-Based Composites

In each of the four jars with screw caps, 1 g of chitosan oligomers of 2000 Da (COS), 0.06 g of stevioside and 0.01 g of one of the polyphenols (either gallic acid, silymarin, ferulic acid or curcumin) were added to 40 mL of hydroalcoholic solution (1:1 *v/v* distilled water and ethanol), followed by treatment in a microwave (Milestone Ethos-One, Sorisole, BG, Italy) at 80 °C for 20 min under stirring. The resulting solutions were centrifuged at 2500 rpm and stored at 4 °C. For the incorporation of the silver nanoparticles to the nanocomposite, 0.1 mL of commercial AgNPs (0.02 mg·mL^{-1}) were added dropwise to 0.9 mL of the previously centrifuged microwave fractions, and the final solution (with pH 7.5) was stirred at room temperature.

4.4. Deep Eutectic Solvent-Based Composites

To prepare the DES-based composites, 20 mL of choline chloride and urea (1:2 *v/v*) DES and 10 mL of glycerol were added to each of the four jars with screw caps, together with 0.03 g of stevioside and 0.01 g of the respective polyphenol (gallic acid, silymarin, ferulic acid or curcumin). The mixture was heated at 80 °C in the microwave, under stirring, for 20 min. Then, 0.1 mL of AgNPs (0.02 mg·mL^{-1}) were added dropwise to 0.9 mL of the microwave fractions (previously centrifuged at 2500 rpm). The mixture, with pH 7.5, was subjected to vigorous stirring for 5 min at room temperature.

4.5. Silver Nanoparticles-Only Treatments

Two additional treatments consisting of 0.1 mL of AgNPs (0.02 mg·mL^{-1}) added dropwise to 0.9 mL of the dispersion medium (either COS or DES), without the polyphenol inclusion compounds, were prepared for control purposes. The mixtures were vigorously stirred for 5 min at room temperature.

4.6. Characterization of the Nanocomposites

The structure and properties of the nanocomposites obtained through the synthetic procedures described above were characterized using Fourier-Transform Infrared spectroscopy (FTIR), scanning electron microscopy (SEM) and transmission electron microscopy (TEM). Results were recently reported in patent P201731489 [36].

4.7. Fungal Isolates and Growth Conditions

The fungal species used in the experiment was *Phytophthora cinnamomi* isolate MYC43, supplied by Centro de Investigaciones Científicas y Tecnológicas de Extremadura—Instituto del Corcho, la Madera y el Carbón Vegetal, Spain. The isolate was maintained in potato-dextrose-agar (PDA) slant tubes, supplied by Merck Millipore (Darmstadt, Germany), stored at 4 °C.

4.8. Efficacy of the Nanocomposites for the Control of Phytophthora cinnamomi

Agar disks (8 mm in diameter) were cut from the margin of a 7-day-old colony growing on PDA and were transferred to a PDA medium supplemented with the nanocomposites at final concentrations of 125 μg·mL^{-1}, 250 μg·mL^{-1} and 500 μg·mL^{-1}. Three replicates were performed for each treatment. For each active ingredient and concentration, inhibition of radial mycelial growth (mm) compared with the untreated control was evaluated after 7 days of incubation at 24 °C, in the dark. The relative growth inhibition (%) of each treatment compared to untreated control was calculated as follows: Growth inhibition (%) = $[(dc - dt)/dc] \times 100$, where dc stands for the average diameter of the fungal colony in the control and dt is the average diameter of the treated colony [63]. Results were expressed as effective concentrations EC$_{50}$ and EC$_{90}$ (i.e., the concentrations which reduced growth inhibition by 50% and 90%) by regressing the inhibition of radial mycelial growth values (% control) against the values of the antifungal nanocomposite concentrations.

5. Conclusions

Composites consisting of silver nanoparticles and polyphenol inclusion compounds were synthesized using two preparation media, one based on chitosan oligomers in an hydroalcoholic solution and the other based on a deep eutectic solvent. Both types of composites showed an increase in the water solubility of the polyphenols and in the in vitro antifungal activity against *Phytophthora cinnamomi* (MYC43 isolate). Nonetheless, the DES-based samples efficacy was remarkably higher than that of their counterparts with a chitosan oligomers-based matrix: Complete inhibition of mycelial growth was attained at concentrations of 250 and 500 μg·mL^{-1}, and even at the lowest dose of 125 μg·mL^{-1}, they resulted in a 90% growth inhibition. As regards the impact of the choice of the different polyphenols, for the DES-based treatments, the highest sensitivity of *P. cinnamomi* corresponded to the composite with gallic acid (EC$_{50}$ = 0.1 μg·mL^{-1}), followed by those with silymarin and ferulic acid, and finally by the one with curcumin, with EC$_{50}$ values of 0.6, 0.6 and 8.9 μg·mL^{-1}, respectively. This may be ascribed to the fact that gallic acid is extremely well absorbed, and very soluble in water as compared with other polyphenols. The reported activities for the composites were remarkably higher than those reported for *Phytophthora* spp. using AgNPs, chitosan or polyphenols separately. This points to the possibility of a successful application of these nanocomposites in agriculture, with the aim of reducing the use of toxic and expensive conventional systemic fungicides. Further research on the ability of the prepared nanocomposites to inhibit growth of *P. cinnamomi* in the context of a plant infection model and on the basic mechanism involved, fungistatic or fungicidal, is underway.

6. Patents

The work reported in this manuscript is related to Spanish patent P201731489.

Author Contributions: P.M.-R., M.T.B.-A., E.P.-L. and J.M.-G. designed the experiments; J.M.-G. contributed reagents and conducted the synthesis; P.M.M. and B.M.I. performed the experiments; P.M.M., J.M.-G., M.T.B.-A. and P.M.-R. analyzed the data and wrote the paper.

Funding: This research was partly funded by Sistemas Biotecnología y Recursos Naturales.

Acknowledgments: P.M.-R. would like to thank Santander Universidades for its financial support through the "Becas Iberoamérica Jóvenes Profesores e Investigadores, España" scholarship program.

References

1. Kim, D.-Y.; Kadam, A.; Shinde, S.; Saratale, R.G.; Patra, J.; Ghodake, G. Recent developments in nanotechnology transforming the agricultural sector: A transition replete with opportunities. *J. Sci. Food Agric.* **2018**, *98*, 849–864. [CrossRef] [PubMed]

2. Berger, T.J.; Spadaro, J.A.; Chapin, S.E.; Becker, R.O. Electrically generated silver ions: Quantitative effects on bacterial and mammalian cells. *Antimicrob. Agents Chemother.* **1976**, *9*, 357–358. [CrossRef] [PubMed]

3. Xia, Z.-K.; Ma, Q.-H.; Li, S.-Y.; Zhang, D.-Q.; Cong, L.; Tian, Y.-L.; Yang, R.-Y. The antifungal effect of silver nanoparticles on *Trichosporon asahii*. *J. Microbiol. Immunol. Infect.* **2016**, *49*, 182–188. [CrossRef] [PubMed]

4. Kim, S.W.; Jung, J.H.; Lamsal, K.; Kim, Y.S.; Min, J.S.; Lee, Y.S. Antifungal effects of silver nanoparticles (AgNPs) against various plant pathogenic fungi. *Mycobiology* **2018**, *40*, 53–58. [CrossRef] [PubMed]

5. Kharissova, O.V.; Dias, H.V.R.; Kharisov, B.I.; Pérez, B.O.; Jiménez Pérez, V.M. The greener synthesis of nanoparticles. *Trends Biotechnol.* **2013**, *31*, 240–248. [CrossRef] [PubMed]

6. Rajan, R.; Chandran, K.; Harper, S.L.; Yun, S.-I.; Kalaichelvan, P.T. Plant extract synthesized silver nanoparticles: An ongoing source of novel biocompatible materials. *Ind. Crop. Prod.* **2015**, *70*, 356–373. [CrossRef]

7. Bujak, T.; Nizioł-Łukaszewska, Z.; Gaweł-Bęben, K.; Seweryn, A.; Kucharek, M.; Rybczyńska-Tkaczyk, K.; Matysiak, M. The application of different *Stevia rebaudiana* leaf extracts in the "green synthesis" of AgNPs. *Green Chem. Lett. Rev.* **2015**, *8*, 78–87. [CrossRef]

8. Sathishkumar, M.; Sneha, K.; Yun, Y.-S. Immobilization of silver nanoparticles synthesized using *Curcuma longa* tuber powder and extract on cotton cloth for bactericidal activity. *Bioresour. Technol.* **2010**, *101*, 7958–7965. [CrossRef] [PubMed]

9. Raut, R.W.; Kolekar, N.S.; Lakkakula, J.R.; Mendhulkar, V.D.; Kashid, S.B. Extracellular synthesis of silver nanoparticles using dried leaves of *Pongamia pinnata* (L.) Pierre. *Nano Micro Lett.* **2010**, *2*, 106–113. [CrossRef]

10. Raut, W.; Lakkakula, R.; Kolekar, S.; Mendhulkar, D.; Kashid, B. Phytosynthesis of silver nanoparticle using *Gliricidia sepium* (Jacq.). *Curr. Nanosci.* **2009**, *5*, 117–122.

11. Rahimi-Nasrabadi, M.; Pourmortazavi, S.M.; Shandiz, S.A.S.; Ahmadi, F.; Batooli, H. Green synthesis of silver nanoparticles using *Eucalyptus leucoxylon* leaves extract and evaluating the antioxidant activities of extract. *Nat. Prod. Res.* **2014**, *28*, 1964–1969. [CrossRef] [PubMed]

12. Safaei, H.R. Eco-friendly method for green recovery of silver nano particles from effluent fixer solution through interacting with *Q. brantii* (oak) peel hydro alcoholic extract. *J. Adv. Eng. Technol.* **2017**, *5*, 1–7.

13. Li, D.; Liu, Z.; Yuan, Y.; Liu, Y.; Niu, F. Green synthesis of gallic acid-coated silver nanoparticles with high antimicrobial activity and low cytotoxicity to normal cells. *Process Biochem.* **2015**, *50*, 357–366. [CrossRef]

14. Safarpoor, M.; Ghaedi, M.; Asfaram, A.; Yousefi-Nejad, M.; Javadian, H.; Zare Khafri, H.; Bagherinasab, M. Ultrasound-assisted extraction of antimicrobial compounds from *Thymus daenensis* and *Silybum marianum*: Antimicrobial activity with and without the presence of natural silver nanoparticles. *Ultrason. Sonochem.* **2018**, *42*, 76–83. [CrossRef] [PubMed]

15. Alves, T.F.; Chaud, M.V.; Grotto, D.; Jozala, A.F.; Pandit, R.; Rai, M.; dos Santos, C.A. Association of silver nanoparticles and curcumin solid dispersion: Antimicrobial and antioxidant properties. *AAPS PharmSciTech* **2017**, *19*, 225–231. [CrossRef] [PubMed]

16. Hussain, Z.; Thu, H.E.; Amjad, M.W.; Hussain, F.; Ahmed, T.A.; Khan, S. Exploring recent developments to improve antioxidant, anti-inflammatory and antimicrobial efficacy of curcumin: A review of new trends and future perspectives. *Mater. Sci. Eng. C* **2017**, *77*, 1316–1326. [CrossRef] [PubMed]

17. Oliver, S.; Vittorio, O.; Cirillo, G.; Boyer, C. Enhancing the therapeutic effects of polyphenols with macromolecules. *Polym. Chem.* **2016**, *7*, 1529–1544. [CrossRef]

18. Liu, J.; Pu, H.; Liu, S.; Kan, J.; Jin, C. Synthesis, characterization, bioactivity and potential application of phenolic acid grafted chitosan: A review. *Carbohydr. Polym.* **2017**, *174*, 999–1017. [CrossRef] [PubMed]

19. Hashemi Gahruie, H.; Niakousari, M. Antioxidant, antimicrobial, cell viability and enzymatic inhibitory of antioxidant polymers as biological macromolecules. *Int. J. Biol. Macromol.* **2017**, *104*, 606–617. [CrossRef] [PubMed]

20. Saranya, T.S.; Rajan, V.K.; Biswas, R.; Jayakumar, R.; Sathianarayanan, S. Synthesis, characterisation and biomedical applications of curcumin conjugated chitosan microspheres. *Int. J. Biol. Macromol.* **2018**, *110*, 227–233. [CrossRef] [PubMed]

21. Bajpai, S.K.; Ahuja, S.; Chand, N.; Bajpai, M. Nano cellulose dispersed chitosan film with Ag NPs/curcumin: An in vivo study on albino rats for wound dressing. *Int. J. Biol. Macromol.* **2017**, *104*, 1012–1019. [CrossRef] [PubMed]

22. Barbinta-Patrascu, M.E.; Badea, N.; Pirvu, C.; Bacalum, M.; Ungureanu, C.; Nadejde, P.L.; Ion, C.; Rau, I. Multifunctional soft hybrid bio-platforms based on nano-silver and natural compounds. *Mater. Sci. Eng. C* **2016**, *69*, 922–932. [CrossRef] [PubMed]

23. Nguyen, T.T.H.; Si, J.; Kang, C.; Chung, B.; Chung, D.; Kim, D. Facile preparation of water soluble curcuminoids extracted from turmeric (*Curcuma longa* L.) powder by using steviol glucosides. *Food Chem.* **2017**, *214*, 366–373. [CrossRef] [PubMed]

24. Kadota, K.; Okamoto, D.; Sato, H.; Onoue, S.; Otsu, S.; Tozuka, Y. Hybridization of polyvinylpyrrolidone to a binary composite of curcumin/α-glucosyl stevia improves both oral absorption and photochemical stability of curcumin. *Food Chem.* **2016**, *213*, 668–674. [CrossRef] [PubMed]

25. Ruesgas-Ramón, M.; Figueroa-Espinoza, M.C.; Durand, E. Application of deep eutectic solvents (DES) for phenolic compounds extraction: Overview, challenges, and opportunities. *J. Agric. Food. Chem.* **2017**, *65*, 3591–3601. [CrossRef] [PubMed]

26. Georgantzi, C.; Lioliou, A.-E.; Paterakis, N.; Makris, D. Combination of lactic acid-based deep eutectic solvents (DES) with β-cyclodextrin: Performance screening using ultrasound-assisted extraction of polyphenols from selected native Greek medicinal plants. *Agronomy* **2017**, *7*, 54. [CrossRef]

27. Almeida, C.M.R.; Magalhães, J.M.C.S.; Souza, H.K.S.; Gonçalves, M.P. The role of choline chloride-based deep eutectic solvent and curcumin on chitosan films properties. *Food Hydrocoll.* **2018**, *81*, 456–466. [CrossRef]

28. Pereira, P.F.; Andrade, C.T. Optimized pH-responsive film based on a eutectic mixture-plasticized chitosan. *Carbohydr. Polym.* **2017**, *165*, 238–246. [CrossRef] [PubMed]

29. Kroon, L.P.N.M.; Brouwer, H.; de Cock, A.W.A.M.; Govers, F. The genus *Phytophthora* anno 2012. *Phytopathology* **2012**, *102*, 348–364. [CrossRef] [PubMed]

30. Munnecke, D.E. Establishment of micro-organisms in fumigated avocado soil to attempt to prevent reinvasion of the soils by *Phytophthora cinnamomi*. *Trans. Br. Mycol. Soc.* **1984**, *83*, 287–294 [CrossRef]

31. Zentmyer, G.A. The world of Phytophthora. In *Phytophthora, Its biology, Taxonomy, Ecology and Pathology*; Erwin, D.C., Bartnicki-García, S., Tsao, P.H., Eds.; American Phytopathological Society: St. Paul, MN, USA, 1983; pp. 1–8.

32. Ali, M.; Kim, B.; Belfield, K.D.; Norman, D.; Brennan, M.; Ali, G.S. Inhibition of *Phytophthora parasitica* and *P. capsici* by silver nanoparticles synthesized using aqueous extract of *Artemisia absinthium*. *Phytopathology* **2015**, *105*, 1183–1190. [CrossRef] [PubMed]

33. Atia, M.M.M.; Buchenauer, H.; Aly, A.Z.; Abou-Zaid, M.I. Antifungal activity of chitosan against *Phytophthora infestans* and activation of defence mechanisms in tomato to late blight. *Biol. Agric. Hortic.* **2005**, *23*, 175–197. [CrossRef]

34. Wang, L.-S.; Wang, C.-Y.; Yang, C.-H.; Hsieh, C.-L.; Chen, S.-Y.; Shen, C.-Y.; Wang, J.-J.; Huang, K.-S. Synthesis and anti-fungal effect of silver nanoparticles-chitosan composite particles. *Int. J. Nanomed.* **2015**, *10*, 2685–2696.

35. Silva-Castro, I.; Martín-García, J.; Diez, J.J.; Flores-Pacheco, J.A.; Martín-Gil, J.; Martín-Ramos, P. Potential control of forest diseases by solutions of chitosan oligomers, propolis and nanosilver. *Eur. J. Plant Pathol.* **2017**, *150*, 401–411. [CrossRef]

36. Martín-Gil, J.; Matei Petruta, M.; Pérez Lebeña, E. Complejo de inclusión para mejorar la biodisponibilidad de compuestos biológicamente activos no hidrosolubles. P201731489, 28 December 2017.

37. Kim, M.K.; Choi, G.J.; Lee, H.S. Fungicidal property of *Curcuma longa* L. rhizome-derived curcumin against phytopathogenic fungi in a greenhouse. *J. Agric. Food Chem.* **2003**, *51*, 1578–1581. [CrossRef] [PubMed]

38. Pompimon, W.; Jomduang, J.; Prawat, U.; Mankhetkorn, S. Anti-*Phytopthora capsici* activities and potential use as antifungal in agriculture of *Alpinia galanga* Swartz, *Curcuma longa* Linn, *Boesenbergia pandurata* Schut and *Chromolaena odorata*: Bioactivities guided isolation of active ingredients. *Am. J. Agric. Biol. Sci.* **2009**, *4*, 83–91. [CrossRef]

39. Hu, L.-F.; Chen, C.-Z.; Yi, X.-H.; Feng, J.-T.; Zhang, X. Inhibition of p-isopropyl benzaldehyde and p-isopropyl benzoic acid extracted from *Cuminum cyminum* against plant pathogens. *Acta Bot. Boreal. Occident. Sin.* **2008**, *11*, 42.

40. Shim, S.-H.; Kim, J.-C.; Jang, K.-S.; Choi, G.-J. Anti-oomycete activity of furanocoumarins from seeds of *Psoralea corylifolia* against *Phytophthora infestans*. *Plant Pathol. J.* **2009**, *25*, 103–107. [CrossRef]

41. Krutyakov, Y.A.; Kudrinskiy, A.A.; Zherebin, P.M.; Yapryntsev, A.D.; Pobedinskaya, M.A.; Elansky, S.N.; Denisov, A.N.; Mikhaylov, D.M.; Lisichkin, G.V. Tallow amphopolycarboxyglycinate-stabilized silver nanoparticles: New frontiers in development of plant protection products with a broad spectrum of action against phytopathogens. *Mater. Res. Express* **2016**, *3*, 1–9. [CrossRef]

42. Banik, S.; Pérez-de-Luque, A. In vitro effects of copper nanoparticles on plant pathogens, beneficial microbes and crop plants. *Span. J. Agric. Res.* **2017**, *15*. [CrossRef]

43. Ostaszewska, T.; Chojnacki, M.; Kamaszewski, M.; Sawosz-Chwalibóg, E. Histopathological effects of silver and copper nanoparticles on the epidermis, gills, and liver of Siberian sturgeon. *Environ. Sci. Pollut. Res.* **2015**, *23*, 1621–1633. [CrossRef] [PubMed]

44. Dobrochna, A.; Jerzy, S.; Teresa, O.; Magda, F.; Malgorzata, R.; Yuichiro, M.; Kacper, M. Effect of copper and silver nanoparticles on trunk muscles in rainbow trout (*Oncorhynchus mykiss*, Walbaum, 1792). *Turk. J. Fish. Aquat. Sci.* **2018**, *18*, 781–788. [CrossRef]

45. Badawy, M.E.I.; Rabea, E.I. Synthesis and antimicrobial activity of *N*-(6-carboxyl cyclohex-3-ene carbonyl) chitosan with different degrees of substitution. *Int. J. Carbohydr. Chem.* **2016**, *2016*, 1–10. [CrossRef]

46. Kim, K.-J.; Sung, W.S.; Suh, B.K.; Moon, S.-K.; Choi, J.-S.; Kim, J.G.; Lee, D.G. Antifungal activity and mode of action of silver nano-particles on *Candida albicans*. *BioMetals* **2008**, *22*, 235–242. [CrossRef] [PubMed]

47. Clement, J.L.; Jarrett, P.S. Antibacterial silver. *Met. Based Drugs* **1994**, *1*, 467–482. [CrossRef] [PubMed]

48. Zhang, J.; Higashi, K.; Ueda, K.; Kadota, K.; Tozuka, Y.; Limwikrant, W.; Yamamoto, K.; Moribe, K. Drug solubilization mechanism of α-glucosyl stevia by NMR spectroscopy. *Int. J. Pharm.* **2014**, *465*, 255–261. [CrossRef] [PubMed]

49. Ferreira, F.D.; Mossini, S.A.G.; Ferreira, F.M.D.; Arrotéia, C.C.; da Costa, C.L.; Nakamura, C.V.; Machinski Junior, M. The inhibitory effects of *Curcuma longa* L. essential oil and curcumin on *Aspergillus flavus* Link growth and morphology. *Sci. World J.* **2013**, *2013*, 1–6. [CrossRef] [PubMed]

50. Lee, W.; Lee, D.G. An antifungal mechanism of curcumin lies in membrane-targeted action within *Candida albicans*. *IUBMB Life* **2014**, *66*, 780–785. [CrossRef] [PubMed]

51. Badhani, B.; Sharma, N.; Kakkar, R. Gallic acid: A versatile antioxidant with promising therapeutic and industrial applications. *RSC Adv.* **2015**, *5*, 27540–27557. [CrossRef]

52. Ing, L.Y.; Zin, N.M.; Sarwar, A.; Katas, H. Antifungal activity of chitosan nanoparticles and correlation with their physical properties. *Int. J. Biomater.* **2012**, *2012*, 1–9. [CrossRef] [PubMed]

53. Juneidi, I.; Hayyan, M.; Mohd Ali, O. Toxicity profile of choline chloride-based deep eutectic solvents for fungi and *Cyprinus carpio* fish. *Environ. Sci. Pollut. Res.* **2016**, *23*, 7648–7659. [CrossRef] [PubMed]

54. Hayyan, M.; Hashim, M.A.; Al-Saadi, M.A.; Hayyan, A.; AlNashef, I.M.; Mirghani, M.E.S. Assessment of cytotoxicity and toxicity for phosphonium-based deep eutectic solvents. *Chemosphere* **2013**, *93*, 455–459. [CrossRef] [PubMed]

55. Radošević, K.; Čanak, I.; Panić, M.; Markov, K.; Bubalo, M.C.; Frece, J.; Srček, V.G.; Redovniković, I.R. Antimicrobial, cytotoxic and antioxidative evaluation of natural deep eutectic solvents. *Environ. Sci. Pollut. Res.* **2018**, *25*, 14188–14196. [CrossRef] [PubMed]

56. Papaccio, G.; Hayyan, M.; Looi, C.Y.; Hayyan, A.; Wong, W.F.; Hashim, M.A. In vitro and in vivo toxicity profiling of ammonium-based deep eutectic solvents. *PLoS ONE* **2015**, *10*. [CrossRef]

57. Mbous, Y.P.; Hayyan, M.; Wong, W.F.; Looi, C.Y.; Hashim, M.A. Unraveling the cytotoxicity and metabolic pathways of binary natural deep eutectic solvent systems. *Sci. Rep.* **2017**, *7*. [CrossRef] [PubMed]

58. Wikene, K.O.; Rukke, H.V.; Bruzell, E.; Tønnesen, H.H. Investigation of the antimicrobial effect of natural deep eutectic solvents (NADES) as solvents in antimicrobial photodynamic therapy. *J. Photochem. Photobiol. B Biol.* **2017**, *171*, 27–33. [CrossRef] [PubMed]

59. Fan, Y.; Li, X.; Yan, L.; Li, J.; Hua, S.; Song, L.; Wang, R.; Sha, S. Enhanced extraction of antioxidants from aqueous solutions by ionic liquids. *Sep. Purif. Technol.* **2017**, *172*, 480–488. [CrossRef]

60. Sun, T.; Zhou, D.; Xie, J.; Mao, F. Preparation of chitosan oligomers and their antioxidant activity. *Eur. Food Res. Technol.* **2007**, *225*, 451–456. [CrossRef]

61. Biswas, A.; Shogren, R.L.; Stevenson, D.G.; Willett, J.L.; Bhowmik, P.K. Ionic liquids as solvents for biopolymers: Acylation of starch and zein protein. *Carbohydr. Polym.* **2006**, *66*, 546–550. [CrossRef]
62. Raja, S.; Murty, V.R.; Thivaharan, V.; Rajasekar, V.; Ramesh, V. Aqueous two phase systems for the recovery of biomolecules—A review. *Sci. Technol.* **2012**, *1*, 7–16. [CrossRef]
63. Şesan, T.E.; Enache, E.; Iacomi, B.M.; Oprea, M.; Oancea, F.; Iacomi, C. In vitro antifungal activity of some plant extracts against *Fusarium oxysporum* in blackcurrant (*Ribes nigrum* L.). *Acta Sci. Pol. Hortorum Cultus* **2017**, *16*, 167–176. [CrossRef]

Article

Microbiome Analysis of Biofilms of Silver Nanoparticle-Dispersed Silane-Based Coated Carbon Steel Using a Next-Generation Sequencing Technique

Akiko Ogawa [1,*], Keito Takakura [1], Katsuhiko Sano [2], Hideyuki Kanematsu [3], Takehiko Yamano [4], Toshikazu Saishin [4] and Satoshi Terada [5]

1 Department of Chemistry and Biochemistry, National Institute of Technology, Suzuka College, Suzuka 510-0294, Japan; h24c17@ed.cc.suzuka-ct.ac.jp
2 D&D Corporation, Yokkaichi 512-1211, Japan; sano@ddcorp.co.jp
3 Department of Material Science and Engineering, National Institute of Technology, Suzuka College, Suzuka 510-0294, Japan; kanemats@mse.suzuka-ct.ac.jp
4 Department of Marine Technology, National Institute of Technology, Toba College, Toba 517-8501, Japan; yamano@toba-cmt.ac.jp (T.Y.); saishin@toba-cmt.ac.jp (T.S.)
5 Department of Applied Chemistry and Biochemistry, University of Fukui, Fukui 910-8507, Japan; terada@u-fukui.ac.jp
* Correspondence: ogawa@chem.suzuka-ct.ac.jp

Received: 16 August 2018; Accepted: 16 October 2018; Published: 22 October 2018

Abstract: Previously, we demonstrated that silver nanoparticle-dispersed silane-based coating could inhibit biofilm formation in conditions where seawater was used as a bacterial source and circulated in a closed laboratory biofilm reactor. However, it is still unclear whether the microbiome of a biofilm of silver nanoparticle-dispersed silane-based coating samples (Ag) differs from that of a biofilm of non-dispersed silane-based coating samples (Non-Ag). This study aimed to perform a microbiome analysis of the biofilms grown on the aforementioned coatings using a next-generation sequencing (NGS) technique. For this, a biofilm formation test was conducted by allowing seawater to flow through a closed laboratory biofilm reactor; subsequently, DNAs extracted from the biofilms of Ag and Non-Ag were used to prepare 16S rRNA amplicon libraries to analyze the microbiomes by NGS. Results of the operational taxonomy unit indicated that the biofilms of Non-Ag and Ag comprised one and no phyla of archaea, respectively, whereas Proteobacteria was the dominant phylum for both biofilms. Additionally, in both biofilms, Non-Ag and Ag, *Marinomonas* was the primary bacterial group involved in early stage biofilm formation, whereas *Anaerospora* was primarily involved in late-stage biofilm formation. These results indicate that silver nanoparticles will be unrelated to the bacterial composition of biofilms on the surface of silane-based coatings, while they control biofilm formation there.

Keywords: biofilm; microbiomes; silver nanoparticles; silane-based coating; *Marinomonas*; *Anaerospora*

---

1. Introduction

A ship's engine room has a plumbing system that includes a power unit, steam pipes, heat exchangers and fuel pipes. Generally, seawater is used in the ship's cooling system; however, this can lead to biofouling. Biofouling is a series of bioprocesses where material surfaces are initially covered by conditioning films of non-organic polymers, followed by biofilm formation by microorganisms, such as bacteria and archaea, and then by macroorganisms, such as algae and balanoids, which adhere to the surface [1]. Biofouling reduces the efficiency of heat exchange in the cooling system, and it may at times result in the destruction of heat exchange pipes [2–4]. The key stage in biofouling is biofilm formation. Biofilms are composed of bacteria and their extracellular polymeric substrates and can cause microbially-influenced corrosion (MIC) of steel [5,6].

Previously, we demonstrated that a combination of silver nanoparticles and silane-based coating inhibited biofilm formation on the surface of pipes of water cooling systems that used seawater as a coolant [7]. However, the microbiome composition of biofilms is still unknown. In this study, we utilized a next-generation sequencing (NGS) technique to study microbiomes of biofilms on steels with silver nanoparticles-dispersed silane-based coating.

2. Materials and Methods

2.1. Specimens

A carbon steel JIS SS400 plate (Sakai Netsu-Giken, Tsu, Japan) was cut into coupons of an area of 10×20 mm^2 (thickness, 1 mm) using a shearing machine (Komatsu Industries, Kanazawa, Japan). Each coupon was washed with acetone and then dried for 24 h in a desiccator before initiating the coating process. Silane-based coating solution was prepared as previously described [7]. Subsequently, silver nanoparticles (diameter, 100 nm; Sigma–Aldrich, St. Louis, MO, USA) were added to the silane-based coating solution. Next, the coating solution was filtered through a nylon mesh #110 (NBC Meshtec Inc., Hino, Japan) to remove any residues or impurities. The filtered Ag solution was then sprayed onto the surface of each SS400 coupon, which were subsequently incubated at 20 °C for 24 h.

2.2. Sampling Seawater

Seawater was collected in a sterile glass bottle (1 L, AGC Techno Glass, Haibara-gun, Japan) at a depth of 2 m on 7 November 2016. The sampling location was 1.6 km offshore in Ise Bay, Japan (34°31.20′ N, 136°48.36′ E). The seawater-filled bottle was covered using double-layered aluminum foil and stored in a refrigerator (4 °C–8 °C) until use.

2.3. Biofilm Formation Test

A laboratory biofilm reactor (LBR) comprising two polycarbonate columns (Sanplatec Corporation, Osaka, Japan), a three-necked, glass culture bottle and a peristaltic pump was used (Figure 1). All coupons were fixed in acryl holders (Sanplatec Corporation) using acryl screw pins (Sanplatec Corporation) with their coated surfaces facing upward. These holders were then placed in LBR columns, which were connected to silicone tubes (As One, Osaka, Japan). The silicone tubes were connected to the three-necked, glass culture bottle. After the assembly was sterilized using an autoclave (Tomy Seiko, Tokyo, Japan) at 121 °C for 15 min, the stored seawater (500 mL) was transferred to the sterile three-necked, glass culture bottle and was circulated by the peristaltic pump at a speed of dial 10 (Thermo Fisher Scientific, Waltham, MA, USA) at 20 °C for 139 h.

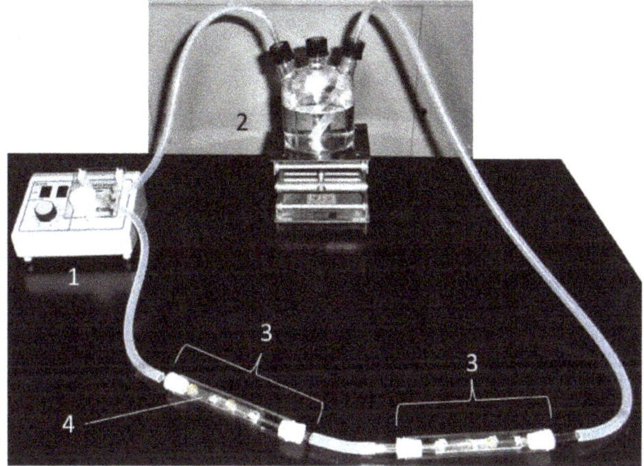

Figure 1. The closed laboratory biofilm reactor. 1: peristaltic pump; 2: culture bottle; 3: column; 4: sample holder. The seawater was circulated in a counterclockwise direction.

2.4. DNA Extraction from Seawater and Biofilms

A PowerSoil® DNA isolation kit (MO BIO Laboratories, Carlsbad, CA, USA) was used for DNA extraction, both from the biofilms on the surface of each coupon and from microbes in the seawater. Stored seawater was filtered through a 0.1-µm polyethersulfone filter (Sartorius Japan, Tokyo, Japan). After filtration, the sediment on the filter was collected in PowerSoil® Bead tubes (MO BIO Laboratories). Biofilms on the surface of each coupon were scratched using a sterile spatula, and the scratched biofilms were collected in PowerSoil® Bead tubes. Then, 60 µL of Solution C1 was added to each tube, which was then inverted several times, secured to a bead crusher (TITEC, Koshigaya, Japan), vortexed at 4600 rpm for 1 min and placed on ice. Vortexing and cooling was repeated nine more times. Any DNA present in tubes was then purified according to a previously reported procedure [7]. The concentration of the purified DNA solution was measured using the Qubit fluorometer (Thermo Fisher Scientific) and a dsDNA high sensitivity (HS) assay kit (Thermo Fisher Scientific).

2.5. 16S rRNA Gene-Based Bacterial Community Analysis

Bacterial and archaebacterial 16S rRNA genes were partially amplified using 16S rRNA V4 region primers: 515f (5′-ACACTCTTTCCCTACACGACGCTCTTCCGATCTGTGCCAG-CMGCCGCGGTAA-3′) and 806r (5′-GTGACTGGAGTTCAGACGTGTGCTCTTCCGATCT-GGACTACHVGGGTWTCTAAT-3′). The first sampled polymerase chain reaction (PCR) solution (total volume, 20 µL) comprised 2.0 µL of PCR buffer (TaKaRa BIO, Kusatsu, Japan), 1 U of ExTaq (TaKaRa BIO), 0.5 µM of primers (515f and 806r each), 0.2 mM of dNTP mixture, 2.0 µL of purified DNA sample and 12.2 µL of deionized distilled water (TaKaRa BIO). For PCR, the samples were initially subjected to 94 °C for 2 min, followed by 94 °C for 30 s, 50 °C for 30 s and 72 °C for 30 s (25 cycles) and finally by 72 °C for 5 min using a thermal cycler (TaKaRa BIO). At the tagging PCR step, each PCR amplicon was tagged for MiSeq Illumina sequencing using Index2 (a unique 8-bp sequence) and Index1 (5′-TCCTCTAC-3′), which was separately inserted into the 2ndF (5′-AATGATACGGCGACCACCGAGATCTACAC-Index2-ACACTCTTTCCCTACACGACGC-3′) and 2ndR primer (5′-CAAGCAGAAGACGGCATACGAGAT-Indexl-GTGACTGGAGTTCAGACGTGTG-3′). The tagging PCR solution (total volume, 20 µL) comprised 2 µL of the first PCR amplicon, 1 U of ExTaq, 2.0 µL of PCR buffer, 0.2 mM of dNTP mixture, 0.5 µM of 2ndF primer, 0.5 µM of 2ndR primer and deionized distilled water. The tagging PCR conditions were as follows: the initial step involved a treatment at 94 °C for 2 min; followed by a treatment at 94 °C for 30 s, 60 °C for 30 s and 72 °C for 30 s (8 cycles); and a final step of 72 °C for 5 min using a thermal cycler (TaKaRa BIO). After reaction completion, the concentration of all tagged PCR products was measured using a dsDNA HS assay kit and Qubit fluorometer. The quality of all tagged PCR products was assessed using a High Sensitivity NGS Fragment Analysis Kit (Advanced Analytical Technology, Ames, IA, USA) and Fragment Analyzer (Advanced Analytical Technology). Finally, all tagged PCR products were pooled in one tube, and NGS was performed using the MiSeq system (Illumina, San Diego, CA, USA). The 16S rRNA gene library was pre-processed and analyzed using a combination of USEARCH [8,9], bioinformatics software package QIIME [10] and FASTX-Toolkit [11], a fast processing tool. Complete tag-matching sequences were extracted from forward- and reverse-sequence raw data files using the fastq_barcode_splitter.pl script, and then, the primer sequences were removed. Extracted sequences were trimmed, and specific sequences, which had a quality score of <20 and length of <40 bp, were discarded using sickle tools [12]. Trimmed forward and reverse sequences with a merged length of 260 bases, reading length of 230 bases and minimum overlap size of 10 bases were merged using the FLASH software [13,14]. Next, merged sequences were filtered from 246 bases to 260 bases, then chimeric sequences were checked using the identify_chimeric_seqs.py script and uchime algorithm of USEARCH against reference sequences of 97% operational taxonomic units (OTU) using the Greengenes database [15]. After removing chimeric sequences from the merged sequences, non-chimeric sequences were clustered based on 97% identity, and an OTU table was created using the pick_de_novo_otus.py script. The Greengenes database was used for taxonomically assigning the bacteria. Raw data files have been deposited in the NCBI Sequence Read Archive and are awaiting an accession number.

3. Results and Discussions

We compared the microbiomes of stored seawater samples (Seawater), of biofilms on the surface of non-dispersed silane-based coating samples (Non-Ag) and of biofilms on the surface of silver nanoparticle-dispersed silane-based coating samples (Ag). At the phylum level, three archaea were detected in Seawater; two archaea were detected on the biofilms of Non-Ag; and no archaea were detected on the biofilms of Ag (Table 1). Crenarchaeota contains hyperthermophilic and acidophilic bacteria [16], whereas Euryarchaeota contains methanogens and halophiles [17]. The content of archaea in Seawater was approximately 2%, whereas that of archaea on the biofilms of Non-Ag was approximately 0.1%. The outstanding difference in the microbiomes of biofilms of Non-Ag compared with that of biofilms of Ag was that no archaea were observed on the biofilms on Ag. These results indicate that marine archaea do not prefer silane-based coating surfaces for attachment and growth; moreover, silver nanoparticles will enhance the effect of silane-based coating killing archaea.

Table 1. Percentage abundance of operational taxonomic units of archaea.

Seawater-1	Seawater-2	Non-Ag-1	Non-Ag-2	Ag-1	Ag-2	Phyla
0.58	0.54	0.12	0.00	0	0	Crenarchaeota
1.20	1.90	0	0	0	0	Euryarchaeota
0.02	0.03	0	0.11	0	0	Parvarchaeota

In all samples, the most common bacterial phylum identified was *Proteobacteria* (#6 in Figure 2), and over 90% of all organisms identified on the biofilms of Non-Ag and Ag belonged to this phylum. This result showed one possibility that MIC may occur in either Non-Ag or Ag because Proteobacteria comprises many MIC-related bacteria such as *Pseudomonas*, sulfate-reducing bacteria, *Acidithiobacillus* and *Gallionella*. Meanwhile, the proportion of Actinobacteria (#1 in Figure 2), Bacteroidetes (#2 in Figure 2), Cyanobacteria (#4 in Figure 2) and Planctomycetes (#5 in Figure 2) had markedly reduced on the biofilms of Non-Ag and Ag compared with Seawater. Huttunen-Saarivirta et al. reported significantly higher biofilm formation on Grade EN 1.4162 stainless steel than that on five other common stainless steels [18]. Moreover, Actinobacteria and Proteobacteria were the main phyla on the biofilms of EN 1.4162 and those of five other common stainless steels. However, the biofouling tendency differed from the corrosion behavior among these six stainless steels [18]. Yu et al. reported that Proteobacteria, Actinobacteria and Bacteroidetes were detected on the biofilms formed on several types of water-distribution pipes and that a plastic-based pipe was suitable for water distribution due to its low biofilm-forming potential and microbial diversity [19]. Considering these reports, both Ag and Non-Ag will have low potential to form biofilms.

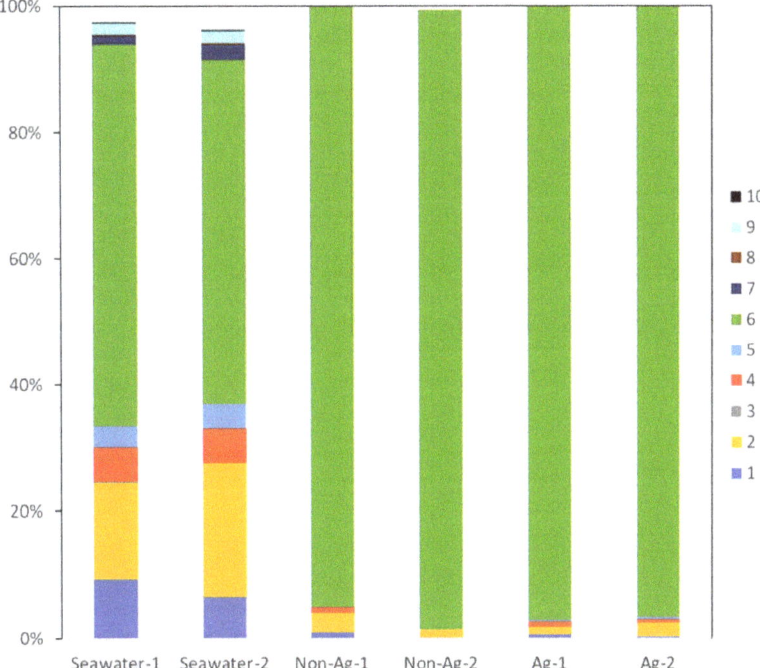

Figure 2. Main bacterial phyla detected in the seawater (Seawater), biofilms of silane-based coating samples (Non-Ag) and biofilms of silver nanoparticles-dispersed silane-based coating samples (Ag). Unassigned and archaeal OTUs were removed. When all samples showed phyla with an abundance of <0.1%, the phyla were excluded from the bacterial percentages. 1: Actinobacteria; 2: Bacteroidetes; 3: Chloroflexi; 4: Cyanobacteria; 5: Planctomycetes; 6: Proteobacteria; 7: SAR406; 8: SBR1093; 9: Verrucomicrobia; 10: ZB3.

At the order level, for Seawater, there were three major bacterial orders: Flavobacteriales of Bacteroidetes (OTU, 14% and 19%, #2 in Figure 3), Rhodobacterales of Alphaproteobacteria (OTU, 14% and 18%, #9 in Figure 3) and Oceanospirillales of Gammaproteobacteria (OTU, 13% and 15%, #16 in Figure 3), respectively. Conversely, biofilms of Non-Ag and Ag had two major bacterial orders: Rhodobacterales, which comprised approximately 50% of the microbiomes, and Oceanospirillales, which comprised approximately 25% of the microbiomes. Bacteroidetes are well known as a main marine bacterial group present across many oceans. They play a vital role in degrading organic compounds in the ocean [20]. In this study, all Seawater, Non-Ag and Ag contained Flavobacteriales of Bacteroidetes; they mainly occupied 14–19% in Seawater; however, Non-Ag and Ag had a low proportion of this order of 1–3%, which means Flavobacteriales had difficulty proliferating on the surface of Non-Ag and Ag. Therefore, the biofilm formation conditions provided by Non-Ag and Ag would not be suitable for Flavobacteriales growth. The order Rhodobacterales contains the family Rhodobacteraceae, whose members are key biofilm formers at the initial stage of biofilm formation in seawater [21]. Besides, Alteromonadales (#15 in Figure 3) and Oceanospirillales have been reported in young biofilms and with potential tolerance toward polysaccharide biodegradation and carbohydrate metabolism in biofilms [22]. In this study, the biofilms of Ag and Non-Ag were rich in Rhodobacterales and Oceanospirillales, indicating an early, not matured, stage of these biofilms. However, Rhodospirillales, whose members are known as marine biocorrosion bacteria [22], was only detected in Seawater (OTU, 0.6% and 0.9%). Additionally, higher proportions of Vibrionales (#18 in Figure 3), which include *Vibrio*, were detected on the biofilms of Non-Ag (OTU, 8.8% and 11.2%) and Ag (OTU, 5.0% and 10.2%) compared with those in Seawater (OTU, 0.5%). The members of *Vibrio* are considered to be corrosion-protective bacteria [22]. These results indicate the possibility that biofilms of Non-Ag and Ag did not contain MIC-related bacteria. As discussed in the previous section, Proteobacteria comprises many MIC-related bacteria, but OTUs of the order level show that the members of Proteobacteria identified on the biofilms of Non-Ag and Ag rarely cause MIC; however, these bacteria are highly related to biofilm formation. The biofilm formation test involved a six-day culture, which was twice the time period used in the previous experiment where Ag was shown to inhibit biofilm formation [7]. Unfortunately, we could not procure sufficient DNA samples to analyze the next generation in the previous three-day culture. Thus, we extended the culture period for the current experiment and obtained sufficient DNA from both the biofilms of Ag and Non-Ag. However, extending the culture period may affect the microbial composition on the biofilm of Ag, i.e., increased number of initial biofilms of Ag may comprise different bacteria.

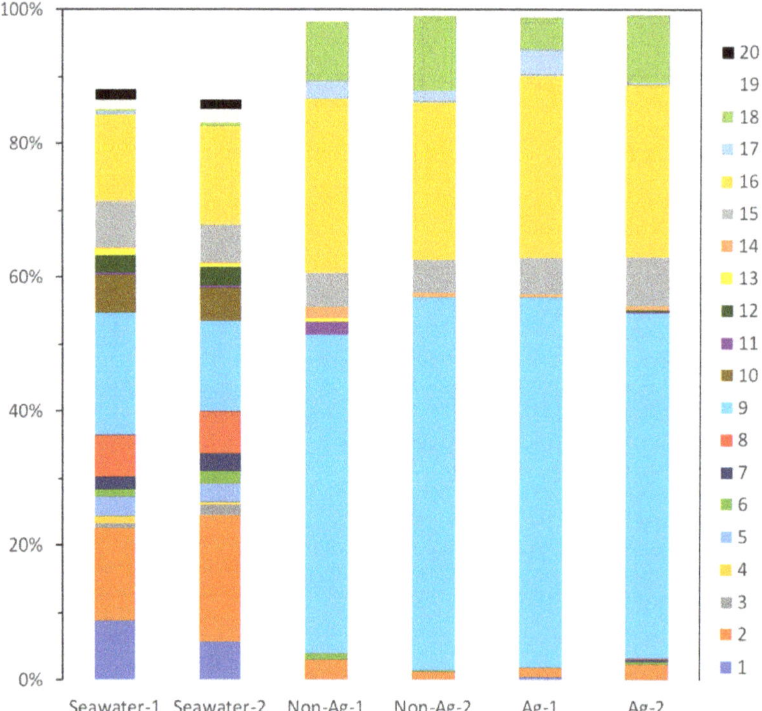

Figure 3. Main bacterial orders detected in the seawater (Seawater), biofilms of silane-based coating samples (Non-Ag) and biofilms of silver nanoparticles-dispersed silane-based coating samples (Ag). Unassigned and archaeal OTUs were removed. When all samples showed orders with an abundance of <1.0%, the orders were excluded from the bacterial percentages. 1: Acidimicrobiales; 2: Flavobacteriales; 3: Rhodothermales; 4: Cryptophyta; 5: Stramenopiles; 6: Synechococcales; 7: unknown order of Pla3; 8: unknown order of Alphaproteobacteria; 9: Rhodobacterales; 10: Rickettsiales; 11: Burkholderiales; 12: Rhodocyclales; 13: Sva0853; 14: Campylobacterales; 15: Alteromonadales; 16: Oceanospirillales; 17: Pseudomonadales; 18: Vibrionales; 19: Arctic96B-7; 20: unknown order of Pedosphaerae.

It is still unclear as to whether the biofilms of Ag and Non-Ag comprise similar biofilm-forming bacteria. In order to estimate the presence of biofilm-forming bacteria in this experiment, three abundant, order-level bacterial groups were summarized (Tables 2–4). On the biofilms of both Non-Ag and Ag, *Anaerospora* of Rhodobacteraceae was the dominant genus, followed by *Marinomonas* of Oceanospirillaceae and an unknown genus of the Vibrionaceae of Vibrionales (not *Vibrio*). *Marinomonas* is an aerobic bacterial genus; Vibrionaceae is a facultative aerobic bacterial family; and *Anaerospora* is an anaerobic bacterial genus, although only *Anaerospora hongkongensis* is registered as a member of *Anaerospora* [23,24]. When considering the oxygen requirement of bacteria, anaerobic bacteria would be more abundant than aerobic bacteria on the biofilms of Non-Ag and Ag. Extracted DNA from the biofilms was derived from living bacterial cells and extracellular DNA [25], which can be released from either living or dead cells that may no longer be present on the biofilms. Hence, we assumed that OTUs would help determine the history of biofilm-related bacteria. Our current biofilm formation test was performed in a closed LBR where dissolved oxygen concentration was estimated to be high during the early culture period, but very poor during the late culture period because stored seawater was used and no additional oxygen was supplied. Based on the presence of Oceanospirillales in the young biofilms, *Marinomonas* were considered to be the dominant biofilm-forming bacteria in early-stage biofilm formation in the LBR. Later, *Anaerospora* dominated as the main biofilm-forming bacteria as the dissolved oxygen concentration decreased during the culture period. In the bacterial phylum section, we considered that MIC might occur in either Non-Ag or Ag because abundant bacteria of biofilms of Non-Ag and Ag were Proteobacteria comprising many MIC-related bacteria such as *Pseudomonas*, sulfate-reducing bacteria, *Acidithiobacillus* and *Gallionella*. Indeed, the abundant members of Proteobacteria of biofilms of Non-Ag and Ag were *Anaerospora* and *Marinomonas*, which were biofilm-forming bacteria, but not reported for MIC. On the other hand, sulfate-reducing bacteria such as *Desulfovibrio*, *Acidithiobacillus* and *Gallionella* had undetectable levels there; however, *Pseudomonas* was slightly detected in the biofilm of Ag (OTU, 0.01% and 0.99%). These results show that both Non-Ag and Ag had biofilms formed on the surface, but they rarely progressed to MIC.

Table 2. Percentage abundance of operational taxonomic units of Rhodobacterales. -: unknown.

Seawater-1	Seawater-2	Non-Ag-1	Non-Ag-2	Ag-1	Ag-2	Families	Genera
5.29	3.79	4.77	4.51	3.86	3.30		Other
5.50	4.11	3.32	2.53	3.67	1.66		-
0.29	0	0	0	0	0		*Amaricoccus*
6.15	4.33	34.86	44.79	47.15	40.30		*Anaerospora*
0.06	0.04	0	0.01	0.01	0		*Loktanella*
0	0.02	0	0	0	0	Rhodobacteraceae	*Octadecabacter*
0.01	0.01	0.02	0.42	0.01	0.03		*Paracoccus*
0.89	1.18	4.42	3.31	0.46	6.05		*Pseudoruegeria*
0.03	0.02	0.01	0	0	0		*Rhodobacter*
0	0	0.01	0.01	0.01	0.01		*Rhodovulum*
0.01	0	0	0.01	0.01	0.01		*Roseivivax*
0.86	0.59	0	0	0	0	Rhodospirillaceae	-
0.04	0	0	0	0	0		*Magnetospirillum*

Table 3. Percentage abundance of operational taxonomic units of Oceanospirillales. -: unknown.

Seawater-1	Seawater-2	Non-Ag-1	Non-Ag-2	Ag-1	Ag-2	Families	Genera
0	0	1.80	0.65	0.81	1.09		Other
1.72	1.70	1.89	2.58	3.72	3.92		-
0	0	0	0.26	0	0.15		*Amphritea*
0.28	0.32	21.76	19.56	18.97	19.60	Oceanospirillaceae	*Marinomonas*
0	0	0	0	0	0.27		*Neptunomonas*
0	0.01	0	0	0	0		*Oceanospirillum*
0	0	0.64	0.30	2.40	0.06		*Oleispira*
0	0	0	0.31	0.56	0.41		-
0.01	0.02	0	0	0	0	SUP05	-

Table 4. Percentage abundance of operational taxonomic units of Vibrionales. -: unknown.

Seawater-1	Seawater-2	Non-Ag-1	Non-Ag-2	Ag-1	Ag-2	Families	Genera
0	0	0	0.01	0	0.01		-
0.03	0.04	1.21	1.68	0.24	0.90	Pseudoalteromonadaceae	*Pseudoalteromonas*
0.41	0.43	7.30	9.47	3.27	9.17		Other
0	0	0.01	0.02	0	0.01		-
0.07	0.01	0.05	0	1.45	0	Vibrionaceae	*Photobacterium*
0.01	0.02	0.21	0.03	0.01	0.08		*Vibrio*

We have already reported that the amount of biofilm of Ag was significantly less than that of Non-Ag [7]; however, what kinds of bacteria formed biofilm was unclear. Therefore, we approached revealing the microbiota of biofilms of Ag and Non-Ag. Comparing the OTU percentages of the biofilm of Ag with that of the biofilm of Non-Ag, Figures 2 and 3 demonstrate that the microbiota composition of biofilm of Ag was almost the same as that of the biofilm of Non-Ag. Therefore, we presumed that the microbiota composition of biofilms formed on silane-based coating did not change whether silver nanoparticles were dispersed in the coating or not; whereas, silver particles in the silane-based coating would control the attachment, growth and quorum sensing of bacteria on the surface, resulting in a small amount of biofilm.

4. Conclusions

In our previous study, we found that a silver nanoparticle-dispersed silane-based coating effectively inhibited biofilm formation. Here, we attempted the analysis of microbiomes on biofilms formed on the coatings using an NGS technique. No archaea phyla were detected on the biofilms of Ag, whereas only one archaea phylum was detected on the biofilms of Non-Ag. Proteobacteria were the dominant bacterial phylum on the biofilms of both Non-Ag and Ag. Comparing OTUs of the biofilms on Ag with those of the biofilms on Non-Ag, no distinct difference was noted in bacterial orders, but biofilm-forming bacterial orders and a biocorrosion-protective bacterial order were found to be present in both biofilms. In addition, members of *Marinomonas* were the main biofilm-forming bacteria under aerobic conditions, but were replaced by those of *Anaerospora* under anaerobic conditions. Moreover, the addition of silver nanoparticles did not affect the microbiome of biofilms on silane-based coating, while it decreased the amount of biofilm on it.

Author Contributions: Conceptualization, A.O.; Methodology, A.O., K.T. and H.K.; Validation, A.O.; Investigation, A.O.; Resources, K.T., H.K, K.S., T.Y. and T.S.; Data Curation, A.O.; Writing-Original Draft Preparation, A.O.; Writing-Review & Editing, A.O. and S.T.; Visualization, A.O. and K.T.; Supervision, A.O.; Project Administration, A.O.; Funding Acquisition, A.O. and H.K.

Funding: This research was partly funded by Suga weathering technology foundation in 2016.

Conflicts of Interest: The authors declare no conflict of interest.

References

1. Flemming, H.C. Biofouling in water systems—Cases, causes and countermeasures. *Appl. Microbiol. Biotechnol.* **2002**, *59*, 629–640. [CrossRef] [PubMed]
2. Bixler, G.D.; Bhushan, B. Biofouling: Lessons from nature. *Philos. Trans. A Math. Phys. Eng. Sci.* **2012**, *370*, 2381–2417. [CrossRef] [PubMed]
3. Kougo, T.; Kuroda, D.; Wada, N.; Ikegai, H.; Kanematsu, H. Biofouling of various metal oxides in marine environment. *J. Phys. Conf. Ser.* **2012**, *352*, 012048. [CrossRef]
4. Flemming, H.C. Why Microorganisms Live in Biofilms and the Problem of Biofouling. In *Marine and Industrial Biofouling*; Flemming, H.C., Murthy, S.P., Venkatesan, R., Coolsey, K.E., Eds.; Springer: Berlin, Germany, 2009; pp. 3–12.
5. Delaunois, F.; Tosar, F.; Vitry, V. Corrosion behavior and biocorrosion of galvanized steel water distribution systems. *Bioelectrochemistry* **2014**, *97*, 110–119. [CrossRef] [PubMed]
6. Beech, I.B.; Sztyler, M.; Gaylarde, C.C.; Smith, W.L.; Sunner, J. Biofilms and biocorrosion. In *Understanding Biocorrosion*; Liengen, T., Féron, D., Basséguy, R., Beech, I.B., Eds.; Woodhead Publishing: Oxford, UK, 2015; pp. 33–56.
7. Ogawa, A.; Kanematsu, H.; Sano, K.; Sakai, Y.; Ishida, K.; Beech, I.; Suzuki, O.; Tanaka, T. Effect of Silver or Copper Nanoparticles-Dispersed Silane Coatings on Biofilm Formation in Cooling Water Systems. *Materials* **2016**, *9*, 632. [CrossRef] [PubMed]
8. Edgar, R.C. Search and clustering orders of magnitude faster than BLAST. *Bioinformatics* **2010**, *26*, 2460–2461. [CrossRef] [PubMed]
9. Edgar, R.C. Drive5. Available online: http://www.drive5.com/ (accessed on 12 August 2018).
10. Caporaso, J.G.; Kuczynski, J.; Stombaugh, J.; Bittinger, K.; Bushman, F.D.; Costello, E.K.; Fierer, N.; Peña, A.G.; Goodrich, J.K.; Gordon, J.I.; et al. QIIME allows analysis of high-throughput community sequencing data. *Nat. Methods* **2010**, *7*, 335–336. [CrossRef] [PubMed]
11. FASTX-Toolkit, FASTQ/A Short-Reads Pre-Processing Tools. Available online: http://hannonlab.cshl.edu/fastx_toolkit/ (accessed on 12 August 2018).
12. Joshi, N.A.; Fass, J.N. Sickle: A Sliding-Window, Adaptive, Quality-Based Trimming Tool for FastQ Files (Version 1.33) [Software]. 2011. Available online: https://github.com/najoshi/sickle (accessed on 12 August 2018).
13. Magoc, T.; Salzberg, S.L. FLASH: Fast length adjustment of short reads to improve genome assemblies. *Bioinformatics* **2011**, *27*, 2957–2963. [CrossRef] [PubMed]
14. FLASh Fast Length Ajustment of Short Reads. Available online: https://ccb.jhu.edu/software/FLASH/ (accessed on 12 August 2018).
15. The Greengenes Database. Available online: http://greengenes.secondgenome.com (accessed on 12 August 2018).
16. Barns, Sue and Siegfried Burggraf: Crenarchaeota. Version 01 January 1997. Available online: http://tolweb.org/Crenarchaeota/9/1997.01.01 (accessed on 15 September 2018).
17. Tree of Life Web Project. Euryarchaeota. Version 01 January 1995 (Temporary). 1995. Available online: http://tolweb.org/Euryarchaeota/10/1995.01.01 (accessed on 15 September 2018).
18. Huttunen-Saarivirta, E.; Rajala, P.; Marja-Aho, M.; Maukonen, J.; Sohlberg, E.; Carpen, L. Ennoblement, corrosion, and biofouling in brackish seawater: Comparison between six stainless steel grades. *Bioelectrochemistry* **2018**, *120*, 27–42. [CrossRef] [PubMed]
19. Yu, J.; Kim, D.; Lee, T. Microbial diversity in biofilms on water distribution pipes of different materials. *Water Sci. Technol.* **2010**, *61*, 163–171. [CrossRef] [PubMed]
20. Wemheuer, B. Diversity and Ecology of the *Roseobacter* Clade and Other Marine Microbes as Revealed by Metagenomic and Metatranscriptomic Approaches. Ph.D. Thesis, Georg-August University School of Science, Gottingen, Germany, 2015.
21. Elifantz, H.; Horn, G.; Ayon, M.; Cohen, Y.; Minz, D. *Rhodobacteraceae* are the key members of the microbial community of the initial biofilm formed in Eastern Mediterranean coastal seawater. *FEMS Microbiol. Ecol.* **2013**, *85*, 348–357. [CrossRef] [PubMed]
22. De Carvalho, C.C.C.R. Marine Biofilms: A Successful Microbial Strategy with Economic Implications. *Front. Mar. Sci.* **2018**, *5*. [CrossRef]

23. Woo, P.C.; Teng, J.L.; Leung, K.W.; Lau, S.K.; Woo, G.K.; Wong, A.C.; Wong, M.K.; Yuen, K.Y. *Anaerospora hongkongensis* gen. nov. sp. nov., a novel genus and species with ribosomal DNA operon heterogeneity isolated from an intravenous drug abuser with pseudobacteremia. *Microbiol. Immunol.* **2005**, *49*, 311–339. [CrossRef]
24. BacDive Strain Identifier BacDive ID: 23337. Available online: https://bacdive.dsmz.de/strain/23337 (accessed on 12 August 2018).
25. Okshevsky, M.; Regina, V.R.; Meyer, R.L. Extracellular DNA as a target for biofilm control. *Curr. Opin. Biotechnol.* **2015**, *33*, 73–80. [CrossRef] [PubMed]

 *antibiotics*

Review

Biogenic Nanosilver against Multidrug-Resistant Bacteria (MDRB)

Caio H. N. Barros †, Stephanie Fulaz †, Danijela Stanisic and Ljubica Tasic *

Laboratory of Chemical Biology, Institute of Chemistry, State University of Campinas, Campinas 13083-970, Brazil; caionasibarros@gmail.com (C.H.N.B.); ste.fulaz@gmail.com (S.F.); danijela.stanisic@iqm.unicamp.br (D.S.)

* Correspondence: ljubica@iqm.unicamp.br

† Present address: School of Bioprocess and Chemical Engineering, University College Dublin, Dublin D04 V1W8, Ireland

Received: 28 June 2018; Accepted: 31 July 2018; Published: 2 August 2018

Abstract: Multidrug-resistant bacteria (MDRB) are extremely dangerous and bring a serious threat to health care systems as they can survive an attack from almost any drug. The bacteria's adaptive way of living with the use of antimicrobials and antibiotics caused them to modify and prevail in hostile conditions by creating resistance to known antibiotics or their combinations. The emergence of nanomaterials as new antimicrobials introduces a new paradigm for antibiotic use in various fields. For example, silver nanoparticles (AgNPs) are the oldest nanomaterial used for bactericide and bacteriostatic purposes. However, for just a few decades these have been produced in a biogenic or bio-based fashion. This review brings the latest reports on biogenic AgNPs in the combat against MDRB. Some antimicrobial mechanisms and possible silver resistance traits acquired by bacteria are also presented. Hopefully, novel AgNPs-containing products might be designed against MDR bacterial infections.

Keywords: silver nanoparticles; biological synthesis; multidrug-resistant bacteria

---

1. Introduction

Antimicrobial resistance refers to the evolutionary capacity developed by microorganisms such as bacteria, fungi, viruses, and parasites to fight and neutralize an antimicrobial agent. According to the World Health Organization (WHO) [1], the intensive use and misuse of antimicrobials has led to an expansion of the number and types of resistant organisms. Moreover, the use of sub-therapeutic antibiotic doses to prevent diseases in animal breeding to improve animal growth can select resistant microorganisms, which can possibly disseminate to humans [2].

The number of pathogens presenting multidrug resistance has had an exponential increase in recent times and is considered an important problem for public health [3]. A wide number of bacteria have been reported as multidrug-resistant (MDR), and they present a high cost of management, including medicines, staff capacity, isolation materials [4], and productivity loss [5]. For instance, in the USA, the cost of conventional tuberculosis treatment for the drug-susceptible bacterium is $17,000 and up to $482,000 for the treatment of the MDR bacterium [5]. In 2017, WHO published the first list of antibiotic-resistant pathogens offering risk to human health and, as such, the development of new drugs is crucial. Priority 1 (critical) microorganisms are carbapenem-resistant *Acinetobacter baumannii*; carbapenem-resistant *Pseudomonas aeruginosa*; and carbapenem-resistant, ESBL-producing *Enterobacteriaceae*. Accounting for priority 2 (high) are vancomycin-resistant *Enterococcus faecium*; methicillin-resistant, vancomycin-intermediate and resistant *Staphylococcus aureus*; clarithromycin-resistant *Helicobacter pylori*; fluoroquinolone-resistant *Campylobacter* spp.; fluoroquinolone-resistant *Salmonellae*; and cephalosporin-resistant, fluoroquinolone-resistant *Neisseria gonorrhoeae*. In priority 3 (medium) are penicillin-non-susceptible *Streptococcus pneumoniae*, ampicillin-resistant *Haemophilus influenzae*, and fluoroquinolone-resistant *Shigella* spp. [6].

The use of drugs combinations, two or more antimicrobial drugs to combat MDRB [7], is already employed in cancer therapy [8], HIV-patients [9], and malaria patients [10]. On the other hand, research groups around the globe are suggesting innovative solutions to treat resistant organisms. Xiao et al. [11] synthesized the block copolymer poly (4-piperidine lactone-b-ω-pentadecalactone) with high antibacterial activity against *E. coli* and *S. aureus*, and low toxicity to NIH-3T3 cells, and suggested that cationic block copolymer biomaterials can be employed in medicine and implants. Zoriasatein et al. [12] showed that a derivative peptide from the snake (*Naja naja*) has an antimicrobial effect against *S. aureus*, *B. subtilis*, *E. coli*, and *P. aeruginosa*. Al-Gbouri and Hamzah [13] reported that an alcoholic extract of *Phyllanthus emblica* exhibits antimicrobial activity against *E. coli*, *S. aureus*, and *P. aeruginosa* and it inhibits biofilm formation of *P. aeruginosa*. Naqvi et al. [14] suggested the combined use of biologically synthesized silver nanoparticles (AgNPs) and antibiotics to combat the MDRB.

The increasing utilization and in-depth studies of nanomaterials have brought new perspectives towards new antimicrobial materials and nanocomposites that could add-in to the MDRB pandemic that we are currently facing. Nanoparticles and nanocomposites comprising zinc oxide [15], copper oxide [16], iron oxide [17], and, especially, silver, have been widely used in textiles [18,19], dental care [20], packaging [21], paints [22], and in a whole myriad of applications. Silver nanoparticles are one of the most exploited nanomaterials for this end, as they have been used for over a century in the healing of wounds and burns. Although chemical methods were successfully employed for AgNPs synthesis, with the need to use sustainable and non-toxic methods in chemistry, a biocompatible modality of AgNPs synthesis came about by using biological routes for nanoparticle synthesis (Figure 1). Biosynthesis or bio-based synthesis of AgNPs may occur through three routes: fungal, bacterial, or by plants, for the reduction of Ag^+ to Ag^0. The saturation of Ag^0 monomers in suspension eventually leads to a burst-nucleation process [23] in which nanoclusters of metallic silver are produced and stabilized by biomolecules from the biological extracts.

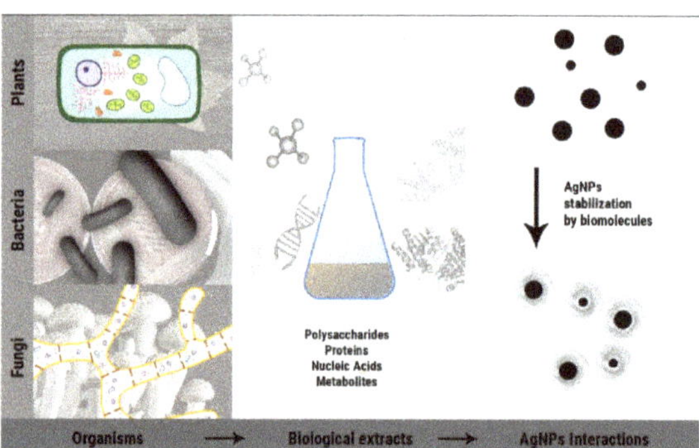

Figure 1. Biological extracts may be prepared from any part of plant material, or via extracellular/intracellular processes using fungi and bacteria cultures. The extracts are rich in biomolecules such as sugars, proteins, nucleic acids, and metabolites that either have a stabilizing potential or reducing and stabilizing potential for the formation of silver nanoparticles.

The demand of products for the combat of MDR bacterial strains such as *Pseudomonas aeruginosa*, methicillin-resistant *Staphylococcus aureus* (MRSA), vancomycin-resistant *Staphylococcus aureus* (VRSA), erythromycin-resistant *Streptococcus pyogenes*, and ampicillin-resistant *Escherichia coli* [24] has led to the design of powerful antimicrobial materials that are reinforced with silver nanoparticles [25]. Today, in medicinal practice, there are wound dressings, contraceptive devices, surgical instruments, bone prostheses, and dental implants which are coated or embedded with nanosilver [26–31]. In daily life, consumers may find nanosilver in room sprays, laundry detergents, water purification devices and paints [26,32,33]. In the final part of this review, some of the recent advances in patented technologies containing AgNPs that establish viable grounds for the development of biogenic AgNPs-containing products for MDRB eradication purposes are cited and discussed.

2. Antibiotics

Antibiotics gained popularity because of their effectiveness or activities against microorganisms, as described by Selman Waksman [34], and refers to an application, and not a class of compound or its function [35]. The first compound with antibacterial activity discovered was arsphenamine, synthesized in 1907 by Alfred Bertheim in Paul Ehrlich's laboratory, with antisyphilitic activity identified in 1909 by Sahachiro Hata [36,37]. Classically, the golden era of antibiotics refers to the period between the 50s and 70s, when the discovery of different classes of antibiotics took place [38]. For a more detailed review of antibiotics and antibacterial drugs, see Bbosa et al., 2014 [39]. Figure 2 illustrates the main antibiotic classes and examples of compounds, with corresponding dates of discovery and resistance as first reported.

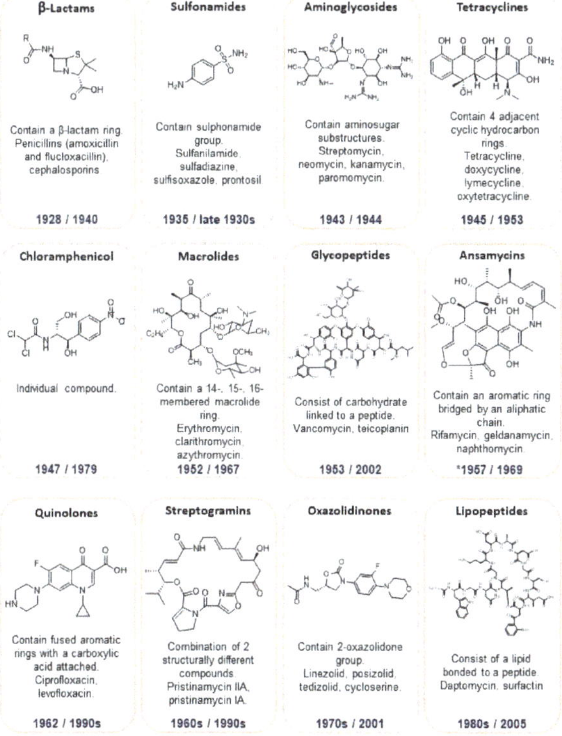

Figure 2. Illustration of different classes of antibiotics. Antibiotics that act as bactericidal agents, i.e., cause cell death, are shown in rectangles with orange borders; antibiotics that act as bacteriostatic agents, i.e., restrict growth and reproduction, are shown in rectangles with dashed line orange borders. Years shown in blue indicate when the antibiotic class was discovered (the first number), and when resistance was first reported (the second number). The structure and years of discovery and resistance refer to the first antibiotic from each class [35,40–46].

3. The Emerging of Antimicrobial Resistance

One of the most famous antibiotics, Penicillin, was discovered in 1928 by Alexander Fleming. In 1940, before its public use, the same group identified a bacterial penicillinase [47], an enzyme able to degrade penicillin. This fact can now be related to the number of antibiotic genes that are naturally present in microbial populations [48]. In Japan, during the 50s, genetically transferable antibiotic resistance was identified. This discovery introduced the concept that antibiotic genes could spread among a population of bacterial pathogens using bacterial conjugation [49,50]. This horizontal gene transfer is important throughout genome evolution and currently presents a serious threat [51]. The bacterial genetic elasticity prompts the acquisition of genetic material, mutational adaptations, or changes in gene expression, leading to the survival of the fittest organism and the generation of resistance to antibiotics [52]. For more details regarding antibiotic resistance development, mechanisms, emergence, and spread see further references [52–58].

Currently, we face a deficiency in the development of new antibiotics to face the growing antimicrobial resistance. The constant increment in the emergence of resistant strains has not been balanced by the availability of new therapeutic agents for many reasons [59,60]. Firstly, policy-makers want to avoid the use of new antibiotics until they are indispensable, because of the resistance development. On the other hand, society needs the pharmaceutical industry to design and develop new drugs, which should not be used. Moreover, antibiotics are used in the short-term, which does not help companies to make a sustained profit. Also, the excessive cost of development and the regulatory onus makes it difficult to attend a demand for cheap antibiotics [61]. Looking at this alarming scenario the design of new therapeutics and/or new approaches is imperative.

4. Biogenic AgNPs as a Weapon against Multidrug-Resistant Bacteria (MDRB)

Traditionally, the synthesis of AgNPs using chemical approaches has been the most explored for a better size and shape control, preparation of nanocomposites and elucidation of electronic properties. However, the necessity of applying the well-known antibacterial activity of AgNPs in biological systems propelled the development of a new synthesis approach. The biological, biogenic, or bio-based methods for AgNPs synthesis present four main advantages: (1) increased biocompatibility, once AgNPs are produced in water and capped with biomolecules such as proteins, sugars or metabolites; (2) diminished toxicity, as the reducing agents are natural compounds that usually have mild reducing strength; (3) easy production, such as preparation of an extract from fungi, bacteria or plants, followed by the addition of a silver salt (typically, silver nitrate); and (4) low cost [62]. Despite positive aspects, the lack of control of shape and size of the nanoparticles is still a challenge for biogenic synthesis methods.

Because every biological synthesis is different from another as a consequence of using distinct species, the capping agents on the surface of the nanoparticles may differ. The concept of "protein corona" [63], for instance, describes the existence and dynamics of a protein shell surrounding nanoparticles in a biological environment or after a biological synthesis [64]. The interaction of biologically synthesized AgNPs with a bacterial cell will inherently involve the contact with the microorganism and the outer biomolecule shell. Thus, this interaction is unique as new joint effects (between biomolecules and the silver itself) can arise and improve the antibacterial action due to a change in toxicity, cell uptake, and bio-distribution [65].

In the case of MDRB, the mechanism of action of AgNPs is distinct from the mechanism by which traditional antibiotics act, and thus resistance does not pose an obstacle that cannot be overcome in most cases. In the following sections, each type of biological synthesis is detailed along with a literature review of biogenic AgNPs being used against MDRB. In most of the papers reviewed, the bacterial strains used for susceptibility and antibacterial tests were clinical isolates from hospital patients, however the list of antibiotics to which the strain is resistant is not always described. Also, in many cases the strain used is standardized (ATCC strains, for example), but no details on the drug resistance capacity are provided. Here, we emphasize the examples where the provenience and description of the bacterial strain are well detailed, along with a robust antibacterial testing methodology.

4.1. Fungal AgNPs against MDRB

The synthesis of AgNPs using fungal cells may be performed outside the cells (extracellular synthesis) or inside the cells (intracellular synthesis) [66]. The former is the most recurrent in the literature, in which a fungal filtrate is obtained after the cultivation of the microorganism and a silver salt solution is added to it. Advantages of extracellular synthesis include ease of purification (as nanoparticles are not inside or attached to the fungus), facilitated downstream processing, and improved size control [67]. Despite usually having high reproducibility, fungal syntheses are time-consuming, as the fungi grow at a slower rate when compared to bacteria or the preparation of a plant extract. Moreover, the reduction of silver ions is also a gradual process, taking up to 96 h for completion. *Fusarium oxysporum* is perhaps the most studied species for AgNPs biosynthesis [19,68]; the mechanism of nanoparticle formation involves the reduction of silver(I) by a nitrate reductase and a shuttle quinone [69]. Scandorieiro et al. [70] demonstrated the synergistic effect of *F. oxysporum* produced AgNPs with oregano essential oil against a range of antibiotic-resistant bacterial strains, including MRSA and beta-lactamase producing strains. Naqvi et al. [14] also showed the effectiveness of a synergistic approach by combining *Aspergillus flavus* produced AgNPs with well-known commercial antibiotics resulting in an increase of up to 7-fold in the area of inhibition against bacterial strains resistant to the same antibiotics. In fact, a combinational therapy is highly desirable taking into consideration the development of AgNPs tolerance in bacteria via genetic evolution [71]. Chowdhury et al. studied the effect of AgNPs synthesized by *Macrophomina phaseolina* against ampicillin and chloramphenicol resistant *E. coli* and noted plasmid fragmentation and a decrease of supercoiled plasmid content upon incubation of the circular DNA with nanoparticles [72]. On the other hand, nanoparticle attachment to the cell wall and leakage of cell components induced by *Penicillium polinicum*-produced AgNPs were observed in transmission micrographs by Neethu et al. [73], which confirms that more than one antibacterial mechanism is possible (this theme is further explored in Section 4.4). Table 1 brings a summary of fungal AgNPs and their activity against MDR bacterial strains.

Table 1. Fungi-mediated AgNPs biosynthesis and their activity against (MDRB).

Fungus	AgNPs Size (nm)	Target MDR Microorganism	Test Type [a]	Test Result [b]	Reference
Aspergillus flavus	5–30	*E. coli*	ZI	15 ± 1.5 mm	[14]
		S. aureus	ZI	16 ± 2 mm	
		M. luteus	ZI	14 ± 1 mm	
		P. aeruginosa	ZI	14 ± 1.5 mm	
		E. faecalis	ZI	15 ± 1.5 mm	
		A. baumanii	ZI	15 ± 1 mm	
		K. pneumoniae	ZI	14 ± 0.6 mm	
		Bacillus spp.	ZI	15 ± 1.5 mm	
Fusarium oxysporum NGD	16.3–70	*Enterobacter* sp.	ZI	31 mm	[74]
		P. aeruginosa	ZI	20 mm	
		K. pneumoniae	ZI	19 mm	
		E. coli	ZI	2 mm	
Trichoderma viride	5–40	*E. coli*	ZI	16–28 mm (*)	[75]
		S. typhi	ZI	19–36 mm (*)	
		S. aureus	ZI	10–19 mm (*)	
		M. luteus	ZI	9–17 mm (*)	
Aspergillus niger	30–40	*S. aureus*	ZI	15 ± 0.23 mm	[76]
		B. cereus	ZI	16 ± 0.32 mm	
		P. vulgaris	ZI	14 ± 0.26 mm	
		E. coli	ZI	14 ± 0.44 mm	
		V. cholerae	ZI	13 ± 0.51 mm	
Tricholoma crassum	5–50	*E. coli* (DH5 α)	ZI	17.5 ± 0.5 (**)	[77]
		A. tumifaciens (LBA4404)	ZI	20.0 ± 0.5 (**)	

Table 1. *Cont.*

Fungus	AgNPs Size (nm)	Target MDR Microorganism	Test Type [a]	Test Result [b]	Reference
Agaricus bisporus	-	*E. coli*	ZI	14 mm	[78]
		Klebsiella sp.	ZI	15 mm	
		Pseudomonas sp.	ZI	-	
		Enterobacter sp.	ZI	18 mm	
		Proteus sp.	ZI	20 mm	
		S. aureus	ZI	17 mm	
		S. typhi	ZI	22 mm	
		S. paratyphi	ZI	17 mm	
Aspergillus clavatus	550–650 (AFM)	*S. aureus*	ZI	20.5 mm	[79]
		S. epidermidis	ZI	19 mm	
Penicilium polonicum	10–15	*A. baumanii*	MIC, MBC, ZI	15.62 µg mL^{-1} (MIC), 31.24 µg mL^{-1} (MBC), 21.2 ± 0.4 mm (ZI)	[73]
Cryphonectria sp.	30–70	*S. aureus* (ATCC-25923)	ZI	16 ± 0.69 mm	[80]
		S. typhi (ATCC-51812)	ZI	12 ± 0.29 mm	
		E. coli (ATCC-39403)	ZI	13 ± 1.54 mm	
Rhizoppus spp.	27–50	*E. coli*	ZI	15–22 mm (***)	[81]
Fusarium oxysporum	77.68	*S. aureus* (MRSA 101)	MIC, MBC	250 µM (MIC), 500 µM (MBC)	[70]
		S. aureus (MRSA 107)	MIC, MBC	250 µM (MIC), 500 µM (MBC)	
		E. coli (ESBL 167)	MIC, MBC	125 µM (MIC), 125 µM (MBC)	
		E. coli (ESBL 169)	MIC, MBC	125 µM (MIC), 125 µM (MBC)	
		E. coli (ESBL 176)	MIC, MBC	125 µM (MIC), 125 µM (MBC)	
		E. coli (ESBL 192)	MIC, MBC	125 µM (MIC), 125 µM (MBC)	
		E. coli (KPC 131)	MIC, MBC	125 µM (MIC), 125 µM (MBC)	
		E. coli (KPC 133)	MIC, MBC	125 µM (MIC), 125 µM (MBC)	
		A. baumannii (CR 01)	MIC, MBC	125 µM (MIC), 125 µM (MBC)	
Aspergillus flavus	5–30	*E. coli*	ZI	15 ± 1.5 mm	[14]
		S. aureus	ZI	16 ± 2 mm	
		M. luteus	ZI	14 ± 1 mm	
		P. aeruginosa	ZI	14 ± 1.5 mm	
		E. faecalis	ZI	15 ± 1.5 mm	
		A. baumannii	ZI	15 ± 1 mm	
		K. pneumoniae	ZI	14 ± 0.6 mm	
		Bacillus spp.	ZI	15 ± 1.5 mm	
Macrophomina phaseolina	5–40	*E. coli* (DH5α-MDR)	ZI	3.0 ± 0.2 mm (**)	[72]
		A. tumefaciens (LBA4404-MDR)	ZI	3.3 ± 0.2 mm (**)	

[a] ZI = zone of inhibition; MIC = Minimum Inhibitory Concentration; MBC = Minimum Bactericidal Concentration.; [b] For tests in which more than one concentration of AgNPs was used, the best results are shown; (*) Values related to a synergistic effect with distinct antibiotics; (**) Values estimated from graphs; (***) More than one bacterial isolate was used.

4.2. Bacterial AgNPs against MDRB

Similarly to fungal biosynthesis, bacterial AgNPs biosynthesis may also be performed extra- or intracellularly [82]. The former can be done by using the cell biomass, where the reducing agents are secreted by the cells and the nanoparticles formed might be attached to the bacterial wall (which can possibly extend the purification process). In contrast, using a bacterial supernatant/cell-free extract has the advantage of facilitating the downstream process and purification procedures by utilizing a sterile biomolecule-rich mixture to synthesize the nanoparticles, often with the aid of microwave [83] or light irradiation [84]. Conversely, the intracellular AgNPs synthesis takes place inside the cell, often in the periplasmic space [85]. This mechanism requires a certain metal resistance from the bacteria [86] or exposure to very low concentrations of the silver salt, as the Ag$^+$ ion must be imported without causing any major damage. The biggest disadvantage of this method is the purification as the nanoparticles must be removed from the interior of the cells. Ultrasonication is usually the most common method used for this end [87].

Singh et al. [88] prepared AgNPs from the culture supernatant of *Aeromonas* sp. THG-FG1.2 extracted from soil and obtained inhibition of several bacterial strains otherwise completely insensitive to erythromycin, lincomycin, novobiocin, penicillin G, vancomycin, and oleandomycin. Desai et al. [89] reported a hydrothermal biosynthesis of AgNPs using a cell-free extract of *Streptomyces* sp. GUT 21 by autoclaving the bacterial extract along with a silver salt solution. The nanoparticles were between 20–50 nm in size and active towards MDRB up to a concentration of 10 μg mL^{-1}. Sunlight exposure is also a good methodology for AgNPs biosynthesis, as demonstrated by Manikprabhu et al. [90]. Nanoparticles were produced from *Sinomonas mesophila* MPKL 26 cell supernatant in contact with silver nitrate upon up to 20 min of sun exposure. Specific secreted extracellular compounds can also be used for AgNPs synthesis. Santos et al. [91] attribute the formation of AgNPs smaller than 10 nm to xanthan gum produced during the growth of *Xanthomonas* spp. The nanoparticles could inhibit, to a certain extent, the growth of MDR *Acinetobacter baumannii* and *Pseudomonas aeruginosa*. Table 2 brings a summary of AgNPs produced by bacteria with activity against MDRB.

Table 2. Bacteria-mediated AgNPs biosynthesis and their activity against MDRB.

Bacteria	AgNPs Size (nm)	Target MDR Microorganism	Test Type [a]	Test Result [b]	Reference
Streptomyces	20–70	*K. pneumoniae* (ATCC 100603)	MIC	4 μg mL^{-1}	[92]
		K. pneumoniae	MIC	1.4 μg mL^{-1}	
		E. coli	MIC	2 μg mL^{-1}	
		Citrobacter	MIC	2 μg mL^{-1}	
Bacillus sp.	14–42	*S. epidermidis* strain 73 (pus)	ZI	15 mm	[93]
		S. epidermidis strain 145 (catheter tips)	ZI	19 mm	
		S. epidermidis strain 152 (blood)	ZI	19 mm	
		S. aureus (MTCC 87)	ZI	18 mm	
		S. typhi	ZI	13 mm	
		S. paratyphi	ZI	15 mm	
		V. cholerae (MTCC 3906)	ZI	18 mm	
Bacillus cereus	24–46	*E. coli*	MIC, ZI	6.25 μg mL^{-1} (MIC), 16 $\pm$ 1 mm (ZI)	[94]
		S. aureus	MIC, ZI	12.5 μg mL^{-1} (MIC), 14 $\pm$ 1 (ZI)	
		K. pneumoniae	MIC, ZI	>3.12 μg mL^{-1} (MIC), 17 $\pm$ 1 mm (ZI)	
		P. aeruginosa	MIC, ZI	3.12 μg mL^{-1} (MIC), 23 $\pm$ 1 mm (ZI)	
Bacillus safensis (LAU 13)	5–95	*E. coli*	ZI	11–19 mm	[95]
		K. granulomatis	ZI	11–19 mm	
		P. vulgaris	ZI	11–19 mm	
		P. aeruginosa	ZI	11–19 mm	
		S. aureus	ZI	11–19 mm	
Aeromonas sp. THG-FG1.2	8–16	*B. cereus* (ATCC 14579)	ZI	13.5 $\pm$ 0.5 mm	[88]
		B. subtilis (KACC 14741)	ZI	13 $\pm$ 1.0 mm	
		S. aureus (ATCC 6538)	ZI	15.5 $\pm$ 0.5 mm	
		E. coli (ATCC 10798)	ZI	13 $\pm$ 0.2 mm	
		P. aeruginosa (ATCC 6538)	ZI	16 $\pm$ 0.1 mm	
		V. parahaemolyticus (ATCC 33844)	ZI	16 $\pm$ 0.1 mm	
		S. enterica (ATCC 13076)	ZI	11 $\pm$ 0.2 mm	
		C. albicans (KACC 30062)	ZI	20 $\pm$ 0.1 mm	
		C. tropicalis (KCTC 7909)	ZI	15 $\pm$ 0.5 mm	
Bacillus thuringiensis	15	*E. coli*	ZI	12 $\pm$ 1 mm (*)	[96]
		P. aeruginosa	ZI	16 $\pm$ 1 mm (*)	
		S. aureus	ZI	9 $\pm$ 1 mm (*)	
Anabaena diololum	10–50	*K. pneumoniae* DF12SA (HQ114261)	ZI	36 $\pm$ 0.82 mm	[97]
	10–50	*E. coli* DF39TA (HQ163793)	ZI	33 $\pm$ 1.63 mm	
	10–50	*S. aureus* DF8TA (JN642261)	ZI	34 $\pm$ 0.81 mm	
Streptomyces sp. GUT 21	23–48	*E. coli* (MTCC 9537)	MIC, ZI	14 μg mL^{-1} (MIC), 27.05 $\pm$ 3.20 mm (ZI)	[89]
		K. pneumoniae (MTCC 109)	MIC, ZI	12 μg mL^{-1} (MIC), 28.50 $\pm$ 2.60 mm (ZI)	
		S. aureus (MTCC 96)	MIC, ZI	15 μg mL^{-1} (MIC), 24.25 $\pm$ 2.09 mm (ZI)	
		P. aeruginosa (MTCC 1688)	MIC, ZI	10 μg mL^{-1} (MIC), 10.05 $\pm$ 3.60 mm (ZI)	

Table 2. *Cont.*

Bacteria	AgNPs Size (nm)	Target MDR Microorganism	Test Type [a]	Test Result [b]	Reference
Bacillus megaterium	80–98.56 (AFM)	*S. pneumoniae* *S. typhi*	ZI ZI	21 mm 18 mm	[98]
Xanthomonas spp.	5–40	*P. aeruginosa* *baumannii*	ZI ZI	10.0 ± 1.0 mm 10.6 ± 0.6 mm	[91]
Sinomonas mesophila MPKL 26	4–50	*S. aureus*	ZI	12 mm	[90]
Bacillus flexus	12–65	*E. coli* *P. aeruginosa* *S. pyogenes* *subtilis*	ZI ZI ZI ZI	11.55 mm 11.05 mm 11.65 mm 11.55 mm	[99]
Bacillus brevis (NCIM 2533)	41–68	*S. aureus* *S. typhi*	ZI ZI	19 mm 7.5 mm	[100]

[a] ZI = zone of inhibition; MIC = Minimum Inhibitory Concentration; MBC = Minimum Bactericidal Concentration;
[b] For tests in which more than one concentration of AgNPs was used, the best results are shown; (*) Values estimated from graphs.

4.3. AgNPs from Plants against MDRB

Production of AgNPs using plant extracts is perhaps the most explored method in biogenic synthesis, probably due to the easiness of the procedure and wide availability of species to work with [101]. The whole plant, the stem, pod, seeds, fruit, flowers, and, most frequently, leaves are used to prepare an extract, which may be done in cold or hot solvent and almost always utilizes water (despite the fact that organic solvent extracts have also been used). The abundance of components such as reducing sugars, ascorbic acid [102], citric acid [103], alkaloids and amino acids [104], along with slightly soluble terpenoids [105], flavonoids [106], and other metabolites in various parts of the plant may easily act as reducing agents, converting Ag^+ to AgNPs in shorter times (when compared to fungal or bacterial syntheses). Due to the lower protein content in most plants, the capping biomolecule shell often has a significant contribution of polysaccharides [107] and other molecules. Most reports on plant biosynthesis are studies of plant species found in the surroundings of the university or city where the laboratory is located, however, in vitro-derived culture of plants can also be used for these purposes [108].

Ma et al. [107] reported on the biosynthesis of 60 nm AgNPs using polysaccharide-rich root extract of *Astragalus membranaceus* and compared the bacterial inhibition against reference strains of *E. coli*, *P. aeruginosa*, *S. aureus*, and *S. epidermidis* with clinically isolated MDR strains of these bacteria. Interestingly, the nanoparticles were slightly more active toward the resistant strains.

The nanoparticle size is known to play an important role in antibacterial activity [24], and this is no different for MDR strains. AgNPs synthesized by *Caesalpinia coriaria* leaf extract, which were 50–53 nm were shown to be more active towards MDR bacterial clinical isolates when compared to 79–99 nm AgNPs [109].

Despite the common belief that biological synthesis implies a lack of control for Ag^+ reduction and poor shape control, Jinu et al. [110] demonstrated the synthesis of cubic and triangular shaped 20 nm AgNPs using *Solanum nigrum* leaf extract. The nanoparticles had a contributing effect along with the antimicrobial plant extract towards six MDRB strains. Moreover, these AgNPs showed antibiofilm activity against *P. aeruginosa* and *S. epidermidis*. Prasannaraj et al. [111] reported an extensive study using ten different plant species for AgNPs biosynthesis, yielding spherical, cubic, and fiber-like nanoparticles. All of them inhibited bacterial growth of clinically isolated MDR pathogens and some also displayed antibiofilm activity against *P. aeruginosa* and *S. epidermidis*. The authors correlate the results with the 3 to 4-fold increase in reactive oxygen species (ROS) by AgNPs.

Intracellular ROS production was also observed by flow cytometry for *Ocimum gratissimum* leaf extract-produced AgNPs [112]; the authors suggest that the membrane damage caused by the nanoparticles could prevent efficient electronic transport in the respiratory chain. This was confirmed by micrographs of MDR *E. coli* and *S. aureus* cells treated with AgNPs, which showed leakage of intracellular content and pits in the membrane.

The antibacterial properties of silver can also be delivered by silver chloride nanoparticles (AgCl-NPs), as shown by Gopinath et al. [113]. AgNPs and AgCl-NPs were produced from *Cissus quadrangularis* leaf extract and were active towards both Gram-negative and Gram-positive MDR strains. In this case, chloride ions were identified in the extract and attributed to the formation of AgCl nanocrystals.

Table 3 presents the gathered data on plant biosynthesis of AgNPs with the corresponding activity against MDRB.

Table 3. Plant-mediated AgNPs biosynthesis and their activity against MDRB.

Plant	Part	AgNPs Size (nm)	Target MDR Microorganism	Test Type [a]	Test Result [b]	Reference
Olive	leaf	20–25	*S. aureus*	ZI	2.4 ± 0.2 cm (*)	[114]
			P. aeruginosa	ZI	2.4 ± 0.2 cm (*)	
			E. coli	ZI	1.8 ± 0.2 cm (*)	
Phyllanthus amarus	Whole plant	24 ± 8	*P. aeruginosa*	MIC, ZI	6.25–12.5 µg mL^{-1} (MIC), 10 ± 0.53 to 21 ± 0.11 mm (ZI)	[115]
Corchorus capsularis	leaf	5–45	*P. aeruginosa*	ZI	17 mm	[116]
			S. aureus	ZI	21 mm	
			Coagulase negative staphylococci	ZI	20 mm	
Tribulus terrestris	fruit	16–28	*S. pyogens*	ZI	10 mm	[117]
			E. coli	ZI	10.75 mm	
			P. aeruginosa	ZI	9.25 mm	
			B. subtilis	ZI	9.25 mm	
			S. aureus	ZI	9.75 mm	
Garcinia mangostana	leaf	35	*E. coli*	ZI	15 mm	[118]
			S. aureus	ZI	20 mm	
Ricinus communis	leaf	29.18 (X-ray diffraction)	*B. fusiformis*	ZI	2.90 cm	[119]
			E.coli	ZI	2.89 cm	
Caesalpinia coriaria	leaf	40–52	*E. coli*	ZI	12.0 ± 0.50 mm	[109]
			P. aeruginosa	ZI	18.3 ± 0.80 mm	
			K. pneumonia	ZI	14.6 ± 1.20 mm	
			S. aureus	ZI	10.3 ± 1.20 mm	
		78–98	*E. coli*	ZI	9.6 ± 0.80 mm	[109]
			P. aeruginosa	ZI	18.3 ± 1.20 mm	
			K. pneumonia	ZI	13.3 ± 0.30 mm	
			S. aureus	ZI	11.0 ± 0.00 mm	
Mimusops elengi	leaf	55–83	*K. pneumoniae*	ZI	18 mm	[120]
			S. aureus	ZI	10 mm	
			M. luteus	ZI	11 mm	
Ocimum gratissimum	leaf	16 ± 2	*E. coli* (MC-2)	MIC, MBC, ZI	4 µg mL^{-1} (MIC), 8 µg mL^{-1} (MBC), 12 ± 0.6 mm (ZI)	[112]
			S. aureus (MMC-20)	MIC, MBC, ZI	8 µg mL^{-1} (MIC), 16 µg mL^{-1} (MBC), 16 ± 1.0 mm (ZI)	
Hydrocotyle sibthorpioides	Whole plant	13.37 ± 10	*K. pneumonia*	ZI	3.0 ± 0.17 mm	[121]
			P. aeruginosa	ZI	2.7 ± 0.32 mm	
			S. aureus	ZI	3.6 ± 0.57 mm	
Vaccinium corymbosum	leaf	10–30	*E. coli* (ATCC 25922)	MIC, MBC, ZI	11.22 ± 0.29 mm	[122]
			S. aureus (ATCC 25923)	MIC, MBC, ZI	13.1 ± 1.1 mm	
			P. aeruginosa (ATCC 27853)	MIC, MBC, ZI	11.6 ± 0.32 mm	
			B. subtilis (ATCC 21332)	MIC, MBC, ZI	12.4 ± 0.40 mm	
Prosopis farcta	leaf	10.8 ± 3.54	*S. aureus* (PTCC 1431)	ZI	9.5 mm	[123]
			B. subtilis (PTCC 1420)	ZI	9 mm	
			E. coli (PTCC 1399)	ZI	9.5 mm	
			P. aeruginosa (PTCC 1074)	ZI	9.5 mm	
Sesbania gradiflora	leaf	10–25	*S. enterica*	ZI	15.67 ± 0.09 mm	[124]
			S. aureus	ZI	10.54 ± 0.23 mm	
Solanum nigrum	leaf	20	*K. pneumoniae*	ZI	21.5 mm	[110]
			P. aeruginosa	ZI	21.3 mm	
			S. epidermidis	ZI	19.6 mm	
			E. coli	ZI	15.3 mm	
			P. vulgaris	ZI	13.3 mm	
			S. aureus	ZI	9.6 mm	

Table 3. *Cont.*

Plant	Part	AgNPs Size (nm)	Target MDR Microorganism	Test Type [a]	Test Result [b]	Reference
Cissus quadrangularis	leaf	15–23 (**)	*S. pyogens*	MIC, ZI	4 μg mL^{-1} (MIC), 7.77 ± 0.25 mm (ZI)	[113]
			S. aureus	MIC, ZI	3 μg mL^{-1} (MIC), 8.83 ± 0.26 mm (ZI)	
			E. coli	MIC, ZI	5 μg mL^{-1} (MIC), 7.9 ± 0.31 mm (ZI)	
			P. vulgaris	MIC, ZI	7 μg mL^{-1} (MIC), 8.4 ± 0.40 mm (ZI)	
Cola nitida	pod	12–80	*E. coli*	ZI	19 ± 0.9 mm	[125]
			K. granulomatis	ZI	11 ± 0.8 mm	
			P. aeruginosa	ZI	28 ± 0.1 mm	
Strychnos potatorum	leaf	28	*S. aureus*	ZI	8 mm	[126]
			K. pneumoniae	ZI	10 mm	
Alstonia scholaris	leaf	80	*E. coli*	ZI	10.0 ± 2.8 mm	[111]
			P. aeruginosa	ZI	8.0 ± 1.4 mm	
			K. pneumoniae	ZI	11.0 ± 1.0 mm	
			S. aureus	ZI	10.0 ± 3.0 mm	
			P. vulgaris	ZI	8.3 ± 0.6 mm	
			S. epidermidis	ZI	10.6 ± 1.2 mm	
Andrographis paniculata	leaf	70	*E. coli*	ZI	8.0 ± 1.4 mm	[111]
			P. aeruginosa	ZI	6.7 ± 0.7 mm	
			K. pneumoniae	ZI	9.3 ± 0.6 mm	
			S. aureus	ZI	8.0 ± 1.0 mm	
			P. vulgaris	ZI	8.3 ± 0.6 mm	
			S. epidermidis	ZI	9.0 ± 1.0 mm	
Aegle marmelos	leaf	70	*E. coli*	ZI	11.0 ± 2.8 mm	[111]
			P. aeruginosa	ZI	9.0 ± 1.4 mm	
			K. pneumoniae	ZI	9.3 ± 1.6 mm	
			S. aureus	ZI	9.7 ± 1.5 mm	
			P. vulgaris	ZI	9.7 ± 0.6 mm	
			S. epidermidis	ZI	8.0 ± 1.0 mm	
Centella asiatica	leaf	90	*E. coli*	ZI	12.7 ± 0.7 mm	[111]
			P. aeruginosa	ZI	8.0 ± 1.4 mm	
			K. pneumoniae	ZI	12.0 ± 1.0 mm	
			S. aureus	ZI	13.0 ± 2.0 mm	
			P. vulgaris	ZI	9.7 ± 0.6 mm	
			S. epidermidis	ZI	14.0 ± 1.0 mm	
Eclipta prostrata	leaf	70	*E. coli*	ZI	10.0 ± 4.0 mm	[111]
			P. aeruginosa	ZI	8.3 ± 2.5 mm	
			K. pneumoniae	ZI	10.0 ± 5.2 mm	
			S. aureus	ZI	12.6 ± 4.9 mm	
			P. vulgaris	ZI	6.6 ± 0.5 mm	
			S. epidermidis	ZI	8.0 ± 0.0 mm	
Moringa oleifera	leaf	50	*E. coli*	ZI	7.7 ± 0.6 mm	[111]
			P. aeruginosa	ZI	8.0 ± 1.7 mm	
			K. pneumoniae	ZI	7.0 ± 1.0 mm	
			S. aureus	ZI	9.0 ± 2.6 mm	
			P. vulgaris	ZI	7.0 ± 2.0 mm	
			S. epidermidis	ZI	7.0 ± 0.0 mm	
Thespesia populnea	bark	70	*E. coli*	ZI	9.0 ± 1.7 mm	[111]
			P. aeruginosa	ZI	10.3 ± 2.1 mm	
			K. pneumoniae	ZI	11.3 ± 1.2 mm	
			S. aureus	ZI	9.3 ± 2.4 mm	
			P. vulgaris	ZI	8.6 ± 1.2 mm	
			S. epidermidis	ZI	8.6 ± 0.7 mm	
Terminalia arjuna	bark	70	*E. coli*	ZI	8.0 ± 0.7 mm	[111]
			P. aeruginosa	ZI	9.0 ± 2.0 mm	
			K. pneumoniae	ZI	14.0 ± 1.0 mm	
			S. aureus	ZI	12.7 ± 1.1 mm	
			P. vulgaris	ZI	8.3 ± 0.6 mm	
			S. epidermidis	ZI	9.0 ± 2.0 mm	
Plumbago zeylanica	Root bark	90	*E. coli*	ZI	8.0 ± 1.4 mm	[111]
			P. aeruginosa	ZI	14.7 ± 0.7 mm	
			K. pneumoniae	ZI	8.3 ± 0.8 mm	
			S. aureus	ZI	7.7 ± 0.6 mm	
			P. vulgaris	ZI	8.3 ± 0.6 mm	
			S. epidermidis	ZI	8.0 ± 1.0 mm	
Semecarpus anacardium	nuts	60	*E. coli*	ZI	10.0 ± 2.0 mm	[111]
			P. aeruginosa	ZI	9.3 ± 1.5 mm	
			K. pneumoniae	ZI	10.0 ± 1.0 mm	
			S. aureus	ZI	7.7 ± 1.1 mm	
			P. vulgaris	ZI	8.3 ± 0.6 mm	
			S. epidermidis	ZI	9.3 ± 1.5 mm	

Table 3. *Cont.*

Plant	Part	AgNPs Size (nm)	Target MDR Microorganism	Test Type [a]	Test Result [b]	Reference
Mukia scabrella	leaf	18–21	*Acinetobacter* sp. *K. pneumoniae* *P. aeruginosa*	ZI ZI ZI	22 mm 19 mm 20 mm	[127]
Phyllanthus amarus	Whole plant	24 ± 8	*P. aeruginosa* (***)	MIC, ZI	6.25–12.5 µg mL^{-1} (MIC), 21 ± 0.11 mm (ZI)	[115]
Ricinodendron heudelotti	Seed kernel	89.0	*E. coli*	MIC, MBC	1.68 µg mL^{-1} (MIC), 6.75 µg mL^{-1} (MBC)	[128]
Gnetum bucholzianum	leaf	67.4	*E. coli*	MIC, MBC	1.687 µg mL^{-1} (MIC), 1.687 µg mL^{-1} (MBC)	[129]
Megaphrynium macrostachyum	leaf	33.7 (Ag), 44.2 (AgCl)	*E. coli*	MIC, MBC	0.515 µg mL^{-1} (MIC), 4.12 µg mL^{-1} (MBC)	[129]
Corchorus olitorus	leaf	30.0 (nm), 37.9 (AgCl)	*E. coli*	MIC, MBC	8.25 µg mL^{-1} (MIC), 16.5 µg mL^{-1} (MBC)	[129]
Ipomoea batatas	leaf	67.3 (Ag), 37.9 (AgCl)	*E. coli*	MIC, MBC	5.3 µg mL^{-1} (MIC), 5.3 µg mL^{-1} (MBC)	[129]
Areca catechu	leaf	22–40	*E. coli* *P. aeruginosa* *S. typhi* *P. vulgaris* *K. pneumoniae*	ZI ZI ZI ZI ZI	20 mm 24 mm 19 mm 23 mm 26 mm	[129]
Cocoa	bean	8.96–54.22	*S. aureus* *K. pneumoniae* (wound) *K. pneumoniae* (urine) *E. coli*	ZI ZI ZI ZI	12 mm (*) 12 mm (*) 13 mm (*) 14 mm (*)	[130]
Cocoa	Pod husk	4–32	*K. pneumoniae* *E. coli*	ZI ZI	10–14 mm 10–14 mm	[131]
Phomis bracteosa	Whole plant	22.41	*E. coli* (ATCC 15224) *S. aureus* (ATCC 6538) *K. pneumoniae* (ATCC 4619)	ZI ZI ZI	13.2 ± 0.12 11.1 ± 0.10 10.3 ± 0.11	[108]
Momordica cymbalaria	fruit	15.5	*E. coli* *M. luteus* *B. cereus* *K. pneumoniae* *S. pneumoniae*	ZI ZI ZI ZI ZI	24.0 ± 1.0 20.0 ± 1.4 22.0 ± 1.0 26.0 ± 1.4 26.0 ± 1.7	[132]
Astragalus membranaceus	root	65.08	*S. aureus* (MRSA) *S. epidermidis* (MRSE) *P. aeruginosa* *E. coli*	MIC, ZI MIC, ZI MIC, ZI MIC, ZI	0.063 mg mL^{-1} (MIC), 12.83 ± 1.04 mm (ZI) 0.063 mg mL^{-1} (MIC), 12.33 ± 0.29 mm (ZI) 0.032 mg mL^{-1} (MIC), 15.17 ± 0.76 mm (ZI) 0.032 mg mL^{-1} (MIC), 14.67 ± 0.76 mm (ZI)	[107]

[a] ZI = zone of inhibition; MIC = Minimum Inhibitory Concentration; MBC = Minimum Bactericidal Concentration; [b] For tests in which more than one concentration of AgNPs was used, the best results are shown; (*) Values estimated from graphs; (**) Silver chloride nanoparticles; (***) 15 strains were tested

4.4. Modes of Action of AgNPs against Bacteria

As stated in previous reviews on the subject [24,133–137], the antibacterial action of silver nanoparticles involves a complex mechanism in which more than one factor can act simultaneously to contribute to an overall effect. Moreover, one must consider the existence of more than one silver species, these being the Ag^0 in the form of nanoparticles and the Ag$^+$ which is released from the surface of the nanoparticles as they are slowly oxidized.

Proteomic analysis of *E. coli* proteins expressed after exposure to AgNPs and Ag$^+$ revealed that both have a similar mode of action, such as overexpressing envelope and heat shock proteins. However, the nanoparticles were effective at inhibiting bacteria in the nanomolar concentration, whereas the Ag$^+$ ions were effective only in the micromolar range [138]. On the other hand, further reports point to the opposite direction. Ag$^+$ release depends on oxidation of metallic silver by oxygen in the air; in a study where *E. coli* was exposed to AgNPs in anaerobic conditions, no bactericidal activity was observed, while in aerobic conditions the usual antimicrobial activity was noticed [139]. This effect can be partially explained by a strong interaction of Ag$^+$ with the cell membrane and cell wall components such as proteins, phospholipids, and thiol-containing groups, as well as by a proton leakage that can induce cell disintegration [140]. As much as the affinity of Ag$^+$ for thiol groups has been known for decades [141], just recently Liao et al. [142] demonstrated how Ag$^+$ can deplete intracellular thiol content of *S. aureus* and bind to cysteine residues of thioredoxin reductase's catalytic site. This enzyme is one of the most important ones related to the antioxidant mechanism and reactive oxygen species (ROS) levels regulation in bacteria. Binding to respiratory chain enzymes is also a factor for intracellular ROS increase [143]. It is worth noting, however, that the protein corona that involves AgNPs has a significant effect on silver ions release. According to a study performed by Wen et al. [144], the binding of cytoskeletal proteins to AgNPs led to a decrease in Ag$^+$ leakage, which could suggest that, similarly, biogenic AgNPs that are capped by biomolecules also have a diminished Ag$^+$ release and thus their antimicrobial action would rely much less on this species.

Regarding the action of the nanoparticles, their size, shape and capping molecules may play significant roles when binding to the cell wall, membrane, and their internalization. In a study performed with silver nanospheres, nanocubes, and nanowires, the latter resulted in diminished antimicrobial activity when compared to the first two due to a smaller effective contact area with the cell membrane [145]. The same explanation applies for truncated octahedral AgNPs outperforming spherical AgNPs [146]. Truncated triangular shaped AgNPs had a better performance than all the other shapes in a study conducted against *E. coli* [147]. Acharya et al. [148] recently reported a study on silver nanospheres and silver nanorods acting against *K. pneumoniae* and attributed the antibacterial activity to the {111} plane shapes, which contain the highest atomic density. Smaller sizes of nanoparticles also lead to an enhanced bactericidal effect [149,150]. This effect is due to a greater surface area in contact with the bacteria that facilitate membrane rupture and internalization [151].

Perhaps one of the most accepted antibacterial mechanisms involves the association of nanoparticles with the cell wall followed by the formation of "pits" [152] and leakage of cellular contents [153]. This corroborates with the fact that AgNPs are usually more active towards Gram-negative bacteria [154], as Gram-positive bacteria have a thicker peptidoglycan cell wall, which could act as an additional physical barrier. Once inside the bacterial cell (a process that is facilitated by sizes smaller than 5 nm [155]), small nanoparticles are able to interfere with the respiratory chain dehydrogenases [156] and also induce generation of intracellular ROS [112,157], which have the ability to cleave DNA [158] and diminish bacterial life. It must be also pointed out that the interaction of AgNPs with the media which they are suspended in has a great influence on AgNPs physicochemical properties and their action on bacterial cells [159]. Figure 3 illustrates all the major mechanisms by which AgNPs display their antibacterial action.

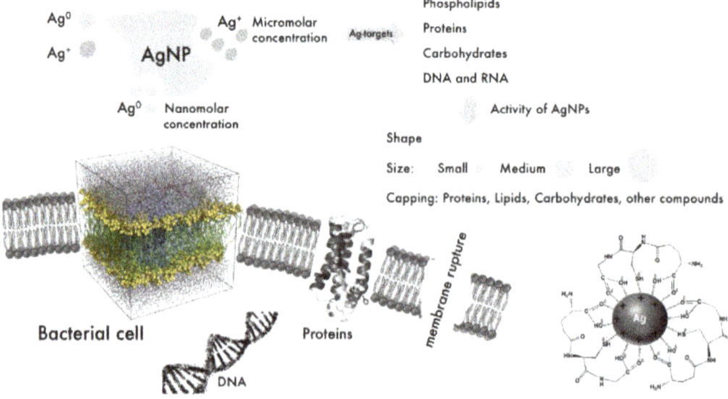

Figure 3. Summary of the factors affecting the antimicrobial capacity of AgNPs and main antibacterial mechanisms. Size, shape and capping agents have a significant influence on the activity against bacterial cells, which are susceptible to nanoparticles because of a strong affinity of the metal with the cell wall and membrane, as well as due to interference in the respiratory chain and generation of reactive oxygen species (ROS).

4.5. Bacterial Resistance to Silver

The increasing application of silver nanomaterials in dressings, packages, and textiles has raised concerns about the development of bacterial resistance to nanosilver, despite the good performance of AgNPs against a range of bacterial strains, as already described. In fact, one of the first reports on resistance to silver was published in 1975, when a strain of *Salmonella typhimurium* resistant to silver nitrate, mercuric chloride, and a range of common antibiotics was identified in three patients in a burn unit [160]. Decades later, this exogenous type of resistance was unveiled by Gupta et al. [161] through the isolation of the plasmid pMG101. This plasmid was identified as the carrier of a silver resistance gene *silE*, which encodes a 143-amino-acid periplasmic Ag^+-specific protein. Upstream of *silE*, a series of genes from the Sil system encode silver efflux-related proteins, such as a protein/cation antiporter system and a P-type cation ATPase (Figure 4). Resistance to silver attributed to *sil* genes was also recently reported for clinical isolates of *Klebsiella pneumonia* and *Enterobacter cloacae* [162]. Endogenous (mutational) silver resistance may also be observed, as reported by Li et al. [163], who observed silver resistance induced in *E. coli* cells by selectively culturing bacterial cells in increasing concentrations of silver nitrate. In this case, mutant cells were deficient in major porins (OmpF and OmpC). Silver efflux is also mediated through a CusCFBA efflux pump system, which has a high amino acid sequence similarity with the Sil system, in spite of being an endogenous type of resistance [164]. Crystal structures of proteins of the CusCFBA system suggest a methionine shuttle efflux mechanism, in which Ag^+ ions are ejected from the bacterial periplasm [165,166]. Nuclear magnetic resonance (NMR) and inductively coupled plasma mass spectrometry (ICP-MS) studies have demonstrated that silver ions may induce a histidine kinase (CuS) dimerization and this conformational change may have a reflex on the upregulation of genes encoding the CusCFBA transport system [167]. The *E. coli* gene *ybdE* belonging to the K38 chromosome was also pointed out as related exclusively to Ag^+ resistance since its deletion in silver-resistant mutant strains had no effect on Cu^+ resistance [168]. Graves et al. [71] recently performed an extensive study using a non-resistant *E. coli* strain for an evolutionary analysis focused on mutations acquired upon exposure to silver nitrate and silver nanoparticles. After 300 generations, the Minimum Inhibitory Concentration (MIC) (using more than one type of AgNPs) of treated bacteria was already between 1.40 and 4.70 times the MIC of control bacteria. Three main mutations were observed: (1) in the *cuS* gene, which encodes the already mentioned histidine kinase which functions as a sensor for the CusCFBA efflux pump; (2) in the *purL* gene, which encodes for an enzyme involved in *de novo* purine nucleotide biosynthesis; and (3) in the *rpoB* gene, responsible for an RNA polymerase beta subunit.

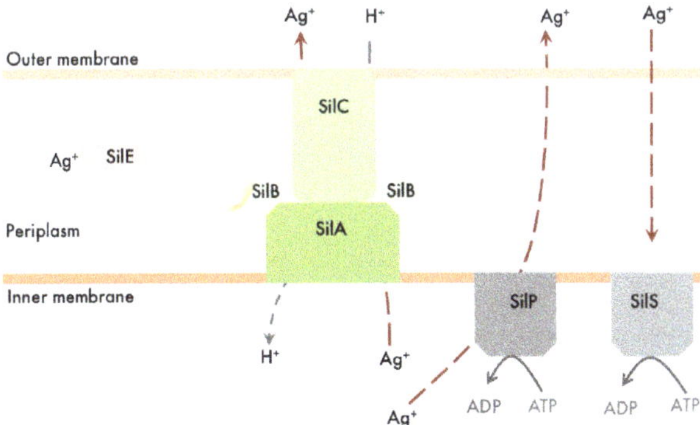

Figure 4. Silver efflux system found in Gram-negative silver-resistant bacteria. SilE is a periplasmic, histidine-rich Ag^+ binding protein; SilS belongs to a two-component (SilRS) transcription regulation system; SilA, SilB, and SilC comprise a three-component chemiosmotic bacterial proton/cation antiporter.

It is worth noting, however, that most of the studies cited are related to exogenous and endogenous Ag$^+$ resistance. The release of Ag$^+$ ions by AgNPs is only one of the forms by which AgNPs might be antimicrobial, as explained in Section 4.4. Few studies have looked at resistance to silver nanoparticles. For instance, Panacek et al. [169] have observed *E. coli* resistance to 28 nm AgNPs in sub-MIC concentrations without any genetic changes noted in *E. coli*. Only a phenotypic change in production of flagellin was noted. Flagellin, an adhesive protein of the flagellum, related to biofilm formation and motility, was found to readily induce nanoparticle aggregation and attenuate their antimicrobial capacity. There is still much to be researched and discovered on outer membrane–metal interactions, especially what accounts for different capping agents, topography, and morphology of AgNPs. Also, other bacterial species and strains must be studied as to map genetic and/or phenotype modifications induced by AgNPs.

5. Nanosilver Applications in Antimicrobial Products

The well-documented antimicrobial activities of AgNPs have attracted great attention from researchers and companies and caused manufacturing of many products which are in everyday use. For instance, dressings, biomedical equipment, paints, packaging materials, and gels containing nanosilver formulations are widely used. However, the number of AgNPs-containing products that are focused on or have been tested against MDRB is still unexpressive and modest. This is even surprisingly true when it comes to biogenically or bio-based synthesized AgNPs. Nevertheless, among many patents of products containing nanosilver, there are some possible applications of patented formulations in the combat against resistant bacteria, which are summarized in Table 4.

Table 4. Patents of AgNPs-based products tested against resistant bacterial strains.

Patent Number	Application	Resistant Bacteria	Reference
WO2006074117A2	Hydrogel	*E. cloacae, K. pneumoniae, E. coli, P. aeruginosa, A. Acinetobacter*	[170]
WO2018010403A1	Pharmaceuticals	*E. cloacae, K. pneumoniae, E. coli, P. aeruginosa, A. Acinetobacter*	[171]
US20100003296A1	Textiles	Methicillin-resistant *S. aureus* (MRSA)	[172]
KR200384433Y1	Apron, perfume	Methicillin-resistant *S. aureus* (MRSA)	[173]
KR100933736B1	Detergent additive	*E. coli*	[174]
CN105412940A	General	Vancomycin-resistant *Enterococcus faecalis*	[175]
WO2005120173A2	General	*P. aeruginosa*	[176]
US7135195B2	General	Methicillin-resistant *S. aureus* (MRSA)	[177]

Despite the controversy that involves the oral use of silver nanoparticles, a recent patent has established a preparation involving AgNPs active towards MDRB suggesting many possible forms of administration, including oral, topical, and intravenous [171]. An invention communicated by Holladay et al. [170] postulates compositions containing AgNPs that may be introduced into a hydrogel for the treatment of various types of infections and inflammations, with activity against MDR *E. cloacae, K. pneumoniae, E. coli, P. aeruginosa,* and *A. Acinetobacter*. In fact, the well-known wound healing capacity of nanosilver is often exploited in dressings and plasters. Liang et al. [178] developed an AgNPs/chitosan composite with amphiphilic properties—a hydrophobic and waterproof surface and a hydrophilic one with a capacity to interact with water and inhibit the growth of the drugs resistant *S. aureus, E. coli,* and *P. aeruginosa*. It is important to point out that these types of dressings with asymmetric wettability properties also enhance re-epithelization and collagen deposition and might be very helpful for wound healing not just because of their antiseptic properties.

Nanocrystalline silver coatings are already available commercially, for example, ACTICOAT™ has been used against MDR *P. aeruginosa* in burn wound infections in rat models [179]. This dressing has also been proven to be effective against methicillin-resistant *S. aureus*, by inhibiting bacterial growth in burn wounds. But it also decreases the secretion and swelling of the damaged tissue areas [180], which speeds up processes of wound healing.

An invention deposited by Paknikar (2006) [176] claims the production of biologically stabilized AgNPs, which were produced from various plants parts, and their incorporation into a variety of possible carriers, such as ointments, sprays, membranes, plasters. The nanoparticles were shown to successfully inhibit MDR strains of *P. aeruginosa* and other highly resistant bacterial strains: *E. coli* ATCC 117, *P. aeruginosa* ATCC 9027, *S. abony* NCTC 6017, *S. typhimurium* ATCC 23564, *K. aerogenes* ATCC 1950, *P. vulgaris* NCBI 4157, *S. aureus* ATCC 6538P, *B. subtilis* ATCC 6633., and *C. albicans*, and, interestingly, were non-cytotoxic towards human leukemic cells (K562), carcinoma cells (HEPG2), and mouse fibroblasts (L929) in the concentrations used against cited MRDB.

Also, there are some reports on materials that contain AgNPs, such as a multipurpose nanocomposite comprising silver nanotriangles and silicon dioxide, which was developed and tested against vancomycin-resistant bacteria *E. Faecalis* (ATCC 51299) [173]. There is a nanocomposite of silver and silver oxide active towards methicillin-resistant *S. aureus* and a broad spectrum of pathogenic bacteria associated with common infections and inflammations in humans [177].

Common household objects can also be enriched with AgNPs to enhance their antimicrobial potential; for example, nanosilver has been used as a detergent additive to enhance the antibiotic effect of the surfactant while not inducing any decrease in the detergent capability of a product [174]. The detergent can be used to disinfect resistant *E. coli* strains. Enhanced hygiene and diminished contamination were also achieved by reinforcing aprons with AgNPs; the material was successful in inhibiting methicillin-resistant *S. aureus*. Cheng and Yan [172] reported and patented the invention on antimicrobial plant fibers enriched with AgNPs that showed strong antimicrobial activity. This material may be applied in various types of linings, clothing, and even for fabricating laboratory or medical coats with improved disinfection properties and thus avoid bacterial contamination.

As stated, there are still much to be discovered and researched until novel fabrics, commodities, and/or pharmaceuticals based on biogenic or bio-based silver nanoparticles became suitable for everyday applications.

6. Conclusions

Some of the main reasons for observing the multidrug resistance in bacteria were discussed along with an introduction of biogenic silver nanoparticles as an alternative or combined technology to overcome this growing health problem. Even though bio-based silver containing nanomaterials are usually not ingested as known antibiotics, mainly due to a lack of understanding of the nanotoxicology associated with nanosilver in the bloodstream or in organs, AgNPs may be incorporated in products such as dressings, sprays, textiles, and paints for MDRB combat to a certain extent. Topical use of ointments and wound dressings have become quite common, as AgNPs not only inhibit bacteria growth but also stimulate epithelial growth and reduce swelling and secretion. Bacterial resistance to silver is a concerning perspective; however, application of bio-based AgNPs may at least postpone it because the extracts used for their synthesis might have natural antimicrobial effects that can act synergistically with the nanosilver. Moreover, combined therapies based on biogenic AgNPs and known antibiotics might be even more effective than the use of only one of them.

The development of biogenic AgNPs-containing products, which are active against MDRB, finds its main obstacle in discovering a systematic, easy to reproduce, and scaled-up process for the production of the uniform nanoparticles with desirable properties that do not vary, which is extremely hard to achieve considering the biological provenience of the extracts. By the time these processes become viable, controlled, and understood, the incorporation of the biologically synthesized nanomaterials as novel biopharmaceuticals or their use as commercial products should find many opportunities in various fields.

Author Contributions: C.H.N.B., S.F. and D.S. performed bibliographic research and wrote the first version of the manuscript. L.T. idealized and revised the manuscript and coordinated the project.

Funding: The authors acknowledge the financial supports received from the Fundação de Amparo à Pesquisa de São Paulo (Fapesp—Projects N°: 2015/12534-5 and 2014/50867-3) and Conselho Nacional de Desenvolvimento Científico e Tecnológico (CNPq—Project N°: 465389/201407).

Conflicts of Interest: The authors declare no conflict of interest.

References

1. World Health Organization. 10 Facts on Antimicrobial Resistance. Available online: http://www.who.int/features/factfiles/antimicrobial_resistance/en/ (accessed on 10 June 2018).
2. Littier, H.M.; Chambers, L.R.; Knowton, K.F. Animal agriculture as a contributor to the global challenge of antibiotic resistance. *CAB Rev.* **2017**, *8*, 1–9. [CrossRef]
3. Roca, I.; Akova, M.; Baquero, F.; Carlet, J.; Cavaleri, M.; Coenen, S.; Cohen, J.; Findlay, D.; Gyssens, I.; Heure, O.E.; et al. The global threat of antimicrobial resistance: Science for intervention. *New Microbe New Infect.* **2015**, *6*, 22–29. [CrossRef] [PubMed]
4. Huebner, C.; Rogellin, M.; Flessa, S. Economic burden of multidrug-resistant bacteria in nursing homes in Germany: A cost analysis based on empirical data. *BJM Open* **2016**, *6*, e008458. [CrossRef] [PubMed]
5. Centers for Disease Control and Prevention. Drug-Resistant TB. Available online: http://www.cdc.gov/tb/topic/drtb/ (accessed on 10 June 2018).
6. World Health Organization. WHO Publishes List of Bacteria for Which New Antibiotics Are Urgently Needed. Available online: http://www.who.int/en/news-room/detail/27-02-2017-who-publishes-list-of-bacteria-for-which-new-antibiotics-are-urgently-needed (accessed on 10 June 2018).
7. Worthington, R.J.; Melander, C. Combination approaches to combat multidrug-resistant bacteria. *Trends Biotechnol.* **2013**, *31*, 177–184. [CrossRef] [PubMed]
8. Lane, D. Designer combination therapy for cancer. *Nat. Biotechnol.* **2006**, *24*, 163–164. [CrossRef] [PubMed]
9. Richman, D.D. HIV chemotherapy. *Nature* **2001**, *410*, 995–1001. [CrossRef] [PubMed]
10. Nosten, F.; White, M.J. Artemisinin-based combination treatment of falciparum malaria. *Am. J. Trop. Med. Hyg.* **2007**, *77*, 191–192.
11. Xiao, Y.; Wang, D.; Heise, A.; Lang, M. Chemo-enzymatic synthesis of poly (4-piperidine lactone-b-ω-pentadecalactone) block copolymers as biomaterials with antibacterial properties. *Biomacromolecules* **2018**, *19*, 2673 2681. [CrossRef] [PubMed]
12. Zoriasatein, M.; Bidhendi, S.M.; Madani, R. Evaluation of antimicrobial properties of derivative peptide of Naja naja snake's venom. *World Fam. Med. J.* **2018**, *16*, 44–62.
13. Al-Gbouri, N.M.; Hamzah, A.M. Evaluation of Phyllanthus emblica extract as antibacterial and antibiofilm against biofilm formation. *TIJAS* **2018**, *49*, 142–151.
14. Naqvi, S.Z.H.; Kiran, U.; Ali, M.I.; Jamal, A.; Hameed, A.; Ahmed, S.; Ali, N. Combined efficacy of biologically synthesized silver nanoparticles and different antibiotics against multidrug-resistant bacteria. *Int. J. Nanomed.* **2013**, *8*, 3187–3195. [CrossRef] [PubMed]
15. Sirelkhatim, A.; Mahmud, S.; Seeni, A.; Kaus, N.H.M.; Ann, L.C.; Bakhori, S.K.M.; Hasan, H.; Mohamad, D. Review on zinc oxide nanoparticles: Antibacterial activity. *Nano-Micro Lett.* **2015**, *7*, 219–242. [CrossRef]
16. Ingle, A.P.; Duran, N.; Rai, M. Bioactivity, mechanism of action, and cytotoxicity of copper-based nanoparticles: A review. *Appl. Microbiol. Biotechnol.* **2014**, *98*, 1001–1009. [CrossRef] [PubMed]
17. Dinali, R.; Ebrahiminezhad, A.; Manley-Harris, M.; Ghasemi, Y.; Berenjian, A. Iron oxide nanoparticles in modern microbiology and biotechnology. *Crit. Rev. Microb.* **2017**, *43*, 493–507. [CrossRef] [PubMed]
18. Dastjerdi, R.; Montazer, M. A review on the application of inorganic nano-structured materials in the modification of textiles: Focus on anti-microbial properties. *Colloids Surf. B Biointerfaces* **2010**, *79*, 5–18. [CrossRef] [PubMed]
19. Ballottin, D.; Fulaz, S.; Cabrini, F.; Tsukamoto, J.; Durán, N.; Alves, O.L.; Tasic, L. Antimicrobial textiles: Biogenic silver nanoparticles against Candida and Xanthomonas. *Mater. Sci. Eng. C* **2017**, *75*, 582–589. [CrossRef] [PubMed]
20. Ertem, E.; Guut, B.; Zuber, F.; Allegri, S.; Le Ouay, B.; Mefti, S.; Formentin, K.; Stellacci, F.; Ren, Q. Core-shell silver nanoparticles in endodontic disinfection solutions enable long-term antimicrobial effect on oral biofilms. *ACS Appl. Mater. Interfaces* **2017**, *9*, 34762–34772. [CrossRef] [PubMed]
21. Nakazato, G.; Kobayashi, R.; Seabra, A.B.; Duran, N. Use of nanoparticles as a potential antimicrobial for food packaging. In *Food Preservation*, 1st ed.; Grumezescu, A., Ed.; Academic Press: Cambridge, MA, USA, 2016.
22. Holtz, R.D.; Lima, B.A.; Filho, A.G.S.; Brocchi, M.; Alves, O.L. Nanostructured silver vanadate as a promising antibacterial additive to water-based paints. *Nanomed. NBM* **2012**, *8*, 935–940. [CrossRef] [PubMed]

23. LaMer, V.K.; Dinegar, R.H. Theory, production and mechanism of formation of monodispersed hydrosols. *J. Am. Chem. Soc.* **1950**, *72*, 4847–4854. [CrossRef]

24. Rai, M.K.; Deshmukh, S.D.; Ingle, A.P.; Gade, A.K. Silver nanoparticles: The powerful nanoweapon against multidrug-resistant bacteria. *J. Appl. Microb.* **2012**, *112*, 841–852. [CrossRef] [PubMed]

25. Radetic, M. Functionalization of textile materials with silver nanoparticles. *J. Mater. Sci.* **2013**, *48*, 95–107. [CrossRef]

26. Maneerung, T.; Tokura, S.; Rujiravanit, R. Impregnation of silver nanoparticles into bacterial cellulose for antimicrobial wound dressing. *Carbohydr. Polym.* **2008**, *72*, 43–51. [CrossRef]

27. Chen, J.; Han, C.M.; Lin, X.W.; Tang, Z.J.; Su, S.J. Effect of silver nanoparticles dressing on second degree burn wound. *Zhonghua Wai Ke Za Zhi* **2006**, *44*, 50–52. [PubMed]

28. Muangman, P.; Chuntrasakul, C.; Silthram, S.; Suvanchote, S.; Benhathanung, R.; Kttidacha, S.; Rueksomtawin, S. Comparison of efficacy of 1% silver sulfadiazine and Acticoat for treatment of partial-thickness burn wounds. *J. Med. Assoc. Thail.* **2006**, *89*, 953–958.

29. Cohen, M.S.; Stern, J.M.; Vanni, A.J.; Kelley, R.S.; Baumgart, E.; Field, D.; Libertino, J.A.; Summerhayes, I.C. In vitro analysis of a nanocrystalline silver-coated surgical mesh. *Surg. Infect.* **2007**, *8*, 397–403. [CrossRef] [PubMed]

30. Lansdown, A.B. Silver in health care: Antimicrobial effects and safety in use. *Curr. Probl. Dermatol.* **2006**, *33*, 17–34. [PubMed]

31. Zhang, Z.; Yang, M.; Huang, M.; Hu, Y.; Xie, J. Study on germicidal efficacy and toxicity of compound disinfectant gel of nanometer silver and chlorhexidine acetate. *Chin. J. Health Lab. Technol.* **2007**, *17*, 1403–1406.

32. Zhang, Y.; Sun, J. A study on the bio-safety for nano-silver as anti-bacterial materials. *Chin. J. Med. Instrum.* **2007**, *31*, 35–38.

33. Nowack, B.; Krug, H.F.; Height, M. 120 Years of nanosilver history: Implications for policy makers. *Environ. Sci. Technol.* **2011**, *45*, 1177–1183. [CrossRef] [PubMed]

34. Waksman, S. History of the word 'antibiotic'. *J. Hist. Med. Allied Sci.* **1973**, *28*, 284–286. [CrossRef] [PubMed]

35. Davies, J.; Davies, D. Origins and evolution of antibiotic resistance. *Microbiol. Mol. Biol. Rev.* **2010**, *74*, 417–433. [CrossRef] [PubMed]

36. Williams, K. The introduction of 'chemotherapy' using arsphenamine—The first magic bullet. *J. R. Soc. Med.* **2009**, *102*, 343–348. [CrossRef] [PubMed]

37. Izumi, Y.; Isozumi, K. Modern Japanese medical history and the European influence. *Keio J. Med.* **2001**, *50*, 91–99. [CrossRef] [PubMed]

38. Amivov, R. A brief history of the antibiotic era: Lessons learned and challenges for the future. *Front. Microbiol.* **2010**, *1*, 134.

39. Bbosa, G.; Mwebaza, N.; Odda, J.; Kyegombe, D.; Ntale, M. Antibiotics/antibacterial drug use, their marketing and promotion during the post-antibiotic golden age and their role in emergence of bacterial resistance. *Health* **2014**, *6*, 410–425. [CrossRef]

40. Thal, L.; Zervos, M. Occurrence and epidemiology of resistance to virginiamycin and streptogramins. *J. Antimicrob. Chemother.* **1999**, *43*, 171–176. [CrossRef] [PubMed]

41. Manten, A.; Van Wijngaarden, L. Development of drug resistance to rifampicin. *Chemotherapy* **1969**, *14*, 93–100. [CrossRef] [PubMed]

42. Chopra, I.; Roberts, M. Tetracycline antibiotics: Mode of action, applications, molecular biology, and epidemiology of bacterial resistance. *Microbiol. Mol. Biol. Rev.* **2001**, *65*, 232–260. [CrossRef] [PubMed]

43. Kirst, H. Introduction to the macrolide antibiotics. In *Macrolide Antibiotics Milestones in Drug Therapy MDT*; Schönfeld, W., Kirst, H., Eds.; Birkhäuser: Basel, Switzerland, 2002; pp. 1–13.

44. Jacoby, G. Mechanisms of resistance to quinolones. *Clin. Infect. Dis.* **2005**, *41*, S120–S126. [CrossRef] [PubMed]

45. Madhavan, H.; Bagyalakshmi, R. Farewell, chloramphenicol? Is this true?: A review. *J. Microbiol. Biotechnol.* **2013**, *3*, 13–26.

46. Mutnick, A.; Enne, V.; Jones, R. Linezolid resistance since 2001: SENTRY Antimicrobial Surveillance Program. *Ann. Pharmacother.* **2003**, *37*, 769–774. [CrossRef] [PubMed]

47. Abraham, E.; Chain, E. An enzyme from bacteria able to destroy penicillin. *Rev. Infect. Dis.* **1940**, *10*, 677–678. [CrossRef]

48. D'Costa, V.; McGrann, M.; Hughes, D.; Wright, G. Sampling the antibiotic resistome. *Science* **2006**, *311*, 374–377. [CrossRef] [PubMed]

49. Davies, J. Vicious circles: Looking back on resistance plasmids. *Genetics* **1995**, *139*, 1465–1468. [PubMed]

50. Helinski, D. Introduction to plasmids: A selective view of their history. In *Plasmid Biology*; Funnell, B., Philips, G., Eds.; ASM Press: Washington, DC, USA, 2004; pp. 1–21.

51. Hacker, J.; Kaper, J. Pathogenicity islands and the evolution of microbes. *Annu. Rev. Microbiol.* **2000**, *54*, 641–679. [CrossRef] [PubMed]

52. Munita, J.M.; Arias, C.A. Mechanisms of Antibiotic Resistance. *Microbiol. Spectr.* **2016**, *4*, 1–37.

53. Steward, P.; Costerton, J. Antibiotic resistance of bacteria in biofilms. *Lancet* **2001**, *358*, 135–138. [CrossRef]

54. Andersson, D. Persistence of antibiotic resistant bacteria. *Curr. Opin. Microbiol.* **2003**, *6*, 452–456. [CrossRef] [PubMed]

55. Nikaido, H. Multidrug Resistance in Bacteria. *Annu. Rev. Biochem.* **2009**, *78*, 119–146. [CrossRef] [PubMed]

56. Kirbis, A.; Krizman, M. Spread of antibiotic resistant bacteria from food of animal origin to humans and vice versa. *Procedia Food Sci.* **2015**, *5*, 148–151. [CrossRef]

57. Lee Ventola, C. The Antibiotic Resistance Crisis—Part 1: Causes and Threats. *Pharm. Ther.* **2015**, *40*, 277–293.

58. Van Duin, D.; Paterson, D. Multidrug resistant bacteria in the community: Trends and lessons learned. *Infect. Dis. Clin. N. Am.* **2016**, *30*, 377–390. [CrossRef] [PubMed]

59. The World Is Running out of Antibiotics, WHO Report Confirms. Available online: http://www.who.int/news-room/detail/20-09-2017-the-world-is-running-out-of-antibiotics-who-report-confirms (accessed on 10 June 2018).

60. Davies, J. Where have all the antibiotics gone? *Can. J. Infect. Dis. Med. Microbiol.* **2006**, *17*, 287–290. [CrossRef] [PubMed]

61. Why Are There So Few Antibiotics in the Research and Development Pipeline? Available online: https://www.pharmaceutical-journal.com/news-and-analysis/features/why-are-there-so-few-antibiotics-in-the-research-and-development-pipeline/11130209.article (accessed on 10 June 2018).

62. Thakkar, K.N.; Mhatre, S.S.; Parikh, R.Y. Biological synthesis of metallic nanoparticles. *Nanomed. NBM* **2010**, *6*, 257–262. [CrossRef] [PubMed]

63. Durán, N.; Silveira, C.P.; Durán, M.; Martinez, D.S.T. Silver nanoparticle protein corona and toxicity: A mini-review. *J. Nanobiotechnol.* **2015**, *13*, 1–17. [CrossRef] [PubMed]

64. Ballottin, D.; Fulaz, S.; Souza, M.L.; Corio, P.; Rodrigues, A.G.; Souza, A.O.; Marcato, P.G.; Gomes, A.F.; Gozzo, F.; Tasic, L. Elucidating protein involvement in the stabilization of the biogenic silver nanoparticles. *Nanoscale Res. Lett.* **2016**, *11*, 1–9. [CrossRef] [PubMed]

65. Shannahan, J.H.; Podila, R.; Aldossari, A.A.; Emerson, H.; Powell, B.A.; Ke, P.C.; Rao, A.M.; Brown, J.M. Formation of a protein corona on silver nanoparticles mediates cellular toxicity via scavenger receptors. *Toxicol. Sci.* **2014**, *143*, 136–146. [CrossRef] [PubMed]

66. Rai, M.; Yadav, A.; Gade, A.K. Myconanotechnology: A new and emerging science. In *Applied Mycology*; Rai, M., Bridge, P.D., Eds.; CABI: Wallingford, UK, 2009; pp. 258–267.

67. Zhao, X.; Zhou, L.; Rajoka, M.S.R.; Yan, L.; Jiang, C.; Shao, D.; Zhu, J.; Shi, J.; Huang, Q.; Yang, H.; et al. Fungal silver nanoparticles: Synthesis, application and challenges. *Crit. Rev. Biotechnol.* **2017**, *38*, 817–835. [CrossRef] [PubMed]

68. Ahmad, A.; Mukherjee, P.; Senapati, S.; Mandal, D.; Khan, M.I.; Kumar, R.; Sastry, M. Extracellular biosynthesis of silver nanoparticles using the fungus Fusarium oxysporum. *Colloids Surf. B* **2003**, *28*, 313–318. [CrossRef]

69. Durán, N.; Marcato, P.D.; Alves, O.L.; De Souza, G.I.H.; Esposito, E. Mechanistic aspects of biosynthesis of silver nanoparticles by several Fusarium oxysporum strains. *J. Nanobiotechnol.* **2005**, *3*, 8. [CrossRef] [PubMed]

70. Scandorieiro, S.; de Camargo, L.C.; Lancheros, C.A.C.; Yamada-Ogatta, S.F.; Nakamura, C.V.; de Oliveira, A.G.; Andrade, C.G.T.J.; Durán, N.; Nakazato, G.; Kobayashi, R.K.T. Synergistic and additive effect of oregano essential oil and biological silver nanoparticles against multidrug-resistant bacterial strains. *Front. Microbiol.* **2016**, *7*, 760. [CrossRef] [PubMed]

71. Graves, J.L., Jr.; Tajkarimi, M.; Cunningham, Q.; Campbell, A.; Nonga, H.; Harrison, S.H.; Barrick, J.E. Rapid evolution of silver nanoparticle resistance in *Escherichia coli*. *Front. Genet.* **2015**, *6*, 42. [CrossRef] [PubMed]

72. Chowdhury, S.; Basu, A.; Kundu, S. Green synthesis of protein capped silver nanoparticles from phytopathogenic fungus *Macrophomina phaseolina* (Tassi) Goid with antimicrobial properties against multidrug-resistant bacteria. *Nanoscale Res. Lett.* **2014**, *9*, 365. [CrossRef] [PubMed]

73. Neethu, S.; Midhun, S.J.; Radhakrishnan, E.K.; Jyothis, M. Green synthesized silver nanoparticles by marine endophytic fungus *Penicillium polonicum* and its antibacterial efficacy against biofilm forming, multidrug-resistant *Acinetobacter baumanii*. *Microb. Pathog.* **2018**, *116*, 263–272. [CrossRef] [PubMed]

74. Gopinath, P.M.; Narchonai, G.; Dhanasekaran, D.; Ranjani, A.; Thajuddin, N. Mycosynthesis, characterization and antibacterial properties of AgNPs against multidrug resistant (MDR) bacterial pathogens of female infertility cases. *Asian J. Pharm. Sci.* **2015**, *10*, 138–145. [CrossRef]

75. Fayaz, A.M.; Balaji, K.; Girilal, M.; Yadav, R.; Kalaichelvan, P.T.; Venketesan, R. Biogenic synthesis of silver nanoparticles and their synergistic effect with antibiotics: A study against gram-positive and gram-negative bacteria. *Nanomed. NBM* **2010**, *6*, 103–109. [CrossRef] [PubMed]

76. Bhat, M.A.; Nayak, B.K.; Nanda, A. Exploitation of filamentous fungi for biosynthesis of silver nanoparticle and its enhanced antibacterial activity. *Int. J. Pharm. Biol. Sci.* **2015**, *6*, 506–515.

77. Ray, S.; Sarkar, S.; Kundu, S. Extracellular biosynthesis of silver nanoparticles using the mycorrhizal mushroom *Tricholoma crassum* (Berk.) SACC: Its antimicrobial activity against pathogenic bacteria and fungus, including multidrug resistant plant and human bacteria. *Dig. J. Nanomater. Biostruct.* **2011**, *6*, 1289–1299.

78. Dhanasekaran, D.; Latha, S.; Saha, S.; Thajuddin, N.; Panneerselvam, A. Extracellular biosynthesis, characterisation and in-vitro antibacterial potential of silver nanoparticles using *Agaricus bisporus*. *J. Exp. Nanosci.* **2013**, *8*, 579–588. [CrossRef]

79. Saravanan, M.; Nanda, A. Extracellular synthesis of silver bionanoparticles from *Aspergillus clavatus* and its antimicrobial activity against MRSA and MRSE. *Colloids Surf. B Biointerfaces* **2010**, *77*, 214–218. [CrossRef] [PubMed]

80. Dar, M.A.; Ingle, A.; Rai, M. Enhanced antimicrobial activity of silver nanoparticles synthesized by *Cryphonectria* sp. evaluated singly and in combination with antibiotics. *Nanomed. NBM* **2013**, *9*, 105–110. [CrossRef] [PubMed]

81. Hiremath, J.; Rathod, V.; Ninganagouda, S.; Singh, D.; Prema, K. Antibacterial activity of silver nanoparticles from *Rhizopus* spp against Gram negative *E. coli*-MDR strains. *J. Pure Appl. Microbiol.* **2014**, *8*, 555–562.

82. Singh, R.; Shedbalkar, U.U.; Wadhwani, S.A.; Chopade, B.A. Bacteriagenic silver nanoparticles: Synthesis, mechanism, and applications. *Appl. Microbiol. Biotechnol.* **2015**, *99*, 4579–4593. [CrossRef] [PubMed]

83. Saifuddin, N.; Wong, C.W.; Yasumira, A.A.N. Rapid biosynthesis of silver nanoparticles using culture supernatant of bacteria with microwave irradiation. *E-J. Chem.* **2009**, *6*, 61–70. [CrossRef]

84. Zhang, X.; Yang, C.; Yu, H.; Sheng, G. Light-induced reduction of silver ions to silver nanoparticles in aquatic environments by microbial extracellular polymeric substances (EPS). *Water Res.* **2016**, *106*, 242–248. [CrossRef] [PubMed]

85. Klaus, T.; Joerger, R.; Olsson, E.; Granqvist, C. Silver-based crystalline nanoparticles, microbially fabricated. *Proc. Natl. Acad. Sci. USA* **1999**, *96*, 13611–13614. [CrossRef] [PubMed]

86. Klaus-Joerger, T.; Joerger, R.; Olsson, E.; Granqvist, C. Bacteria as workers in the living factory: Metal-accumulating bacteria and their potential for materials science. *Trends Biotechnol.* **2001**, *19*, 15–20. [CrossRef]

87. Kalishwaralal, K.; Deepak, V.; Pandian, S.R.K.; Kottaisamy, M.; BarathManiKanth, S.; Kartukeyan, B.; Gurunathan, S. Biosynthesis of silver and gold nanoparticles using *Brevibacterium casei*. *Colloids Surf. B* **2010**, *77*, 257–262. [CrossRef] [PubMed]

88. Singh, H.; Du, J.; Yi, T. Biosynthesis of silver nanoparticles using *Aeromonas* sp. THG-FG1.2 and its antibacterial activity against pathogenic microbes. *Artif. Cells Nanomed. Biotechnol.* **2017**, *45*, 584–590. [CrossRef] [PubMed]

89. Desai, P.P.; Prabhurajeshwar, C.; Chandrakanth, K.R. Hydrothermal assisted biosynthesis of silver nanoparticles from Streptomyces sp. GUT 21 (KU500633) and its therapeutic antimicrobial activity. *J. Nanostruct. Chem.* **2016**, *6*, 235–246. [CrossRef]

90. Manikprabhu, D.; Cheng, J.; Chen, W.; Sunkara, A.K.; Mane, S.B.; Kumar, R.; Das, M.; Hozzein, W.N.; Duan, Y.; Li, W. Sunlight mediated synthesis of silver nanoparticles by a novel actinobacterium (*Sinomonas mesophila* MPKL 26) and its antimicrobial activity against multi drug resistant *Staphylococcus aureus*. *J. Photochem. Photobiol.* **2016**, *158*, 202–205. [CrossRef] [PubMed]

91. Santos, K.S.; Barbosa, A.M.; Costa, L.P.; Pinheiro, M.S.; Oliveira, M.B.P.P.; Padilha, F.F. Silver nanocomposite biosynthesis: Antibacterial activity against multidrug-resistant strains of *Pseudomonas aeruginosa* and *Acinetobacter baumannii*. *Molecules* **2016**, *21*, 1255. [CrossRef] [PubMed]

92. Subashini, J.; Khanna, V.G.; Kannabiran, K. Anti-ESBL activity of silver nanoparticles biosynthesized using soil *Streptomyces* species. *Bioprocess Biosyst. Eng.* **2014**, *37*, 999–1006. [CrossRef] [PubMed]

93. Thomas, R.; Nair, A.P.; Mathew, J.; Ek, R. Antibacterial activity and synergistic effect of biosynthesized AgNPs with antibiotics against multidrug-resistant biofilm-forming coagulase-negative Staphylococci isolated from clinical samples. *Appl. Biochem. Biotechnol.* **2014**, *173*, 449–460. [CrossRef] [PubMed]

94. Arul, D.; Balasubramani, G.; Balasubramanian, V.; Natarajan, T.; Perumal, P. Antibacterial efficacy of silver nanoparticles and ethyl acetate's metabolites of the potent halophilic (marine) bacterium, *Bacillus cereus* A30 on multidrug resistant bacteria. *Pathog. Glob. Health* **2017**, *111*, 367–382. [CrossRef] [PubMed]

95. Lateef, A.; Ojo, S.A.; Akinwale, A.S.; Azeez, L.; Gueguim-Kana, E.B.; Beukes, L.S. Biogenic synthesis of silver nanoparticles using cell-free extract of Bacillus safensis LAU 13: Antimicrobial, free radical scavenging and larvicidal activities. *Biologia* **2015**, *70*, 1295–1306. [CrossRef]

96. Jain, D.; Kachhwaha, S.; Jain, R.; Srivastava, G.; Kothari, S.L. Novel microbial route to synthesize silver nanoparticles using spore crystal mixture of *Bacillus thuringiensis. Indian J. Exp. Biol.* **2010**, *48*, 1152–1156. [PubMed]

97. Singh, G.; Babele, P.K.; Shahi, S.K.; Sinha, R.P.; Tyagi, M.B.; Kumar, A. Green synthesis of silver nanoparticles using cell extracts of *Anabaena doliolum* and screening of its antibacterial and antitumor activity. *J. Microbiol. Biotechnol.* **2014**, *24*, 1354–1367. [CrossRef] [PubMed]

98. Saravanan, M.; Vemu, A.K.; Barik, S.K. Rapid biosynthesis of silver nanoparticles from Bacillus megaterium (NCIM 2326) and their antibacterial activity on multi drug resistant clinical pathogens. *Colloids Surf. B Biointerfaces* **2011**, *88*, 325–331. [CrossRef] [PubMed]

99. Priyadarshini, S.; Gopinath, V.; Priyadharsshini, N.M.; MubarakAli, D.; Velusamy, P. Synthesis of anisotropic silver nanoparticles using novel strain, *Bacillus flexus* and its biomedical application. *Colloids Surf. B Biointerfaces* **2013**, *102*, 232–237. [CrossRef] [PubMed]

100. Saravanan, M.; Barik, S.K.; MubarakAli, D.; Prakash, P.; Pugazhendhi, A. Synthesis of silver nanoparticles from *Bacillus brevis* (NCIM 2533) and their antibacterial activity against pathogenic bacteria. *Microb. Pathog.* **2018**, *116*, 221–226. [CrossRef] [PubMed]

101. Ahmed, S.; Ahmad, M.; Swami, B.L.; Ikram, S. A review on plants extract mediated synthesis of silver nanoparticles for antimicrobial applications: A green expertise. *J. Adv. Res.* **2016**, *7*, 17–28. [CrossRef] [PubMed]

102. Prathna, T.C.; Chandrasekaran, N.; Raichur, A.M.; Mukherjee, A. Biomimetic synthesis of silver nanoparticles by *Citrus limon* (lemon) aqueous extract and theoretical prediction of particle size. *Colloids Surf. B Biointerfaces* **2011**, *82*, 152–159. [CrossRef] [PubMed]

103. Jiang, X.C.; Chen, C.Y.; Chen, W.M.; Yu, A.B. Role of citric acid in the formation of silver nanoplates through a synergistic reduction approach. *Langmuir* **2010**, *26*, 4400–4408. [CrossRef] [PubMed]

104. Makarov, V.; Love, A.; Sinitsyna, O.; Yaminsky, S.M.; Taliansky, M.; Kalinina, N. Green nanotechnologies: Synthesis of metal nanoparticles using plants. *Acta Naturae* **2014**, *6*, 35–44. [PubMed]

105. Singh, A.K.; Talat, M.; Singh, D.P.; Srivastava, O.N. Biosynthesis of gold and silver nanoparticles by natural precursor clove and their functionalization with amine group. *J. Nanopart. Res.* **2010**, *12*, 1667–1675. [CrossRef]

106. Barros, C.H.N.; Cruz, G.C.F.; Mayrink, M.; Tasic, L. Bio-based synthesis of silver nanoparticles from orange waste: Effects of distinct biomolecule coatings on size, morphology, and antimicrobial activity. *Nanotechnol. Sci. Appl.* **2018**, *11*, 1–14. [CrossRef] [PubMed]

107. Ma, Y.; Liu, C.; Qu, D.; Chen, Y.; Huang, M.; Liu, Y. Antibacterial evaluation of silver nanoparticles synthesized by polysaccharides from *Astragalus membranaceus* roots. *Biomed. Pharmacother.* **2017**, *89*, 351–357. [CrossRef] [PubMed]

108. Anjum, S.; Abbasi, B.H. Biomimetic synthesis of antimicrobial silver nanoparticles using in vitro-propagated plantlets of a medicinally important endangered species: *Phlomis bracteosa. Int. J. Nanomed.* **2016**, *11*, 1663–1675.

109. Jeeva, K.; Thiyagarajan, M.; Elangovan, V.; Geetha, N.; Venkatachalam, P. *Caesalpinia coriaria* leaf extracts mediated biosynthesis of metallic silver nanoparticles and their antibacterial activity against clinically isolated pathogens. *Ind. Crops Prod.* **2012**, *52*, 714–720. [CrossRef]

110. Jinu, U.; Jayalakshmi, N.; Anbu, A.S.; Mahendran, D.; Sahi, S.; Venkatachalam, P. Biofabrication of cubic phase silver nanoparticles loaded with phytochemicals from *Solanum nigrum* leaf extracts for potential antibacterial, antibiofilm and antioxidant activities against MDR human pathogens. *J. Clust. Sci.* **2017**, *28*, 489–505. [CrossRef]

111. Prasannaraj, G.; Venkatachalam, P. Enhanced antibacterial, anti-biofilm and antioxidant (ROS) activities of biomolecules engineered silver nanoparticles against clinically isolated Gram positive and Gram negative microbial pathogens. *J. Clust. Sci.* **2017**, *28*, 645–664. [CrossRef]

112. Das, B.; Dash, S.K.; Mandal, D.; Ghosh, T.; Chattopadhyay, S.; Tripathy, S.; Das, S.; Dey, S.K.; Das, D.; Roy, S. Green synthesized silver nanoparticles destroy multidrug resistant bacteria via reactive oxygen species mediated membrane damage. *Arab. J. Chem.* **2017**, *10*, 862–876. [CrossRef]

113. Gopinath, V.; Priyadarshini, S.; Priyadharsshini, N.M.; Pandian, K.; Velusamy, P. Biogenic synthesis of antibacterial silver chloride nanoparticles using leaf extracts of *Cissus quadrangularis* Linn. *Mater. Lett.* **2013**, *91*, 224–227. [CrossRef]

114. Khalil, M.M.H.; Ismail, E.H.; El-Baghdady, K.Z.; Mohamed, D. Green synthesis of silver nanoparticles using olive leaf extract and its antibacterial activity. *Arab. J. Chem.* **2014**, *7*, 1131–1139. [CrossRef]

115. Singh, K.; Panghal, M.; Kadyan, S.; Chaudhary, U.; Yadav, J.P. Green silver nanoparticles of *Phyllanthus amarus*: As an antibacterial agent against multi drug resistant clinical isolates of *Pseudomonas aeruginosa*. *J. Nanobiotechnol.* **2014**, *12*, 40. [CrossRef] [PubMed]

116. Kasithevar, M.; Periakaruppan, P.; Muthupandian, S.; Mohan, M. Antibacterial efficacy of silver nanoparticles against multi-drug resistant clinical isolates from post-surgical wound infections. *Microb. Pathog.* **2017**, *107*, 327–334. [CrossRef] [PubMed]

117. Gopinath, V.; MubarakAli, D.; Priyadarshini, S.; Priyadharsshini, N.M.; Thajuddin, N.; Velusamy, P. Biosynthesis of silver nanoparticles from *Tribulus terrestris* and its antimicrobial activity: A novel biological approach. *Colloids Surf. B Biointerfaces* **2012**, *96*, 69–74. [CrossRef] [PubMed]

118. Veerasamy, R.; Xin, T.Z.; Gunasagaran, S.; Xiang, T.F.W.; Yang, E.F.C.; Jeyakumar, N.; Dhanaraj, S.A. Biosynthesis of silver nanoparticles using mangosteen leaf extract and evaluation of their antimicrobial activities. *J. Saudi Chem. Soc.* **2011**, *15*, 113–120. [CrossRef]

119. Singh, A.; Mittal, S.; Shrivastav, R.; Dass, S.; Srivastava, J.N. Biosynthesis of silver nanoparticles using *Ricinus communis* L. leaf extract and its antibacterial activity. *Dig. J. Nanomater. Biostruct.* **2012**, *7*, 1157–1163.

120. Prakash, P.; Gnanaprakasam, P.; Emmanuel, R.; Arokiyaraj, S.; Saravanan, M. Green synthesis of silver nanoparticles from leaf extract of *Mimusops elengi*, Linn. for enhanced antibacterial activity against multi drug resistant clinical isolates. *Colloids Surf. B Biointerfaces* **2013**, *108*, 255–259. [CrossRef] [PubMed]

121. Garg, M.; Devi, B.; Devi, R. In vitro antibacterial activity of biosynthesized silver nanoparticles from ethyl acetate extract of *Hydrocotyle sibthorpioides* against multidrug resistant microbes. *Asian J. Pharm. Clin. Res.* **2017**, *10*, 263–266. [CrossRef]

122. Li, K.; Ma, C.; Jian, T.; Sun, H.; Wang, L.; Xu, H.; Li, W.; Su, H.; Cheng, X. Making good use of the byproducts of cultivation: Green synthesis and antibacterial effects of silver nanoparticles using the leaf extract of blueberry. *J. Food Sci. Technol.* **2017**, *54*, 3569–3576. [CrossRef] [PubMed]

123. Miri, A.; Sarani, M.; Bazaz, M.R.; Darroudi, M. Plant-mediated biosynthesis of silver nanoparticles using *Prosopis farcta* extract and its antibacterial properties. *Spectrochim. Acta A Mol. Biomol. Spectrosc.* **2015**, *141*, 287–291. [CrossRef] [PubMed]

124. Das, J.; Das, M.P.; Velusamy, P. *Sesbania grandiflora* leaf extract mediated green synthesis of antibacterial silver nanoparticles against selected human pathogens. *Spectrochim. Acta A Mol. Biomol. Spectrosc.* **2013**, *104*, 265–270. [CrossRef] [PubMed]

125. Lateef, A.; Azeez, M.A.; Asafa, T.B.; Yekeen, T.A.; Akinboro, A.; Oladipo, I.C.; Azeez, L.; Ajibade, S.E.; Ojo, S.A.; Gueguim-Kana, E.B.; et al. Biogenic synthesis of silver nanoparticles using a pod extract of *Cola nitida*: Antibacterial and antioxidant activities and application as a paint additive. *J. Taibah Univ. Sci.* **2016**, *10*, 551–562. [CrossRef]

126. Kagithoju, S.; Godishala, V.; Nanna, R.S. Eco-friendly and green synthesis of silver nanoparticles using leaf extract of *Strychnos potatorum* Linn.F. and their bactericidal activities. *3 Biotech* **2015**, *5*, 709–714. [CrossRef] [PubMed]

127. Prabakar, K.; Sivalingam, P.; Rabeek, S.I.M.; Muthuselvam, M.; Devarajan, N.; Arjunan, A.; Karthick, R.; Suresh, M.M.; Wembonyama, J.P. Evaluation of antibacterial efficacy of phyto fabricated silver nanoparticles using *Mukia scabrella* (Musumusukkai) against drug resistance nosocomial gram negative bacterial pathogens. *Colloids Surf. B Biointerfaces* **2013**, *104*, 282–288. [CrossRef] [PubMed]

128. Meva, F.E.; Ebongue, C.O.; Fannang, S.V.; Segnou, M.L.; Ntoumba, A.A.; Kedi, P.B.E.; Loudang, R.N.; Wanlao, A.Y.; Mang, E.R.; Mpondo, E.A.M. Natural substances for the synthesis of silver nanoparticles against Escherichia coli: The case of *Megaphrynium macrostachyum* (Marantaceae), Corchorus olitorus (Tiliaceae), *Ricinodendron heudelotii* (Euphorbiaceae), *Gnetum bucholzianum* (Gnetaceae), and Ipomoea batatas (Convolvulaceae). *J. Nanomater.* **2017**, *2017*, 6834726.

129. Shruthi, G.; Prasad, K.S.; Vinod, T.P.; Balamurugan, V.; Shivamallu, C. Green synthesis of biologically active silver nanoparticles through a phyto-mediated approach using Areca catechu leaf extract. *ChemistrySelect* **2017**, *2*, 10354–10359. [CrossRef]
130. Azeez, M.A.; Lateef, A.; Asafa, T.B.; Yekeen, T.A.; Akinboro, A.; Oladipo, I.C.; Gueguim-Kana, E.B.; Beukes, L.S. Biomedical applications of cocoa bean extract-mediated silver nanoparticles as antimicrobial, larvicidal and anticoagulant agents. *J. Clust. Sci.* **2017**, *28*, 149–164. [CrossRef]
131. Lateef, A.; Azeez, M.A.; Asafa, T.B.; Yekeen, T.A.; Akinboro, A.; Oladipo, I.C.; Azeez, L.; Ojo, S.A.; Gueguim-Kana, E.B.; Beukes, L.S. Cocoa pod husk extract-mediated biosynthesis of silver nanoparticles: Its antimicrobial, antioxidant and larvicidal activities. *J. Nanostruct. Chem.* **2016**, *6*, 159–169. [CrossRef]
132. Swamy, M.K.; Akhtar, M.S.; Mohanty, S.K.; Sinniah, U.R. Synthesis and characterization of silver nanoparticles using fruit extract of *Momordica cymbalaria* and assessment of their in vitro antimicrobial, antioxidant and cytotoxicity activities. *Spectrochim. Acta A Mol. Biomol. Spectrosc.* **2015**, *151*, 939–944. [CrossRef] [PubMed]
133. Durán, N.; Durán, M.; Jesus, M.B.; Seabra, A.B.; Fávaro, W.J.; Nakazato, G. Silver nanoparticles: A new view on mechanistic aspects on antimicrobial activity. *Nanomed. NBM* **2016**, *12*, 789–799. [CrossRef] [PubMed]
134. Li, Q.; Mahendra, S.; Lyon, D.Y.; Brunet, L.; Liga, M.V.; Li, D.; Alvarez, P.J.J. Antimicrobial nanomaterials for water disinfection and microbial control: Potential applications and implications. *Water Res.* **2008**, *42*, 4591–4602. [CrossRef] [PubMed]
135. Manke, A.; Wang, L.; Rojanasakul, Y. Mechanisms of nanoparticle-induced oxidative stress and toxicity. *BioMed Res. Int.* **2013**, *2013*, 942916. [CrossRef] [PubMed]
136. Siddiqi, K.S.; Husen, A.; Rao, R.A.K. A review on biosynthesis of silver nanoparticles and their biocidal properties. *J. Nanobiotechnol.* **2018**, *16*, 14. [CrossRef] [PubMed]
137. Zheng, K.; Setyawati, M.I.; Leong, D.T.; Xie, J. Antimicrobial silver nanomaterials. *Coord. Chem. Rev.* **2018**, *357*, 1–17. [CrossRef]
138. Lok, C.; Ho, C.M.; Chen, R.; He, Q.Y.; Yu, W.Y.; Sun, H.; Tam, P.K.; Chiu, J.F.; Che, C.M. Proteomic analysis of the mode of antibacterial action of silver nanoparticles. *J. Proteome Res.* **2006**, *5*, 916–924. [CrossRef] [PubMed]
139. Xiu, Z.; Zhang, Q.; Puppala, H.L.; Colvin, V.L.; Alvarez, P.J.J. Negligible particle-specific antibacterial activity of silver nanoparticles. *Nano Lett.* **2012**, *12*, 4271–4275. [CrossRef] [PubMed]
140. Dibrov, P.; Dzioba, J.; Gosink, K.K.; Hase, C.C. Chemiosmotic mechanism of antimicrobial activity of Ag^+ in *Vibrio cholerae*. *Antimicrob. Agents Chemother.* **2002**, *46*, 2668–2670. [CrossRef] [PubMed]
141. Liau, S.Y.; Read, D.C.; Pugh, W.J.; Furr, J.R.; Russell, A.D. Interaction of silver nitrate with readily identifiable groups: Relationship to the antibacterial action of silver ions. *Lett. Appl. Microbiol.* **1997**, *25*, 279–283. [CrossRef] [PubMed]
142. Liao, X.; Yang, F.; Li, H.; So, P.K.; Yao, Z.; Wia, W.; Sun, H. Targeting the thioredoxin reductase–thioredoxin system from *Staphylococcus aureus* by silver ions. *Inorg. Chem.* **2017**, *56*, 14823–14830. [CrossRef] [PubMed]
143. Holt, K.B.; Bard, A.J. Interaction of silver(I) ions with the respiratory chain of *Escherichia coli*: An electrochemical and scanning electrochemical microscopy study of the antimicrobial mechanism of micromolar Ag^+. *Biochemistry* **2005**, *44*, 13214–13223. [CrossRef] [PubMed]
144. Wen, Y.; Geitner, N.K.; Chen, R.; Ding, F.; Chen, P.; Andorfer, R.E.; Govindan, P.N.; Ke, P.C. Binding of cytoskeletal proteins with silver nanoparticles. *RSC Adv.* **2013**, *3*, 22002–22007. [CrossRef]
145. Hong, X.; Wen, J.; Xiong, X.; Hu, Y. Shape effect on the antibacterial activity of silver nanoparticles synthesized via a microwave-assisted method. *Environ. Sci. Pollut. Res.* **2016**, *23*, 4489–4497. [CrossRef] [PubMed]
146. Alshareef, A.; Laird, K.; Cross, R.B.M. Shape-dependent antibacterial activity of silver nanoparticles on *Escherichia coli* and *Enterococcus faecium* bacterium. *Appl. Surf. Sci.* **2017**, *424*, 310–315. [CrossRef]
147. Pal, S.; Tak, Y.K.; Song, J.M. Does the antibacterial activity of silver nanoparticles depend on the shape of the nanoparticle? A study of the Gram-negative bacterium Escherichia coli. *Appl. Environ. Microb.* **2007**, *73*, 1712–1720. [CrossRef] [PubMed]
148. Acharya, D.; Singha, K.M.; Pandey, P.; Mohanta, B.; Rajkumari, J.; Singha, L.P. Shape dependent physical mutilation and lethal effects of silver nanoparticles on bacteria. *Sci. Rep.* **2018**, *8*, 201. [CrossRef] [PubMed]
149. Ivask, A.; Kurvet, I.; Kasemets, K.; Blinova, I.; Aruoja, V.; Suppi, S.; Vija, H.; Kakinen, A.; Titma, T.; Heinlaan, M.; et al. Size-dependent toxicity of silver nanoparticles to bacteria, yeast, algae, crustaceans and mammalian cells in vitro. *PLoS ONE* **2014**, *9*, e102108. [CrossRef] [PubMed]

150. Kumari, M.; Pandey, S.; Giri, V.P.; Bhattacharya, A.; Shukla, R.; Mishra, A.; Nautiyal, C.S. Tailoring shape and size of biogenic silver nanoparticles to enhance antimicrobial efficacy against MDR bacteria. *Microb. Pathog.* **2017**, *105*, 346–355. [CrossRef] [PubMed]

151. Lu, Z.; Rong, K.; Li, J.; Yang, H.; Chen, R. Size-dependent antibacterial activities of silver nanoparticles against oral anaerobic pathogenic bacteria. *J. Mater. Sci. Mater. Med.* **2013**, *24*, 1465–1471. [CrossRef] [PubMed]

152. Chen, M.; Yang, Z.; Wu, H.; Pan, X.; Xie, X.; Wu, C. Antimicrobial activity and the mechanism of silver. *Int. J. Nanomed.* **2011**, *6*, 2873–2877.

153. Kora, A.J.; Sashidhar, R.B. Biogenic silver nanoparticles synthesized with rhamnogalacturonan gum: Antibacterial activity, cytotoxicity and its mode of action. *Arab. J. Chem.* **2018**, *11*, 313–323. [CrossRef]

154. Kim, J.S.; Kuk, E.; Yu, N.K.; Kim, J.; Park, S.J.; Lee, H.J.; Kim, S.H.; Park, Y.K.; Park, Y.H.; Hwang, C.; et al. Antimicrobial effects of silver nanoparticles. *Nanomed. NBM* **2007**, *3*, 95–101. [CrossRef] [PubMed]

155. Choi, O.; Hu, Z. Size dependent and reactive oxygen species related nanosilver toxicity to nitrifying bacteria. *Environ. Sci. Technol.* **2008**, *42*, 4583–4588. [CrossRef] [PubMed]

156. Li, W.; Xie, X.; Shi, Q.; Zeng, H.; OU-Yang, Y.; Chen, Y. Antibacterial activity and mechanism of silver nanoparticles on *Escherichia coli*. *Appl. Microbiol. Biotechnol.* **2010**, *85*, 1115–1122. [CrossRef] [PubMed]

157. Ahmad, A.; Wei, Y.; Syed, F.; Rehman, A.U.; Khan, A.; Ullah, S.; Yuan, Q. The effects of bacteria-nanoparticles interface on the antibacterial activity of green synthesized silver nanoparticles. *Microb. Pathog.* **2017**, *102*, 133–142. [CrossRef] [PubMed]

158. Qayyum, S.; Oves, M.; Khan, A.U. Obliteration of bacterial growth and biofilm through ROS generation by facilely synthesized green silver nanoparticles. *PLoS ONE* **2017**, *12*, e0181363. [CrossRef] [PubMed]

159. Pareek, V.; Gupta, R.; Panwar, J. Do physico-chemical properties of silver nanoparticles decide their interaction with biological media and bactericidal action? A review. *Mater. Sci. Eng. C* **2018**, *90*, 739–749. [CrossRef] [PubMed]

160. Mchugh, G.L.; Moellering, R.C.; Hopkins, C.C.; Swartz, M.N. Salmonella typhimurium resistant to silver nitrate, chloramphenicol, and ampicillin. *Lancet* **1975**, *1*, 235–240. [CrossRef]

161. Gupta, A.; Matsui, K.; Lo, J.; Silver, S. Molecular basis for resistance to silver cations in *Salmonella*. *Nat. Med.* **1999**, *5*, 183–185. [CrossRef] [PubMed]

162. Finley, P.J.; Norton, R.; Austin, C.; Mitchell, A.; Zank, S.; Durham, P. Unprecedented silver resistance in clinically isolated Enterobacteriaceae: Major implications for burn and wound management. *Antimicrob. Agents Chemother.* **2015**, *59*, 4734–4741. [CrossRef] [PubMed]

163. Li, X.; Nikaido, H.; Williams, K.E. Silver-resistant mutants of Escherichia coli display active efflux of Ag$^+$ and are deficient in porins. *J. Bacteriol.* **1997**, *179*, 6127–6132. [CrossRef] [PubMed]

164. Randall, C.P.; Gupta, A.; Jackson, N.; Busse, D.; O'Neill, A.J. Silver resistance in Gram-negative bacteria: A dissection of endogenous and exogenous mechanisms. *J. Antimicrob. Chemother.* **2015**, *70*, 1037–1046. [CrossRef] [PubMed]

165. Su, C.; Long, F.; Zimmermann, M.T.; Rajashankar, K.R.; Jernigan, R.L.; Yu, E.W. Crystal structure of the CusBA heavy-metal efflux complex of *Escherichia coli*. *Nature* **2011**, *470*, 558–563. [CrossRef] [PubMed]

166. Xue, Y.; Davis, A.V.; Balakrishnan, G.; Stasser, J.P.; Staehlin, B.M.; Focia, P.; Spiro, T.G.; Penner-Hahn, J.E.; O'Halloran, T.V. Cu(I) recognition via cation-π and methionine interactions in CusF. *Nat. Chem. Biol.* **2008**, *4*, 107–109. [CrossRef] [PubMed]

167. Gudipaty, S.A.; McEvoy, M.M. The histidine kinase CusS senses silver ions through direct binding by its sensor domain. *Biochim. Biophys. Acta* **2014**, *1844*, 1656–1661. [CrossRef] [PubMed]

168. Franke, S.; Grass, G.; Nies, D.H. The product of the ybdE gene of the Escherichia coli chromosome is involved in detoxification of silver ions. *Microbiology* **2001**, *147*, 965–972. [CrossRef] [PubMed]

169. Panáček, A.; Kvítek, L.; Smékalová, M.; Večeřová, R.; Kolář, M.; Röderová, M.; Dyčka, F.; Šebela, M.; Prucek, R.; Tomanec, O.; et al. Bacterial resistance to silver nanoparticles and how to overcome it. *Nat. Nanotechnol.* **2018**, *13*, 65–71. [CrossRef] [PubMed]

170. Holladay, R.; Moeller, W.; Mehta, D.; Brooks, J.H.J.; Roy, R.; Mortenson, M. Silver/Water, Silver Gels and Silver-Based Compositions; and Methods for Making and Using the Same. World Intellectual Property Organization 2006074117A2, 5 January 2005.

171. Jinjun, L.; Qiangbai, L.; Jianchao, S. Use of Medicinal Nanomaterial Composition Dg-5 Applied to Anti-Drug Resistant Bacteria. World Intellectual Property Organization 2018010403A1, 13 July 2016.

172. Jiachong, C.; Jixiong, Y. Manufacturing Methods and Applications of Antimicrobial Plant Fibers Having Silver Particles. U.S. Patent 20100003296, 7 January 2010.
173. Nano Silver and Perfume Contain an Apron. Korean Patent 200384433Y1, 16 May 2005.
174. Method for Preparing Nano-Silver Particle and Detergent Composition by Using Them. Korean Patent 100933736B1, 26 June 2008.
175. Composite Nanometer Antibacterial Material Used for Treating Vancomycin Drug Resistant Pathogenic Bacteria. Chinese Patent 105412940A, 2 December 2015.
176. Paknikar, K.M. Anti-Microbial Activity of Biologically Stabilized Silver Nano Particles. World Intellectual Property Organization 2005120173A2, 22 December 2005.
177. Holladay, R.J.; Christensen, H.; Moeller, W.D. Treatment of Humans with Colloidal Silver Composition. U.S. Patent 7135195B2, 14 November 2006.
178. Liang, D.; Lu, Z.; Yang, H.; Gao, J.; Chen, R. Novel asymmetric wettable AgNPs/chitosan wound dressing: In vitro and in vivo evaluation. *ACS Appl. Mater. Interfaces* **2016**, *8*, 3958–3968. [CrossRef] [PubMed]
179. Yabanoglu, H.; Basaran, O.; Aydogan, C.; Azap, O.K.; Karakayali, F.; Moray, G. Assessment of the effectiveness of silver-coated dressing, chlorhexidine acetate (0.5%), citric acid (3%), and silver sulfadiazine (1%) for topical antibacterial effects against the multi-drug resistant *Pseudomonas aeruginosa* infecting full-skin thickness burn wounds on rats. *Int. Surg.* **2013**, *98*, 416–423. [PubMed]
180. Huang, Y.; Li, X.; Liao, Z.; Zhang, G.; Liu, Q.; Tang, J.; Peng, Y.; Liu, X.; Luo, Q. A randomized comparative trial between Acticoat and SD-Ag in the treatment of residual burn wounds, including safety analysis. *Burns* **2007**, *33*, 161–166. [CrossRef] [PubMed]

Article

Microwave-Assisted Green Synthesis of Silver Nanoparticles Using *Juglans regia* Leaf Extract and Evaluation of Their Physico-Chemical and Antibacterial Properties

Mahsa Eshghi [1], Hamideh Vaghari [1,2], Yahya Najian [2], Mohammad Javad Najian [2], Hoda Jafarizadeh-Malmiri [1,*] and Aydin Berenjian [3,*]

[1]   Faculty of Chemical Engineering, Sahand University of Technology, Tabriz 513351996, Iran; mahsaeshghii@gmail.com (M.E.); vaghari_h@yahoo.com (H.V.)

[2]   Research and Development Department, Najian Herbal Group, East Azarbaijan, Tabriz 5159193858, Iran; h_jafarizadeh@yahoo.com (Y.N.); najian_herbal_group@yahoo.com (M.J.N.)

[3]   School of Engineering, Faculty of Science and Engineering, University of Waikato, Hamilton 3240, New Zealand

*   Correspondence: h_jafarizadeh@sut.ac.ir or h_jafarizadeh@yahoo.com (H.J.-M.); aydin.berenjian@waikato.ac.nz (A.B.); Tel.: +98-413-3459099 (H.J.-M.); +64-7-858-5119 (A.B.)

Received: 4 July 2018; Accepted: 27 July 2018; Published: 30 July 2018

Abstract: Silver nanoparticles (Ag NPs) were synthesized using *Juglans regia* (*J. regia*) leaf extract, as both reducing and stabilizing agents through microwave irradiation method. The effects of a 1% (w/v) amount of leaf extract (0.1–0.9 mL) and an amount of 1 mM $AgNO_3$ solution (15–25 mL) on the broad emission peak (λ_{max}) and concentration of the synthesized Ag NPs solution were investigated using response surface methodology (RSM). Fourier transform infrared analysis indicated the main functional groups existing in the *J. regia* leaf extract. Dynamic light scattering, UV-Vis spectroscopy and transmission electron microscopy were used to characterize the synthesized Ag NPs. Fabricated Ag NPs with the mean particle size and polydispersity index and maximum concentration and zeta potential of 168 nm, 0.419, 135.16 ppm and −15.6 mV, respectively, were obtained using 0.1 mL of *J. regia* leaf extract and 15 mL of $AgNO_3$. The antibacterial activity of the fabricated Ag NPs was assessed against both Gram negative (*Escherichia coli*) and positive (*Staphylococcus aureus*) bacteria and was found to possess high bactericidal effects.

Keywords: green synthesis; silver nanoparticles; microwave irradiation; *Juglans regia*; antibacterial activity

1. Introduction

Silver is known as strong antimicrobial agent due to its toxicity against most microorganisms including bacteria, fungi, and viruses [1]. As compared to the silver element in bulk state, silver nanoparticles (Ag NPs) exhibit unusual physico-chemical and antimicrobial properties due to their lower particle size (less than 100 nm) and higher surface area to volume ratio, which increase the proportion of high-energy surface atoms [1–3]. Because of the marvelous antimicrobial properties of Ag NPs, they have been widely utilized in numerous fields such as medicine, electronics, biotechnology, water disinfection, air treatment, and packaging [3].

Green chemistry is the use of chemistry principles to reduce or eliminate the use of toxic reagents, leading to lower amounts of undesirable residues, which in turn are harmful to human health or the environment [4,5]. Incorporation of green chemistry and nanotechnology is of great interest and has gained much attention over the past decade. Green synthesis of Ag NPs, using plants and their derivative extracts as an alternative method for the conventional physical and chemical NP fabrication methods, has attracted considerable attention in recent years [1,6]. In fact, various metabolites existing in the plants including sugars, alkaloids, phenolic acids, terpenoids, polyphenols, and proteins play an important role in the bioreduction of silver ions to Ag NPs and their stabilization [7]. Several studies have been completed on green synthesis of Ag NPs using *Aloe vera* and *Pelargonium zonali* leaf extract and chitosan [8–10].

Walnut belongs to *Juglandaceae* family and has the scientific name *Juglans regia* (*J. regia*). Numerous health benefits of walnut for human have been reported, some of which are related to the existence of its main bioactive compounds including polyphenolic compounds, flavonoids, proteins, carotenoids, lipids, and alkaloids. Leaf of walnut contains malic acid, sucrose, α-tocopherol, 3-*O*-caffeoylquinic acids, and quercetin *O*-pentoside as the most abundant organic acids, disaccharide, tocopherol isomer and phenolic compounds, respectively [11–13].

The present work focuses on (i) the potential of walnut leaf extract in the fabrication of Ag NPs; (ii) Ag NPs synthesized parameter optimization using microwave irradiation to achieve Ag NPs with more desirable physico-chemical properties and (iii) an antibacterial activity assessment of the produced Ag NPs.

2. Results

2.1. Formation of Ag NPs

During the synthesis of Ag NPs, the color of the mixture solution containing silver salt and walnut leaf extract changed from colorless into brown and dark. In fact, the synthesized Ag NPs, due to their surface plasmon resonance (SPR), changed the color of the mixture solution and this color change verified the formation of Ag NPs using walnut leaf extract and microwave irradiation (Figure 1). As clearly observed in Table 1, λ_{max} of the synthesized Ag NPs was varied 424–429 nm, which was in a favorable range for Ag NPs [8,10]. The particle size of the synthesized Ag NPs could be correlated with their broad emission peaks (λ_{max}), where longer wavelengths in λ_{max} of Ag NPs were associated with bigger size [9]. This indicated that *J. regia* leaf extract successfully reduced silver ions and formed Ag NPs.

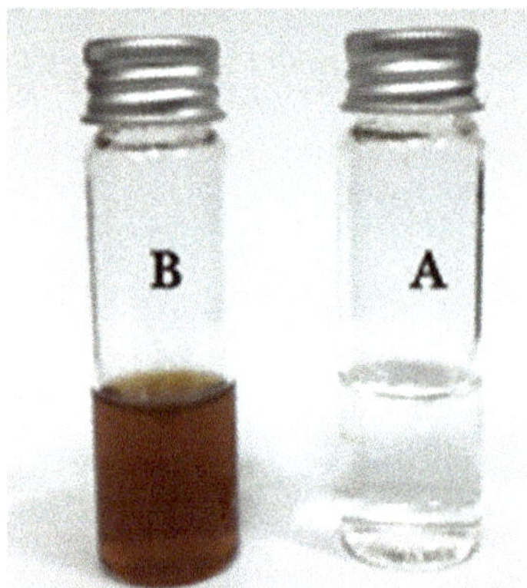

Figure 1. Color and appearance changes during synthesis of silver nanoparticles (Ag NPs) using *J. regia* leaf extract. *J. regia* leaf extract containing silver nitrate before (**A**) and after (**B**) exposure to microwave irradiation.

Figure 2 shows the Fourier transform-infrared (FT-IR) spectrum of *J. regia* leaf extract which was in the region range of 400–4000 cm^{-1}. The IR spectrum of leaf extract absorption bands at 3454.87 and 2072.86 cm^{-1} represent the phenolic–OH and N=C groups. Absorption bands at 1637.48 cm^{-1} are related to the amino group, bending vibrations, and the –OH group. The obtained results indicated that the phenolic compounds and proteins were the two main components of the *J. regia* leaf extract which had key roles in the formation of the stabilized Ag NPs [11,12].

Table 1. Experimental runs according to the central composite design (CCD) and response variables for synthesis of Ag NPs.

Sample No.	Amount of Leaf Extract (mL)	Amount of Silver Salt (mL)	λ_{max} (nm)		Concentration (ppm)	
			Exp	Pre	Exp	Pre
1	0.10	20.00	425	424.896	104.813	101.702
2	0.78	16.46	429	429.104	100.003	97.802
3	0.78	23.53	424	424.00	93.503	96.614
4	0.50	15.00	427	426.854	72.443	74.643
5	0.50	25.00	*	*	48.393	43.081
6	0.50	20.00	426	427.250	56.453	61.835
7	0.50	20.00	426	427.250	56.453	61.835
8	0.217	23.53	426	426.043	50.343	55.654
9	0.50	20.00	428	427.250	66.073	61.835
10	0.90	20.00	427	426.957	*	*
11	0.217	16.46	424	424.146	*	*
12	0.50	20.00	429	427.250	66.723	61.835
13	0.50	20.00	*	*	63.473	61.835

Exp, Experimental values of studied responses; Pre, Predicted values of studied responses. *, Out of range.

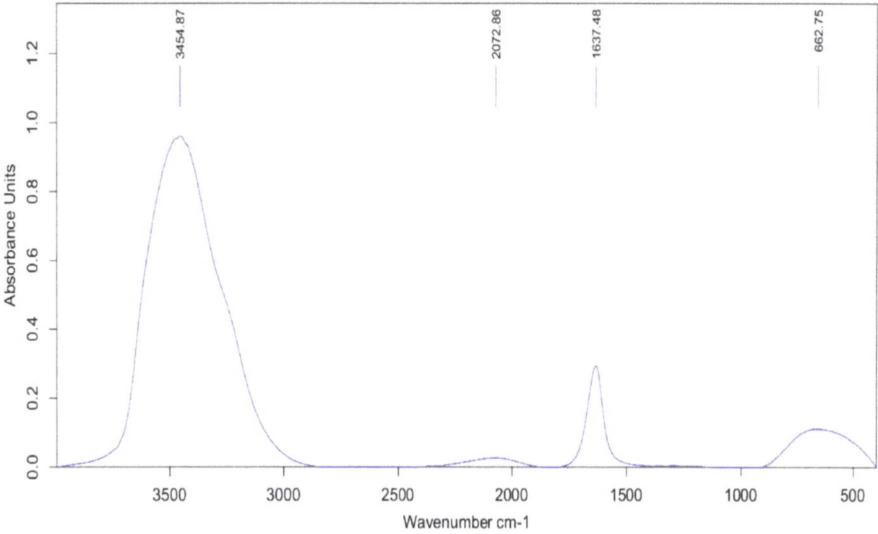

Figure 2. Fourier transform-infrared (FT-IR) spectrum of *J. regia* leaf extract.

2.2. Models Generation and Synthesis Conditions Optimization

Based on experimental runs, the values of broad emission peaks (λ_{max}, nm) and concentration (ppm) for the fabricated Ag NPs were achieved (Table 1) and according to these obtained data, models were created. Table 2 shows coefficients of the model terms and models accuracy based on R square and its adjusted (R^2, R^2-adj) and lack of fit. The high values of R^2 (>77.80) and R^2-adj (>85.61) and lack of fit ($p > 0.05$) verified higher suitability of the generated models to predict the synthesis response parameters. p-Values of the generated model terms are also described in Table 3. As can be seen, the interaction of the amounts of leaf extract and silver salt solution had significant ($p < 0.05$) effects on λ_{max} and concentration of the formed Ag NPs.

Table 2. Regression coefficients, R^2, R^2-adj, and probability values for the fitted models.

Regression Coefficient	λ_{max} (nm)	Concentration (ppm)
β_0 (constant)	386.44	249.7
β_1 (main effect)	−45,084	−513.04
β_2 (main effect)	3.09	−3.68
β_{11} (quadratic effect)	−8.27	336.8
β_{22} (quadratic effect)	−0.061	−0.119
β_{12} (interaction effect)	−1.75	10.56
R^2	77.81%	95.25%
R^2-adj	85.62%	90.49%
Lack of fit (p-value)	0.985	0.142

β_0 is a constant, β_1, β_{11} and β_{12} are the linear, quadratic, and interaction coefficients of the quadratic polynomial equation, respectively.

Table 3. p-Values of the regression coefficients in the obtained models.

Main Effects	Main Effects		Quadratic Effects		Interacted Effect
	X_1	X_2	X_{11}	X_{22}	$X_1 X_2$
λ_{max} (nm) (p-value)	0.003	0.693	0.00	0.580	0.045
Concentration (ppm) (p-value)	0.018	0.153	0.224	0.247	0.030

Figure 3A,B indicates the effects of the amount of *J. regia* leaf extract and amount of $AgNO_3$ on the λ_{max} of the synthesized Ag NPs. As clearly observed in Figure 3, the minimum λ_{max} (particle size) was obtained at both minimum amount of *J. regia* leaf extract and of $AgNO_3$ solution and the maximum amount of *J. regia* leaf extract and of $AgNO_3$ solution. The experimental value of the concentration for the synthesized Ag NPs ranged 19–80 ppm (Table 1). The effects of the amount of *J. regia* leaf extract and of $AgNO_3$ on the concentration of the fabricated Ag NPs are shown in Figure 4A,B.

As clearly observed in Figure 4, the maximum concentration of the synthesized Ag NPs was obtained using the highest amount of the *J. regia* leaf extract. The obtained results were in line with the findings of other research [8–10]. They found that by increasing the amount of plant extract, the concentration of the bioreductants increased in the extract, which in turn increased the concentration of the formed Ag NPs.

In the synthesis of Ag NPs, the main objective is formation of NPs with desirable physico-chemical properties, including minimum particle size (λ_{max}) and maximum concentration. This synthesis process is known as optimized procedure. According to the generated models for the synthesis of Ag NPs, an overlaid contour plot, as graphical optimization (Figure 5), was plotted to better visualize of optimum area (white colored zone).

The result of a numerical optimization also demonstrated that by using 0.1 mL of *J. regia* leaf extract and 15 mL of $AgNO_3$, Ag NPs with minimum λ_{max} of 421 nm and maximum concentration of 135.64 ppm were produced. A verification test using optimum synthesis parameters also indicated insignificant ($p > 0.05$) differences between the values of predicted and experimental of λ_{max} and concentration of the fabricated Ag NPs and verified the suitability of the models.

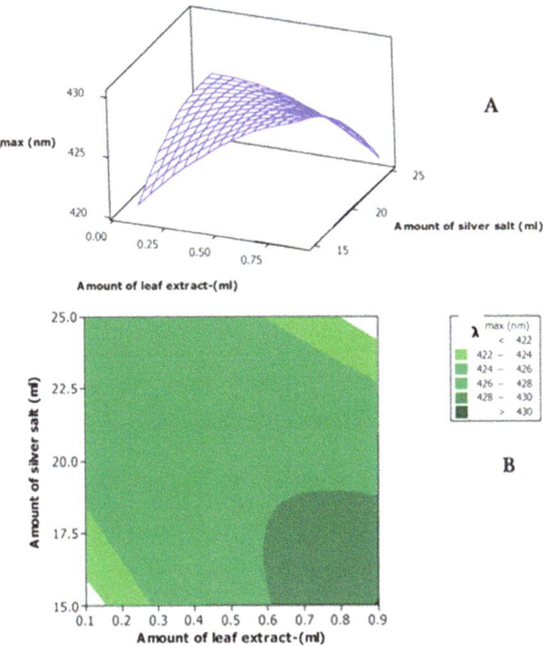

Figure 3. Response surface (**A**) and contour plots (**B**) for λ_{max} of the synthesized Ag NP solution as function of the amount of *J. regia* leaf extract and amount of $AgNO_3$ solution.

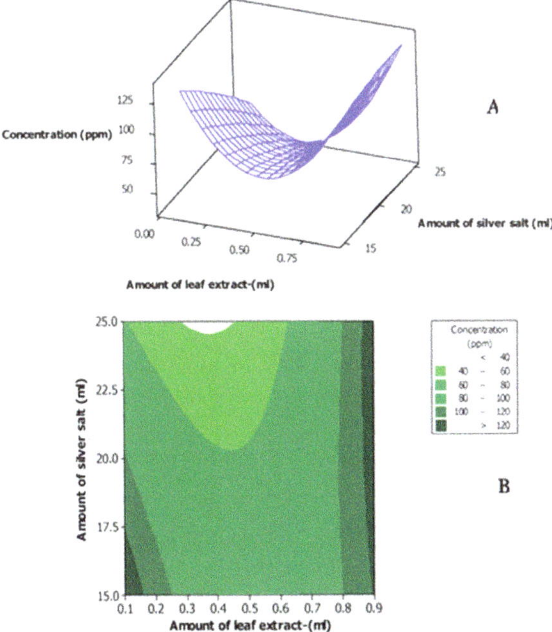

Figure 4. Response surface (**A**) and contour plots (**B**) for the concentration of the synthesized Ag NP solution as function of amount of *J. regia* leaf extract and amount of $AgNO_3$ solution.

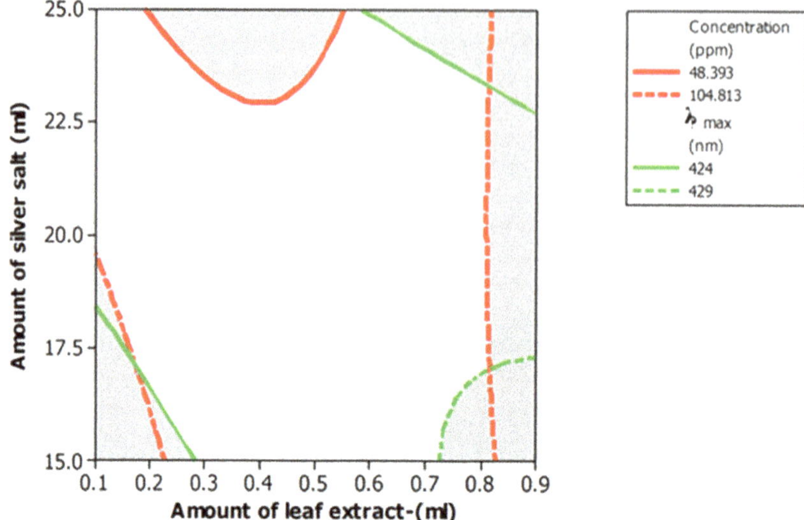

Figure 5. Overlaid contour plot of Ag NPs λ_{max} and concentration with acceptable levels as a function of amount of *J. regia* leaf extract and amount of AgNO$_3$ solution.

2.3. Physico-Chemical Characteristics of the Synthesized Ag NPs at Obtained Optimum Conditions

Formation of Ag NPs using *J. regia* leaf extract at obtained optimum conditions was confirmed by changes in the color of the mixture solution. Dynamic light scattering (DLS) analysis also indicated that the synthesized Ag NPs had particle size, polydispersity index (PDI), and zeta potential values of 168 nm, 0.419 and −15.6 mV, respectively. The particle size distributions (PSD) of the sample are also shown in Figure 6.

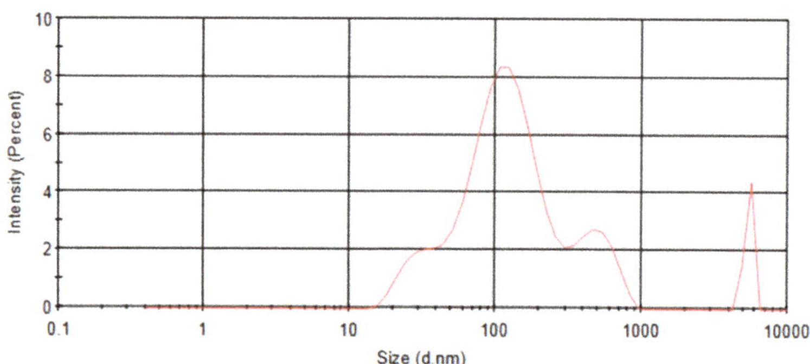

Figure 6. Particle size distribution of synthesized Ag NPs at obtained optimum synthesis conditions using *J. regia* leaf extract.

2.4. Antibacterial Activity

The antibacterial activity of synthesized Ag NPs on the growth of Gram-positive (*S. aureus*) and Gram-negative (*E. coli*) bacteria during incubation indicated that the diameter of the clear zone for synthesized Ag NPs in the plate containing *S. aureus* and *E. coli* were 16 and 10 mm, respectively (Figure 7). The results also indicated that the mean diameter of formed clear zone around the Ampicillin disc in the plates containing *S. aureus* and *E. coli* were 35 and 33 mm, respectively.

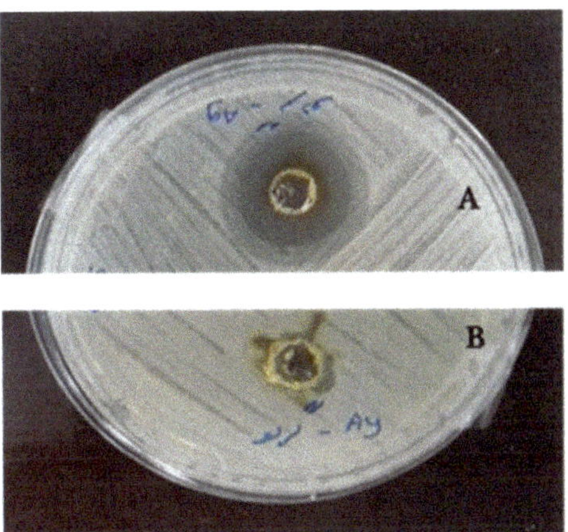

Figure 7. Created zones of inhibition with *S. aureus* (**A**) and *E. coli* (**B**) incubated at 37 °C for 24 h for synthesized Ag NPs using *J. regia* leaf extract.

The obtained results indicated that the fabricated Ag NPs had higher antibacterial activity against Gram-positive bacteria compared to the Gram-negative bacteria. The obtained results were in agreement with findings of Ahmadi et al., Mohammadlu et al., and Torabfam and Jafarizadeh-Malmiri [8–10]. The main reason behind the bactericidal activity of the Ag NPs against the bacteria strains is related to their effects on the permeability of the cell wall and membrane. In fact, the released silver ions (Ag$^+$) from the Ag NPs were attached into the anionic groups of the cell wall, such as polycyclic aromatic hydrocarbon and teichoic acids. They also formed polyelectrolyte complexes, which limited the transference of nutrients and provided metabolites into and outside the cell [9]. Furthermore, caffeoylquinic acid is the main phenylpropanoids in *J. regia* leaf extract, which has various bioactivities such as antioxidant, antibacterial, anticancer, antihistamic, and other biological effects. The direct antimicrobial activity of caffeoylquinic acid implies an array of possibilities, including effects in the cell envelope [13].

3. Materials and Methods

3.1. Materials

J. regia leaves were picked from walnut trees in Tabriz, Iran. AgNO$_3$, as silver salt, was bought from Dr. Mojallali (Dr. Mojallali Chemical Complex Co., Tehran, Iran). Standard solution of Ag NPs (with particle size of 10 nm and concentration of 1000 ppm) was purchased from Tecnan-Nanomat (Spain). *Escherichia coli* (PTCC 1270) and *Staphylococcus aureus* (PTCC 1112) were attained from microbial Persian-type culture collection (PTCC, Tehran, Iran). There is no ATCC (American type culture collection) number for *E. coli* and this bacteria is clinical isolate. However, the ATCC number of *S. aureus* is 6538. Nutrient agar was bought from Biolife (Biolife Co., Milan, Italy).

3.2. Preparation of J. regia Leaf Extract

The *J. regia* leaf was washed, dried (in dark room), powdered, and 1 g of the prepared powder was added into 100 mL of boiling distilled water for 5 min. After cooling the solution, it was filtered (Whatman No. 1 filter paper) using a Buchner funnel under vacuum pressure and the clear walnut leaf extract was stored at a cold temperature (4 °C).

3.3. Ag NPs Synthesis Using J. regia Leaf Extract

The Ag NPs solution was obtained by a domestic microwave-assisted synthetic approach. $AgNO_3$ solution (1 mM) was made by dissolving 0.017 g of its powder in 100 mL of deionized double-distilled water. In a typical synthesis, different amounts of $AgNO_3$ solution (15–25 mL) were mixed with different amounts of *J. regia* leaf extract (0.1–0.9 mL) and the mixture solutions were put into a microwave oven (MG-2312W, LG Co., Seoul, Korea) at a constant power of 800 W and microwave exposure time (180 s).

3.4. Physico-Chemical Assay

3.4.1. Fourier Transform-Infrared (FT-IR) Spectra Analysis

In order to identify the possible reducing and stabilizing biomolecules of *J. regia* leaf extract, FT-IR measurements were carried out. The FT-IR spectrum of the extract was recorded on a Bruker Tensor27 spectrometer (Bruker Co., Karlsruhe, Germany) using KBr pellets in the 4000–400 cm^{-1} region.

3.4.2. Surface Plasmon Resonance

Ag NPs, due to their SPR, have a strong absorption of light, which is shown as broad emission peaks (λ_{max}) in the wavelength ranging from 380 to 450 nm [8–10]. Therefore, formation of Ag NPs using *J. regia* leaf extract can be confirmed by the absorption spectrum of the mixture solutions containing fabricated Ag NPs by using a Jenway UV-Vis spectrophotometer 6705 (Cole-Parmer Co., Staffordshire, UK). Furthermore, by preparing several serial dilute Ag NPs solutions (10–1000 ppm) and establishing standard curves based on the defined concentrations of the Ag NPs solutions and their absorbance unit values, it is possible to determine the concentration of the formed Ag NPs using *J. regia* leaf extract [10].

3.4.3. Particle Size, Particle Size Distribution, Polydispersity Index and Zeta Potential of the Synthesized Ag NPs

In order to measure values of the mean particle size (nm), PDI (ranging 0–1) and zeta potential (mV) of the fabricated Ag NPs and their PSD, a DLS particle size analyzer (Nanotrac Wave, Microtrac, Montgomeryville, PA, USA) was utilized. The DLS technique scatters a laser light beam at the surface of dispersed NPs, which results in the detection of the backscattered light. PDI is a dimensionless value which shows that the uniformity of the synthesized NPs and their surface electric charge are related to the PDI and zeta potential values of the synthesized NPs [14,15].

3.5. Antibacterial Assay

Bactericidal activity of the formed Ag NPs was evaluated using the well diffusion method. For this reason, bacterial suspensions containing 1.5×10^8 colony-forming units of bacteria was prepared based on a 0.5 McFarland standard and 0.1 mL of that amount was spread on the surface of solid nutrient agar in the plates and a hole, 5 mm in diameter, was made in the solid agar. 10 μL of the produced Ag NPs solution was then poured into the created well and put in incubator at 37 °C for 24 h. The antibacterial activity of the synthesized Ag NPs correlated to the diameter of created clear zones around the holes. An Ampicillin disc 5mm in diameter (Oxoid, 10 μg/disc) was used as a positive control for both Gram-positive and Gram-negative bacteria strains and the antibacterial activity of the synthesized Ag NPs was compared to the Ampicillin.

3.6. Experimental Design, Statistical Analysis and Optimization Procedure

The experiment was planned using a central composite design (CCD) and response surface methodology (RSM) was used to evaluate the effects of two independent parameters, namely amount of leaf extract (X_1) and amount of $AgNO_3$ solution (X_2), on the prepared Ag NPs. The studied response variables were broad emission peak (λ_{max}) (Y_1, nm) and concentration (Y_2, ppm) of the synthesized Ag NPs. As clearly observed in Table 1, thirteen experimental treatments were assigned with five different levels for each independent parameter using Minitab software (v.16 statistical package, Minitab Inc., Pennsylvania State, PA, USA). In order to correlate the λ_{max} (Y_1) and concentration (Y_2) of the synthesized Ag NPs to the studied synthesis variables, a second order polynomial equation (Equation (1)) was used. Where β_0 is a constant, β_1, β_{11}, and β_{12} correspond to the linear, quadratic and interaction effects, respectively [16]. The suitability of the model was studied, accounting for the coefficient of determination (R^2) and adjusted coefficient of determination (R^2-adj) [17]. Analysis of variance (ANOVA) was also used to provide the significance determinations of the resulted models in terms of p-value. Small p-values (lower than 0.05) were considered as statistically significant [18].

$$Y = \beta_0 + \beta_1 X_1 + \beta_2 X_2 + \beta_{11} X_1{}^2 + \beta_{22} X_2{}^2 + \beta_{12} X_1 X_2 \tag{1}$$

Numerical optimization was carried out to determine exact amounts of silver salt and *J. regia* leaf extract in the Ag NPs synthesis to produce NPs with minimum λ_{max} (particle size) and maximum concentration. Graphical optimization was also used to better visualize the effects of the synthesis parameters on the response variables (λ_{max} and concentration) [19]. Suitability and accuracy of the generated models in predicting the response variables in the defined range of the Ag NPs synthesis parameters were validated by synthesis of Ag NPs using obtained optimum synthesized conditions and comparison of the experimental values of the response variables and their predicted values [20].

4. Conclusions

Fabricated Ag NPs using a rapid, one-step green approach, based on microwave irradiation showed desirable physico-chemical properties with antibacterial activity against *E. coli* and *S. aureus*. However, their antibacterial activity was significantly ($p < 0.05$) higher toward Gram-positive bacteria strains. The synthesized Ag NPs using the rapid developed synthesis method can be used as a favorable antibacterial agent in different areas such as medicine and food packaging.

Author Contributions: M.E. performed the Ag NPs synthesis and physico-chemical analyses. H.V. performed strain cultivation and antibacterial assay. Y.N. and M.J.N. contributed with valuable discussions and scientific advice. H.J.-M. designed the experiments and wrote the draft manuscript. A.B. contributed helpful discussion and statistical analysis during the preparation of manuscript.

Funding: This research received no external funding.

Acknowledgments: This research was undertaken with material support from the Najian Herbal Group (Tabriz, Iran). The authors appreciate this support. This research received no specific grant from any funding agency in the public, commercial, or not-for-profit sectors.

References

1. Mohammadlu, M.; Jafarizadeh-Malmiri, H.; Maghsoudi, H. A review on green silver nanoparticles based on plants: Synthesis, potential applications and eco-friendly approach. *Int. Food Res. J.* **2016**, *23*, 446–463.
2. Kon, K.; Rai, M. Metallic nanoparticles: Mechanism of antibacterial action and influencing factors. *J. Comp. Clin. Pathol. Res.* **2013**, *2*, 160–174.
3. Zhou, Y.; Kong, Y.; Kundu, S.; Cirillo, J.D.; Liang, H. Antibacterial activities of gold and silver nanoparticles against *Escherichia coli* and *Bacillus calmette-Guérin*. *J. Nanobiotechnol.* **2012**, *10*, 19–28. [CrossRef] [PubMed]
4. Ingale, A.G.; Chaudhari, A. Biogenic synthesis of nanoparticles and potential applications: An eco-friendly approach. *J. Nanomed. Nanotechnol.* **2013**, *4*, 165–167. [CrossRef]
5. Hebbalalu, D.; Lalley, J.; Nadagouda, M.N.; Varma, R.S. Greener techniques for the synthesis of silver nanoparticles using plant extracts, enzymes, bacteria, biodegradable polymers, and microwaves. *ACS Sustain. Chem. Eng.* **2013**, *1*, 703–712. [CrossRef]

6. Vamanu, E.; Ene, M.; Bita, B.; Ionescu, C.; Craciun, L.; Sarbu, I. In vitro human microbiota response to exposure to silver nanoparticles biosynthesized with mushroom extract. *Nutrients* **2018**, *10*, 607. [CrossRef] [PubMed]

7. Zhang, Y.; Cheng, X.; Zhang, Y.; Xue, X.; Fu, Y. Biosynthesis of silver nanoparticles at room temperature using aqueous Aloe leaf extract and antibacterial properties. *Colloids Surf. A Physicochem. Eng. Asp.* **2013**, *423*, 63–68. [CrossRef]

8. Mohammadlu, M.; Jafarizadeh-Malmiri, H.; Maghsoudi, H. Hydrothermal green silver nanoparticles synthesis using *Pelargonium/Geranium* leaf extract and evaluation of their antifungal activity. *Green Process. Synth.* **2017**, *6*, 31–42. [CrossRef]

9. Ahmadi, O.; Jafarizadeh-Malmiri, H.; Jodeiri, N. Eco-friendly microwave enhanced green silver nanoparticles synthesis using Aloe vera leaf extract and their physico-chemical and antibacterial studies. *Green Process. Synth.* **2018**, *7*, 231–240. [CrossRef]

10. Torabfam, M.; Jafarizadeh-Malmiri, H. Microwave–enhanced silver nanoparticles synthesis using chitosan biopolymer–Optimization of the process conditions and evaluation of their characteristics. *Green Process. Synth.* **2017**. [CrossRef]

11. Jaiswal, B.S.; Tailang, M. *Juglans regia*: A review of its traditional uses phytochemistry and pharmacology. *Indo Am. J. Pharm. Res.* **2017**, *7*, 390–398.

12. Amaral, J.S.; Seabra, R.M.; Andrade, P.B. Phenolic profile in the quality control of walnut (*Juglans regia* L.) leaf. *Food Chem.* **2004**, *88*, 373–379. [CrossRef]

13. Pereira, J.A.; Oliveira, I.; Sousa, A. Walnut (*Juglans regia* L.) leaf: Phenolic compounds, antimicrobial activity and antioxidant potential of different cultivars. *Food Chem. Toxicol.* **2007**, *45*, 2287–2295. [CrossRef] [PubMed]

14. Eskandari-Nojedehi, M.; Jafarizadeh-Malmiri, H.; Rahbar-Shahrouzi, J. Optimization of processing parameters in green synthesis of gold nanoparticles using microwave and edible mushroom (*Agaricus bisporus*) extract and evaluation of their antibacterial activity. *Nanotechnol. Rev.* **2016**, *5*, 537–548. [CrossRef]

15. Eskandari-Nojedehi, M.; Jafarizadeh-Malmiri, H.; Jafarizad, A. Microwave accelerated green synthesis of gold nanoparticles using gum Arabic and their physico-chemical properties assessments. *Z. Phys. Chem.* **2018**, *232*, 325–343. [CrossRef]

16. Amirkhani, L.; Moghaddas, J.; Jafarizadeh-Malmiri, H. *Candida rugosa* lipase immobilization on magnetic silica aerogel nanodispersion. *RSC Adv.* **2016**, *6*, 12676–12687. [CrossRef]

17. Anarjan, N.; Jaberi, N.; Yeganeh-Zare, S.; Banafshehchin, E.; Rahimirad, A.; Jafarizadeh-Malmiri, H. Optimization of mixing parameters for α-tocopherol nanodispersions prepared using solvent displacement method. *J. Am. Oil Chem. Soc.* **2014**, *91*, 1397–1405. [CrossRef]

18. Eskandari-Nojedehi, M.; Jafarizadeh-Malmiri, H.; Rahbar-Shahrouzi, J. Hydrothermal biosynthesis of gold nanoparticle using mushroom (*Agaricus bisporous*) extract: Physico-chemical characteristics and antifungal activity studies. *Green Process. Synth.* **2018**, *7*, 38–47. [CrossRef]

19. Anarjan, N.; Jafarizadeh-Malmiri, H.; Nehdi, I.A.; Sbihi, H.M.; Al-Resayes, S.I.; Tan, C.P. Effects of homogenization process parameters on physicochemical properties of astaxanthin nanodispersions prepared using a solvent-diffusion technique. *Int. J. Nanomed.* **2015**, *10*, 1109–1118.

20. Ahdno, H.; Jafarizadeh-Malmiri, H. Development of a sequenced enzymatically pre-treatment and filter pre-coating process to clarify date syrup. *Food Bioprod. Process.* **2017**, *101*, 193–204. [CrossRef]

Article

Antimicrobial Potential and Cytotoxicity of Silver Nanoparticles Phytosynthesized by Pomegranate Peel Extract

Renan Aparecido Fernandes [1,2], Andresa Aparecida Berretta [3], Elina Cassia Torres [3], Andrei Felipe Moreira Buszinski [3], Gabriela Lopes Fernandes [2], Carla Corrêa Mendes-Gouvêa [2], Francisco Nunes de Souza-Neto [4], Luiz Fernando Gorup [4,5], Emerson Rodrigues de Camargo [4] and Debora Barros Barbosa [2,*]

1   Department of Dentistry, University Center of Adamantina (UNIFAI), Adamantina 17800-000, São Paulo, Brazil; renanfernandes@fai.com.br
2   Department of Dental Materials and Prosthodontics, São Paulo State University (UNESP), School of Dentistry, Araçatuba 16015-050, São Paulo, Brazil; fernandesgabriela@hotmail.com (G.L.F.); carla_cmendes@hotmail.com (C.C.M.-G.)
3   Laboratory of Research, Development & Innovation, Apis Flora Industrial e Comercial Ltda., Ribeirão Preto 14020-670, São Paulo, Brazil; andresa.berretta@apisflora.com.br (A.A.B.); elinacassia@hotmail.com (E.C.T.); andrei.buszinski@apisflora.com.br (A.F.M.B.)
4   Department of Chemistry, Federal University of São Carlos, São Carlos 13565-905, São Paulo, Brazil; francisconsn29@gmail.com (F.N.S.-N.); lfgorup@gmail.com (L.F.G.); camargo@ufscar.br (E.R.C.)
5   FACET-Department of Chemistry, Federal University of Grande Dourados, Dourados 79804-970, Mato Grosso do Sul, Brazil
*   Correspondence: debora@foa.unesp.br; Tel.: +55-18-3636-3284

Received: 27 April 2018; Accepted: 25 June 2018; Published: 26 June 2018

Abstract: The phytosynthesis of metal nanoparticles is nowadays attracting the increased attention of researchers and is much needed given the worldwide matter related to environmental contamination. The antimicrobial activity of colloidal and spray formulation of silver nanoparticles (AgNPs) synthesized by pomegranate peel extract against *Candida albicans* and *Staphylococcus aureus*, and their cytotoxicity in mammalian cells were tested in the present study. Dry matter, pH, total phenolics, and ellagic acid in the extract were determined. Then, AgNPs were phytosynthesized and characterized by X-ray diffraction, electron transmission microscopy, dynamic light scattering, zeta potential, and Ag^+ dosage. Spray formulations and respective chemical-AgNP controls were prepared and tested. The peel extract reduced more than 99% of Ag^+, and produced nanoparticles with irregular forms and an 89-nm mean size. All AgNP presented antimicrobial activity, and the spray formulation of green-AgNP increased by 255 and 4 times the effectiveness against *S. aureus* and *C. albicans*, respectively. The cytotoxicity of colloidal and spray green-AgNP was expressively lower than the respective chemical controls. Pomegranate peel extract produced stable AgNP with antimicrobial action and low cytotoxicity, stimulating its use in the biomedical field.

Keywords: silver; nanoparticles; *Candida albicans*; *Staphylococcus aureus*; herbal medicine; Punicaceae

---

1. Introduction

Recently, a state of alert on a topic that affects people globally, antimicrobial resistance, has received much attention. This has led to the deaths of more than 700,000 people a year worldwide and this number has risen every year [1]. It is estimated that there will be a reduction in the world population of 11–444 million people in 2050 if antimicrobial resistance is not bypassed [1].

As an alternative against antimicrobial resistance, one approach gaining in strength is the use of inorganic particles at the nanoscale. The most prominent metals in the group of inorganic nanoparticles are copper, zinc, titanium, magnesium, gold, and silver [2–4]. In this context, silver nanoparticles have been the most exploited as they have a wide range of toxicity against several microorganisms such as *Staphylococcus aureus*, *Escherichia coli*, *Candida albicans*, and others [5].

The incorporation and use of silver nanoparticles has been observed in sundry sectors, for instance, in the food industry as an attempt to produce packaging with antimicrobial activity [6]. Its use in the area of cosmetics has also received prominence, as has its use in housecleaning, antiseptics, sunscreens, soap, and shampoo [7–9] as well as in textile manufacturing [10].

Considering the synthesis of silver nanoparticles, many routes have been presented such as electrochemical [11], radiation [12], photochemistry [13], and by biological methods [14]. Phytochemical synthesis has been noteworthy since the use of chemical compounds may result in undesirable toxic effects not only for the human organism but also for the environment. Its effectiveness in the production of silver nanoparticles has been demonstrated by the use of compounds of different plants in the ion reduction, being characterized as rapid, low cost, and environmentally friendly synthesis [15]. Furthermore, green-silver nanoparticles are usually less cytotoxic when compared to those reduced by conventional chemical agents [16]. It is believed that silver nanoparticles reduced by plant extracts do not carry on their surface chemical compounds used for the reduction and stabilization of chemically produced silver nanoparticles that are toxic to human cells. It is still believed that the phytochemicals present in the extracts are carried on the surface of the silver nanoparticles, reducing their cytotoxic effect, aside from presenting different forms of chemically produced silver nanoparticles [16]. Important aspects in green-synthesis should be taken into account including the choice of plant to be used, being the plants which grow in different regions of the world more eligible for this [16]. The previously known potential of the plant including antioxidant, anti-inflammatory, and antimicrobial such as the case of *Punica granatum* (pomegranate) should also be considered [17–19]. Some studies have also used *Punica granatum* to reduce silver ions to silver nanoparticles [19–21]. Silver nanoparticles were green-synthesized and showed significantly lower cytotoxicity when compared to the silver nanoparticles synthesized by a chemical pathway. This fact has stimulated the search for the use of reduced silver nanoparticles by means of plant extracts for biological purposes such as the treatment of contagious infectious diseases, especially those in need of topical treatment.

Thus, taking together the benefits of pomegranate and the antimicrobial applicability of silver nanoparticles, the present study aimed to synthesize silver nanoparticles using pomegranate peel extract, and to produce spray formulations containing the previously green-synthesized silver nanoparticles. Their antimicrobial activity against *Staphylococcus aureus* and *Candida albicans*, and their cytotoxicity effect on fibroblast cells were investigated.

2. Results

2.1. Characterization of Peel Extract, Silver Nanoparticles and Formulations

The pH and the dry matter of the peel extract obtained by maceration followed by percolation were 3.13 and 86.39 ($\pm$0.96) % *w/w*, and the total phenolics expressed in gallic acid and the ellagic acid were 392.0 ($\pm$9) and 3.64 ($\pm$0.03) mg/g, respectively.

The formation of silver nanoparticles was confirmed by comparing the XRD patterns and the corresponding standard patterns of cubic of silver nanoparticles (Figure 1), according to the diffraction standard (JCPDS file No. 04-0783). The reflection peak (2 2 2) is characteristic of the substrate (Si), where silver particles were deposited as a thin film. TEM images (Figure 2) showed different forms and sizes of silver nanoparticles produced by green and conventional chemical routes as well as in their respective formulations. In general, green-synthesis produced particles with a larger size than those obtained by conventional synthesis. Dynamic Laser Scanning (DLS) analyses of the formulations prepared with green or conventional silver nanoparticles demonstrated different particle sizes, being the mean values of 89 ± 21 and 19 ± 4 nm for the green and conventional formulation, respectively. The values of zeta potential of green and conventional silver nanoparticles were lower than −30 mV (−46.2 ± 6.06 mV green, and −67.5 ± 3.69 mV conventional), indicating the stability of both colloidal silver nanoparticles.

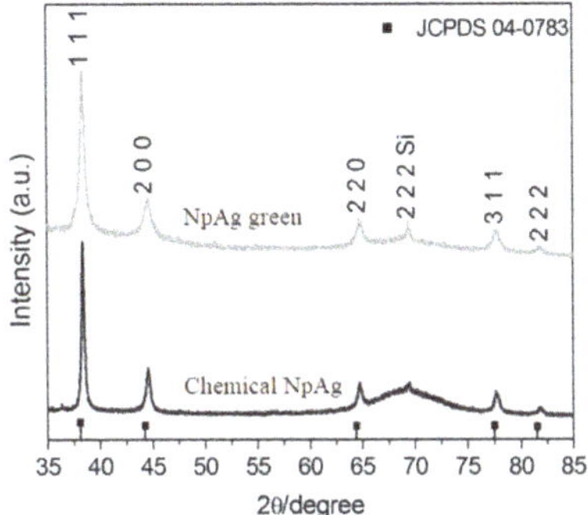

Figure 1. X-ray diffraction (XRD) of the green and chemical silver nanoparticles.

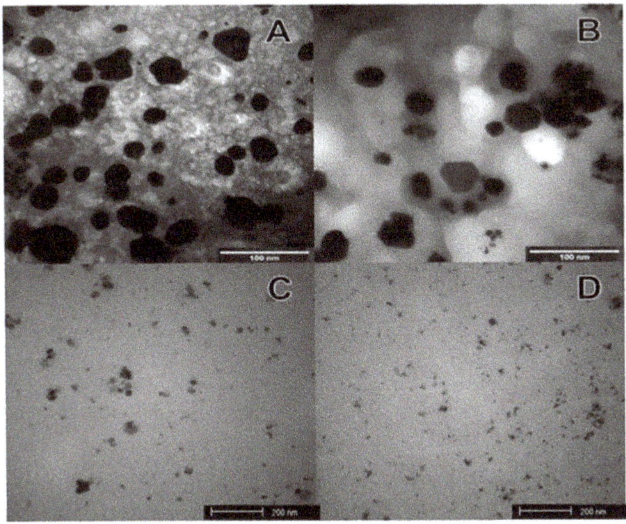

Figure 2. Images of transmission electron microscopy (TEM): (**A**) Green silver nanoparticles; (**B**) Silver nanoparticles green formulation; (**C**) Chemical silver nanoparticles; (**D**) Silver nanoparticles chemical formulation.

Almost 100% of the Ag^+ ions coming from $AgNO_3$ were reduced by the pomegranate peel extract (99.89%) and sodium citrate (99.51%). However, in the spray formulation containing chemical-silver nanoparticles, the percentage of reduction was diminished to 68.18% although the formulation maintained stable regarding Ag^+ ions concentration for 28 days (Table 1). Zeta potential data confirmed the stability of the spray formulations regardless of the method used to obtain the silver nanoparticles (Table 2). The total phenolics in the spray formulations with or without silver nanoparticles were quantified at 0, 7, 14, and 28 days after having been prepared (Figure 3), and it has been significantly reduced in the green-synthesized silver nanoparticle formulation after 14 days with values ranging from 0.405 to 0.295 mg/g.

Table 1. Values of the silver ionic reduction and zeta potential for green and chemical silver nanoparticle formulations in different periods.

Time	Silver Nanoparticles Green Formulation			Silver Nanoparticles Chemical Formulation		
	µgAg⁺/mL	% of Reduction	Zeta Potential	µgAg⁺/mL	% of Reduction	Zeta Potential
T0	0.249	99.93%	-73.7 ± 6.49	1.769	68.15%	-78.2 ± 3.06
T7	0.178	99.95%	-68.3 ± 4.92	1.927	65.31%	-72.9 ± 3.10
T14	0.220	99.94%	-72.8 ± 6.49	1.543	72.22%	-85.5 ± 3.36
T28	0.186	99.95%	-68.6 ± 5.62	1.846	66.77%	-76.5 ± 4.05

Table 2. Silver ion concentration (µgAg⁺/mL) and percentage of silver ions reduction after the reactions, AgNP percentage, and values of minimum inhibitory concentration (MIC) of silver nanoparticles and pomegranate peel extract found for *Staphylococcus aureus* and *Candida albicans*.

Samples	Silver Ions Concentration	Silver Ions Remaining %	Ag NP %	MIC (µg/mL)	
				S. aureus	*C. albicans*
Control *	10,303.26	95.52	4.48	4.13	4.59
Pomegranate peel extract	-	-	-	391	781
Silver nanoparticles green	10.89	0.11	99.89	67.50	68.75
Silver nanoparticles chemical	130.40	1.21	98.79	0.50	0.25
Pomegranate peel extract formulation	-	-	-	0.37	0.18
Silver nanoparticles green formulation	0.249	0.01	99.99	0.26	16.87
Silver nanoparticles chemical formulation	1.769	31.85	68.15	0.56	1.12

* Control = Carboxymethylcellulose, propylene glycol, silver nitrate.

Silver nanoparticles green formulation   **Pomegranate peel extract formulation**

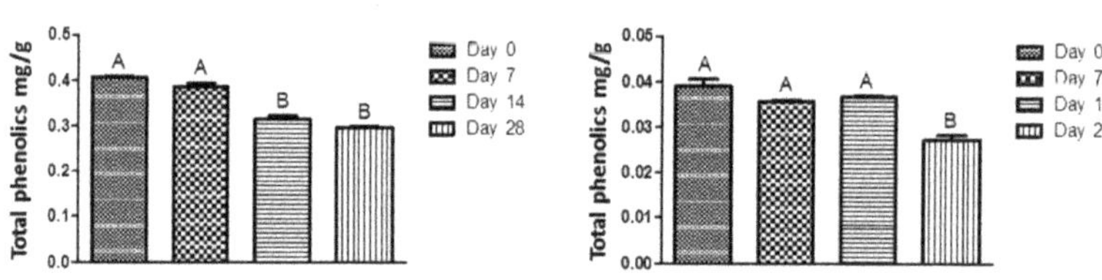

Different capital letters denote significant differences ($p < 0.05$; one-way ANOVA followed by Tukey's Multiple Comparison Test) among the groups.

Figure 3. Total phenolics concentration for the silver nanoparticles green formulation and pomegranate peel extract formulation in different periods. Different capital letters denote significant difference ($p < 0.05$; one-way ANOVA followed by Tukey's multiple comparison test) among the groups.

2.2. Antimicrobial Activity

The antimicrobial activity expressed as MIC values of silver nanoparticles and pomegranate peel extract (µg/mL) (Table 1) was, in general, considerably lower for the spray formulations than the active inputs regardless of the microorganisms tested. MIC values against *C. albicans* for active inputs and spray formulations were 781 and 0.18 for the peel extract, 68.75 and 16.87 for the green-, and 0.25 and 1.12 for the chemical-silver nanoparticles. While for *S. aureus*, the values were 391 and 0.37, 67.5, and 0.26, and 0.5 and 0.56 for pomegranate peel extract, green-, and chemical-silver nanoparticles in the active inputs and spray formulations, respectively. In addition, different conditions of humidity and temperature did not affect the effectiveness of the spray formulations against both microorganisms.

2.3. Cytotoxicity

Figure 4 shows the fibroblast L929 cells viability in view of different concentrations of silver nanoparticles (green and conventional route). Green silver nanoparticles presented lower cytotoxicity than conventional ones. A dosage of 50 µg/mL was necessary to initiate the toxicity, but the cell viability was nearly 80%, while conventional-silver nanoparticles were quite toxic at very low concentration (6.25 µg/mL) and was similar to the negative control (DMSO) with viability lower than 20%. Furthermore, the addition of the reagents to prepare the formulations did not interfere in the toxicity of the conventional-silver nanoparticles, whereas the cytotoxicity for the green-silver nanoparticles formulation as well as for the extract formulation was considerably increased.

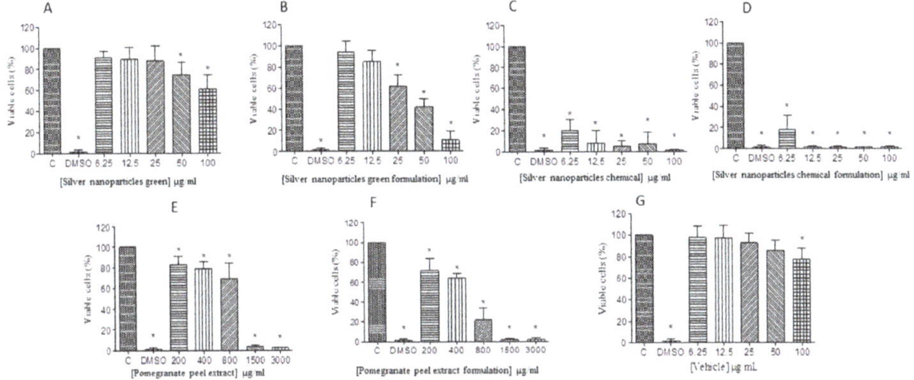

Figure 4. Cytotoxicity evaluation of respective active input (green and chemical silver nanoparticles and pomegranate peel extract), their respective formulations, and the vehicle (compounds of spray-formulation without the active inputs). (**A**) Silver nanoparticles green; (**B**) Silver nanoparticles green formulation; (**C**) Silver nanoparticles chemical; (**D**) Silver nanoparticles chemical formulation; (**E**) Pomegranate peel extract; (**F**) Pomegranate peel extract formulation and (**G**) Vehicle.

3. Discussion

For future reproducibility of the experiment, the extract obtained by maceration followed by percolation was duly characterized in relation to dry matter, total phenolics content, ellagic acid, and pH. Total phenolics were determined only in samples that contained the pomegranate peel extract, and then the chemical formulation did not present any phenolic content in its composition. Polyphenols are effective hydrogen donors and are correlated to the number and position of hydroxyl groups and conjugations as well as the presence of donor electrons in the aromatic ring B, because of the ability of this aromatic ring to withstand the electron depletion located in the π electron system [22]. The antimicrobial activity of various polyphenols and plant extracts have been evaluated in pharmaceutical and food studies [23,24]. Some phenolic compounds present in sage, rosemary, thyme, hops, coriander, tea, cloves, and basil are known to exhibit antimicrobial effects against foodborne pathogens. Their mechanisms of action need to be further elucidated, and might be due to a plethora of phenolic compounds present in a very single plant extract. Furthermore, as the bioactive compounds in the extract presented antioxidant and anti-inflammatory activities, the antimicrobial potential of the pomegranate peel extract in the in vivo trials could show better results, and should be strongly stimulated in further studies. Regarding the multi conceptual nature of the term antioxidant and bringing it into the context of this study, some polyphenols present in low concentrations could prevent or reduce the extent of oxidative damage in mammalian cells. Taken together, these natural biomolecules could indirectly protect the cells and reduce the cytotoxicity of silver nanoparticles.

The correct selection of the plant and the standardization of the methods to obtain the extracts to be used as reducing or capping agent in the nanosynthesis of metal particles should be preponderant when the green process is elected for the production of products in large scale. Additionally, a plethora of plants used in the phytosynthesis of metal nanoparticles [25–27] and the lack of information of the extraction techniques used in the articles has hindered the comparison of the present results with those found in the literature. For instance, different values and methods of total phenolics quantification can be observed in the literature as described by Kalaycioglu et al. (2017) [28]. Similarly, other factors can interfere in the evaluation and comparisons of the extracts such as the chemical and genotypic composition of the plant, the variety and the soil type, the place of the plant origin, the harvest season, maturation method, aside from the solvent and the process used for the obtention of the pomegranate extract, among others [29].

Scanning electron microscopy (SEM) and transmission electron microscopy (TEM) images showed the smallest particles obtained by conventional chemical synthesis, and DLS data confirmed these findings with mean sizes of 89 and 19 nm for green and chemical nanoparticles, respectively. The fission of colloidal particles of different sizes and shapes may be related to additives (salts, polymers), solvent properties (boiling temperature, affinity with created surfaces), the addition of nucleation, among others [30]. The reagents used in the chemical synthesis would produce particles with more predictable characteristics than the several substances and compounds present in the plant extract and used in the phytosynthesis route, which would interfere with the size and form of the nanoparticles and make phytosynthesis a challenge in controlling the reaction process and the morphological aspects of the particles. Moreover, the presence of different bioactive substances in the extract would reduce only a fraction of the silver ions present in the solution. The remaining silver ions would form other nuclei and further the growth of the previously formed silver nanoparticles [31]. This process is called Ostwald Ripening, where the largest particles consume the smaller ones and grow larger, where the dissolution of the smaller ones and deposition of ions on the surface of larger ones occur [32].

Almost 100% of ions reduction was observed for both synthesis routes. However, when the chemical silver nanoparticles were added to the formulation, a dissociation of ions from nearly 30% was observed when compared to chemical silver nanoparticles alone. This fact could be due to the presence of the components as carboxymethylcellulose and propylene glycol in the spray formulation which possibly favored the silver ion dissociation into the system [33]. The presence of oxygen or ligands for Ag^+ in the formulations may increase the dissolution rate of AgNP and lead to increased dissolution through the formation of Ag^+ complexes [34]. Ag^+ in solution will interact with various ions and molecules that are present in aqueous media. Important ligands to be considered for Ag^+ are sulfide and organic ligands such as the carboxylic acids group which are used as Ag coatings (e.g., citrate, lactate). Carboxyl ligands such as carboxymethylcellulose strongly bind Ag^+, which may affect the dissolution of AgNP and the bioavailability of Ag^+ [35].

Furthermore, the size of the Ag in the NP affects the extent and kinetics of the AgNP dissolution as the smallest nanoparticles dissolve faster and to a greater extent [36]. This would explain the difference in the dissolution of the nanoparticles in the formulations. Their dissolution is of high relevance for the possible toxic effects of AgNP as Ag^+ appears in many cases to determine their toxicity [37]. This fact was not observed when green-synthesis was carried out. This could be related to several compounds present in the extract which would readily react with the released silver ions, or the encapsulation of the silver nanoparticles promoted by those phytocompounds may have avoided the silver ions dissociation from the silver nanoparticles and its release to the solution.

Zeta potential test demonstrated the stability of the silver nanoparticles, most notedly in the spray formulations. Electrical charges on the surface of the nanoparticles prevent agglomeration, and thus afford the stability of the nanoparticles [38,39]. Indeed, silver nanoparticles and spray formulations presented a mean of 70 mV, which indicates their high stability of silver nanoparticles [40].

Antimicrobial results are also promising for the silver nanoparticles as well as the pomegranate extract obtained. The formulations notably showed better results when compared with the input active only. This fact could be explained for the proper dispersion of the active inputs (silver nanoparticles and pomegranate peel extract) in the spray formulation. Additionally, a synergistic effect could have occurred between those active inputs and the methylparaben present in the formulation. In the literature, studies with an antimicrobial effect of pomegranate extract were conducted against

Staphylococcus aureus, Enterobacter aerogene, Salmonella typhi, and *Klebsiella pneumonia* [41]. The MIC values obtained in this study for pomegranate extract were in accordance with Bakkiyaraj et al. (2013) [42] for both the microorganisms studied, and a difference was observed in *C. albicans*, but this fact may be explained by the difference between the *C. albicans* strains used in the studies.

Chemical-silver nanoparticles, in formulation or not, produced MIC values against *S. aureus* about 10-fold lower than those produced by Prema et al. (2017) [33] (60 µg/mL), who also produced silver nanoparticles stabilized with CMC. Indeed, the antimicrobial activity of chemical silver nanoparticles was also determined by Monteiro et al. (2011) [43] with MIC values for *C. albicans* (0.5 µg/mL) in accordance with this present study.

Noteworthy is the difference found in the present study in respect of cytotoxicity between the chemical and green routes to obtain silver nanoparticles. Studies have shown that silver nanoparticles produced with *Protium serratum* and *Nyctanthes arbortristis* extracts were biocompatible when tested in L929 fibroblasts [44,45]. It is believed that what makes the silver nanoparticle toxic to human cells is the type of reducing agent used such as sodium citrate or sodium borohydride [46]. Even in conventional syntheses of silver nanoparticles, reagents are used that prevent the aggregation of these nanoparticles [47], which may further favor their cytotoxicity.

In the case of phytosynthesis of metal nanoparticles, plant extracts, aside from acting as reducing agents, would act to stabilize the particles against dissolution, hence reducing the toxicity of the silver nanoparticles solution. Furthermore, it is possible that some compounds in the extracts may have a synergistic effect with the silver nanoparticles [48], making them less toxic to human cells. Furthermore, extracts of *Punica granatum* have exhibited antioxidant [49] and anti-inflammatory [50] activity, and may have contributed to reducing the cytotoxicity of green- in comparison with chemical-silver nanoparticles.

In general, the stability assay (silver ions dosage, zeta potential, and antimicrobial activity) showed a high stabilizing capacity of the formulations. However, the spray formulations of green silver nanoparticles and pomegranate peel extract showed a significant reduction in the content of total phenolics in 14 and 28 days. The decrease in the content of total phenolics may have occurred due to the temperature variations inherent in the stability test, as occurred in the study of [51] where the temperature affected the total phenolics content in the roselle-mango juice blends. Moreover, in formulations containing green-silver nanoparticles, the components of the extract may have been degraded or associated with the nanoparticles, explaining the faster decrease of the total phenolics content when compared to the pomegranate extract formulation. Interestingly, ion dosage, zeta potential, and antimicrobial activity were not affected by different conditions of temperature, time, and humidity of the stability test.

Altogether, the reported results suggest that the plant extract mediated syntheses of AgNP showed a pronounced lower cytotoxic effect in mouse fibroblast cells (L929) than the syntheses of AgNP by the chemical method. Of note is the implication that different sizes between the green- chemical-AgNP as well as the expected impurities sedimented on both obtained nanoparticles could have had on their toxicity. Although it is quite tricky to obtain AgNP with a well-defined form and size and prevent the particles aggregation [52], it is of importance to complement and support our findings, then strongly recommend an eco-friendly approach to producing green-AgNP and prototype wound-care sprays containing these particles.

4. Materials and Methods

4.1. Plant Material and Preparation of Pomegranate Peel Extract

Pomegranate samples were collected from a crop cultivated in Eixo (21°08′01″ S, 51°06′06″ W), Mirandópolis, São Paulo, Brazil, during May 2015. Pomegranate peels were separated and stove-dried at 50 °C, ground, and sieved to a granulometry lower than 2 mm. Peels were submitted to alcohol extraction using 70% ethanol by maceration, followed by the percolation process [53]. The extract was characterized in relation to pH, dry matter, and total phenolics expressed as gallic acid. The chemical marker of pomegranate, ellagic acid, was also identified and quantified.

4.1.1. Determination of Total Phenolics, pH, and Dry Matter

To determine the total phenolics, an analytical curve of gallic acid (Sigma-Aldrich Chemical Co., St. Louis, MO, USA) was carried out [54]. All extracts obtained and the standard solution of gallic acid were prepared in 50 mL volumetric flasks using water as the solvent. The samples were homogenized and, the flasks were brought to the ultrasonic bath for 30 min. A 0.5 mL aliquot was transferred to another 50 mL flask where 2.5 mL of Folin-Denis reagent (Qhemis-High Purity, Hexis, São Paulo, Brazil) and 5.0 mL of 29% sodium carbonate (Cinética, São Paulo, Brazil) were added. The samples were protected from light and the readings were performed after 30 min in a UV-Vis spectrophotometer at 760 nm [53]. The pH was measured direct from a solution of 1% extract, using a pH kit (Merck KGaA, Darmstadt, Germany) and dry matter was calculated after drying on a sample stove at 105 °C and was expressed in percentage w/w. All data were analyzed in triplicate.

4.1.2. Determination of the Ellagic Acid Content

A Shimadzu liquid chromatograph and a Shimpack ODS C18 (Shimadzu Corporation, Kyoto, Japan) reverse phase column (100 mm × 2.6 mm) were used to determine the ellagic acid content by high performance liquid chromatography (HPLC). Analytical conditions were optimized based on de Sousa et al. (2007) [55] with modifications. As the mobile phase, HPLC grade methanol and a 2% aqueous acetic acid solution with gradient elution (0–7 min, 20–72.5% v/v methanol, 7–7.5 min, 72.5–95% v/v methanol, 7.5–8.5 min 95% v/v methanol, 8.5–9 min 95–20% v/v methanol, 9–10 min 20% v/v methanol) were used. The flow rate was 1.0 mL/min, and the separation was achieved at 25 °C. The injection volume was 5 μL and the wavelength used was 254 nm. Peaks were determined by comparison with an authenticated ellagic acid standard. Briefly, the sample was transferred to a 20 mL volumetric flask which was diluted with HPLC grade methanol. Extraction was undertaken using a vortex for 5 min and ultrasonic bath for 1 h. For the extracts, samples were transferred to volumetric flasks of 10 mL, using methanol HPLC as the solvent. All samples were vortexed for 5 min and sonicated for 30 min. Samples were filtered through 0.45 μm filter. All samples were prepared in triplicate.

4.2. Synthesis of Green-Silver Nanoparticles

The protocols described by Gorup et al. (2011) [56] and Das et al. 2015 [57] with modifications were used to produce silver nanoparticles. Briefly, 3.5% of carboxymethylcellulose (CMC) (Labsynth, Diadema, Brazil), 20% of propylene glycol (PG) (Labsynth, Diadema, Brazil), 100 mM of silver nitrate (SN) (Merck KGaA, Darmstadt, Germany), pomegranate peel extract at 30 mg/mL, and water to make up 100% of the samples were used. Silver nanoparticles were not purified relative to the excess reagents. The reaction was carried out at 50 °C for 12 min, and it was selected based on previous results.

4.3. Synthesis of Chemical-Silver Nanoparticles

Chemical-silver nanoparticles were produced according to Gorup et al. [53]. AgNO$_3$ (Merck KGaA, Darmstadt, Hesse, Germany) was dissolved in water, and brought to boiling at 90 °C. After 2 min of boiling, an aqueous solution of sodium citrate (Na$_3$C$_6$H$_5$O$_7$) (Merck KGaA, Darmstadt, Hesse, Germany) was added, and kept boiling for another 6 min until the solution reached a yellow amber color. The stoichiometric ratio was 1:3, respectively for AgNO$_3$ and Na$_3$C$_6$H$_5$O$_7$. Silver nanoparticles were not purified relative to the excess reagents.

4.4. Preparation of the Spray Formulations

The reagents used were CMC (Labsynth, Diadema, Brazil), PG (Labsynth, Diadema, Brazil), and methylparaben (Labsynth, Diadema, Brazil) in a proportion of 0.1%, 7%, and 0.1%, respectively. The active inputs (green- or chemical-silver nanoparticles and pomegranate peel extract) concentrations were based on the minimum inhibitory concentration and cytotoxicity. Therefore, the final concentrations of active inputs in the spray formulations were: 337.5 μg/mL of green-silver nanoparticles, 5.55 μg/mL chemical-silver nanoparticles, and 94 μg/mL of crude peel extract dry matter.

4.5. Characterization of the Silver Nanoparticles and the Spray Formulations

4.5.1. X-ray Diffraction (XRD), Dynamic Light Scattering (DLS), and Zeta Potential Analysis

A Shimadzu XRD diffractometer with a Cu Kα radiation operating at 30 kV and 30 mA and 2θ range from 35° to 85° with step scan of 0.02° and scan speed $0.2° \cdot min^{-1}$ was used to perform XRD analysis. To collect silver nanoparticles patterns, the nanoparticles were deposited on the surface of a silicon substrate (Si) by dripping the aqueous colloidal dispersion on the substrate at room temperature until the solvent had evaporated.

DLS experiments were performed at room temperature and at a fixed angle of 173° on a Zetasizer Nano ZS (Malvern Instruments Ltd., Malvern, UK) equipped with 50 mW 533 nm laser and a digital auto correlator. The number-average values obtained were compared to the size distributions of the silver nanoparticles. For the zeta potential test a Zetasizer (Malvern instruments, Malvern, UK) with an MPT-2 titrator was used. Aliquots from each test suspension were obtained to conduct zeta potential, and mean values were obtained from three independent measurements.

4.5.2. TEM Analyzes

The nanocompounds morphology was characterized by TEM images in a Jeol JEM-100 CXII (JEOL USA Inc., Peabody, St. Louis, MO, USA) microscope equipped with Hamamatsu ORCA-HR digital camera.

4.6. Silver Ions Dosage

The dosages of free silver ions (Ag^+) present in the compounds and spray formulations were performed to observe if the total amount of Ag added in the synthesis reaction was successfully reduced. A specific electrode 9616 BNWP (Thermo Scientific, Beverly, MA, USA) coupled to an ion analyzer (Orion 720 A^+, Thermo Scientific, Beverly, MA, USA) was used. A 1000 μg Ag/mL standard was prepared by adding 1.57 g of AgNO3 to 1 L of deionized water. The combined electrode was calibrated with standards containing 6.25 to 100 μg Ag/mL to achieve equivalent silver concentrations in the compounds. A silver ionic strength adjuster solution (ISA, Cat. No. 940011) that provided a constant background ionic strength was used (1 mL of each sample/standard: 0.02 mL ISA).

4.7. Stability Test of the Spray Formulations

The spray formulations were submitted to a stability test with controlled conditions of temperature and time. This test was based on Anvisa protocols (Cosmetics stability guide ISBN 85-88233-15-0; Copyright© Anvisa, 2005) and the guide to stability studies (Ordinance No. 593 of 25 August 2000). Briefly, samples of each spray formulation were submitted to alternating cycles of temperature daily ranging from 40 to −5 °C for 28 days. The tests selected to evaluate the stability of the samples were ion dosage, total phenolics content, zeta potential, and minimal inhibitory concentration (MIC). All tests were done in the same conditions as described before, and were carried out at 0, 7, 14, and 28 days.

4.8. Antimicrobial Activity of the Silver Nanoparticles and the Spray Formulations

Minimal inhibitory concentration of the silver nanoparticles samples were determined following the instructions of the Clinical Laboratory Standards Institute with some modifications. The samples were first diluted in water and subsequently in culture medium specific for each microorganism, Mueller Hinton broth (BD Difco, Franklin Lakes, USA) for *Staphylococcus aureus* (ATCC 25923), and RPMI (Sigma-Aldrich, St. Louis, MO, USA) for *Candida albicans* SC 5314) [58]. The microorganisms were adjusted to 5×10^5 cells/mL for *S. aureus* and 5×10^3 cells/mL for *C. albicans*, and the plates were incubated for 24 h and 48 h in aerobiosis at 37 °C for *S. aureus* and *C. albicans*, respectively. After incubation, the plates were visually read. The assays were performed in triplicate.

4.9. Cytotoxicity Analysis

For the evaluation of cytotoxicity, fibroblast cells of the L929 lineage were used. Cells were cultured in DMEM culture supplemented with 10% fetal bovine serum (FBS), penicillin G (100 U/mL) (Gibco®, Carlsbad, USA), streptomycin (100 µg/mL), amphotericin B (25 µg/mL) and incubated in a stove at 37 °C with 5% CO_2. Cells were subcultured (5–7 days), using 0.9% saline to wash them and 0.25% trypsin to disintegrate them from the vial. After disruption, these cells were centrifuged at 1000 rpm for 10 min at 10 °C, resuspended in complete DMEM medium (supplemented with FBS), and cell counted in a Neubauer's chamber.

The sub-cultured third to eighth passage fibroblasts were inoculated into 96-well microplates at a density of 0.5×10^5 cells/well. They were then incubated at 37 °C with 5% CO_2. After 24 h, 20 µL of different dilutions of each sample were added to the wells of the plate containing the cells in medium not supplemented with SBF (incomplete medium) and incubated. Twenty-four hours post-treatment, the medium was withdrawn, cells were washed with saline and 20 µL of resazurin (Sigma-Aldrich) 0.01% *w/v* in deionized H_2O was added to each well containing 180 µL of DMEM medium supplemented with 10% Of SFB. The plates were then incubated for 4 h at 37 °C and fluorescence was measured at 540 and 590 nm for excitation and emission, respectively [59]. Cell viability was expressed as a percentage of viable cells when compared to the control group without treatment.

4.10. Statistical Analysis

GraphPad Prism software (GraphPad Software, Inc., La Jolla, CA, USA) was employed for the statistical analysis with a confidence level of 95%. Parametric statistical analyses were conducted with one-way ANOVA followed by Tukey's multiple comparison test for total phenols and zeta potential. For the ion test the statistical analyses was Dunn's multiple comparison test.

5. Conclusions

In light of the results obtained and the limitations of the present study, it was concluded that the use of pomegranate peel extract showed it to be an efficient reducing agent for the production of silver nanoparticles. Moreover, the antimicrobial potential and the low cytotoxicity demonstrated by green-silver nanoparticles have stimulated the search for improvements in the bio-nanotechnology field. Furthermore, the anti-inflammatory and antioxidant properties of pomegranate have encouraged further studies to use nanosystems with future application in prophylaxis or treatment of biofilm-dependent diseases.

Author Contributions: R.A.F., D.B.B., and A.A.B. conceived and designed the experiments; R.A.F., E.C.T., A.F.M.B., G.L.F., and C.C.M.-G. performed the experiments; R.A.F., D.B.B., and A.A.B. analyzed the data; F.N. d.S.-N., L.F.G., and E.R.d.C. contributed reagents/materials/analysis tools; R.A.F. and D.B.B. wrote the paper.

Funding: This study was supported by São Paulo Research Foundation (FAPESP), Brazil, (Process n° 2016/04230-9).

Acknowledgments: The authors would like to thank the company Apis Flora Indl. Coml. Ltda. for the facilities and for the production of the spray formulations containing pomegranate peel extract, and the Laboratory of photochemioprotection of the Faculty of Pharmaceutical Sciences of Ribeirão Preto for the facilities in performing some of the laboratory assays. We thank the Brazilian Agricultural Research Corporation (EMBRAPA) to allow for some tests on the research on its premises and also the Federal University of São Carlos for disposal of its dependencies and technologies. This research work was supported by the São Paulo Research Foundation (FAPESP), Brazil, (Process n° 2016/04230-9).

Conflicts of Interest: The authors declare no conflict of interest. The founding sponsors had no role in the design of the study; in the collection, analyses, or interpretation of data; in the writing of the manuscript, and in the decision to publish the results.

References

1. Raut, S.; Adhikari, B. The need to focus China's national plan to combat antimicrobial resistance. *Lancet Infect. Dis.* **2017**, *17*, 137–138. [CrossRef]
2. Guzman, M.; Dille, J.; Godet, S. Synthesis and antibacterial activity of silver nanoparticles against gram-positive and gram-negative bacteria. *Nanomed. Nanotechnol. Biol. Med.* **2012**, *8*, 37–45. [CrossRef] [PubMed]
3. He, G.; Qiao, M.; Li, W.; Lu, Y.; Zhao, T.; Zou, R.; Li, B.; Darr, J.A.; Hu, J.; Titirici, M.M.; et al. S, N-Co-Doped Graphene-Nickel Cobalt Sulfide Aerogel: Improved Energy Storage and Electrocatalytic Performance. *Adv. Sci.* **2017**, *4*, 1600214. [CrossRef] [PubMed]
4. Jankun, J.; Landeta, P.; Pretorius, E.; Skrzypczak-Jankun, E.; Lipinski, B. Unusual clotting dynamics of plasma supplemented with iron(III). *Int. J. Mol. Med.* **2014**, *33*, 367–372. [CrossRef] [PubMed]
5. Hebeish, A.; El-Rafie, M.H.; El-Sheikh, M.A.; Seleem, A.A.; El-Naggar, M.E. Antimicrobial wound dressing and anti-inflammatory efficacy of silver nanoparticles. *Int. J. Biol. Macromol.* **2014**, *65*, 509–515. [CrossRef] [PubMed]
6. Kuorwel, K.K.; Cran, M.J.; Sonneveld, K.; Miltz, J.; Bigger, S.W. Antimicrobial activity of biodegradable polysaccharide and protein-based films containing active agents. *J. Food Sci.* **2011**, *76*, R90–R102. [CrossRef] [PubMed]
7. Bansod, S.D.; Bawaskar, M.S.; Gade, A.K.; Rai, M.K. Development of shampoo, soap and ointment formulated by green-synthesised silver nanoparticles functionalised with antimicrobial plants oils in veterinary dermatology: Treatment and prevention strategies. *IET Nanobiotechnol.* **2015**, *9*, 165–171. [CrossRef] [PubMed]
8. Tulve, N.S.; Stefaniak, A.B.; Vance, M.E.; Rogers, K.; Mwilu, S.; LeBouf, R.F.; Schwegler-Berry, D.; Willis, R.; Thomas, T.A.; Marr, L.C. Characterization of silver nanoparticles in selected consumer products and its relevance for predicting children's potential exposures. *Int. J. Hyg. Environ. Health* **2015**, *218*, 345–357. [CrossRef] [PubMed]
9. Benn, T.; Cavanagh, B.; Hristovski, K.; Posner, J.D.; Westerhoff, P. The release of nanosilver from consumer products used in the home. *J. Environ. Qual.* **2010**, *39*, 1875–1882. [CrossRef] [PubMed]
10. Velazquez-Velazquez, J.L.; Santos-Flores, A.; Araujo-Melendez, J.; Sánchez-Sánchezc, R.; Velasquilloc, C.; Gonzálezd, C.; Martínez-Castañone, G.; Martinez-Gutierreza, F. Anti-biofilm and cytotoxicity activity of impregnated dressings with silver nanoparticles. *Mater. Sci. Eng. C Mater. Biol. Appl.* **2015**, *49*, 604–611. [CrossRef] [PubMed]
11. Treshchalov, A.; Erikson, H.; Puust, L.; Tsarenko, S.; Saar, R.; Vanetsev, A.; Tammeveski, K.; Sildos, I. Stabilizer-free silver nanoparticles as efficient catalysts for electrochemical reduction of oxygen. *J. Colloid Interface Sci.* **2017**, *491*, 358–366. [CrossRef] [PubMed]
12. Malkar, V.V.; Mukherjee, T.; Kapoor, S. Synthesis of silver nanoparticles in aqueous aminopolycarboxylic acid solutions via gamma-irradiation and hydrogen reduction. *Mater. Sci. Eng. C Mater. Biol. Appl.* **2014**, *44*, 87–91. [CrossRef] [PubMed]
13. Lombardo, P.C.; Poli, A.L.; Castro, L.F.; Perussi, J.R.; Schmitt, C.C. Photochemical Deposition of Silver Nanoparticles on Clays and Exploring Their Antibacterial Activity. *ACS Appl. Mater. Interfaces* **2016**, *8*, 21640–21647. [CrossRef] [PubMed]
14. Rafique, M.; Sadaf, I.; Rafique, M.S.; Tahir, M.B. A review on green-synthesis of silver nanoparticles and their applications. *Artif. Cells Nanomed. Biotechnol.* **2016**, *8*, 1–20. [CrossRef]
15. Roy, N.; Gaur, A.; Jain, A.; Bhattacharya, S.; Rani, V. Green-synthesis of silver nanoparticles: An approach to overcome toxicity. *Environ. Toxicol. Pharmacol.* **2013**, *36*, 807–812. [CrossRef] [PubMed]
16. Das, R.K.; Brar, S.K. Plant mediated green-synthesis: Modified approaches. *Nanoscale* **2013**, *5*, 10155–101562. [CrossRef] [PubMed]

17. Lansky, E.P.; Newman, R.A. *Punica granatum* (pomegranate) and its potential for prevention and treatment of inflammation and cancer. *J. Ethnopharmacol.* **2007**, *109*, 177–206. [CrossRef] [PubMed]
18. Pande, G.; Akoh, C.C. Antioxidant capacity and lipid characterization of six Georgia-grown pomegranate cultivars. *J. Agric. Food Chem.* **2009**, *57*, 9427–9436. [CrossRef] [PubMed]
19. Edison, T.J.; Sethuraman, M.G. Biogenic robust synthesis of silver nanoparticles using *Punica granatum* peel and its application as a green catalyst for the reduction of an anthropogenic pollutant 4-nitrophenol. *Spectrochim. Acta Part A Mol. Biomol. Spectrosc.* **2013**, *104*, 262–264. [CrossRef] [PubMed]
20. Ahmad, N.S.; Rai, R. Rapid green-synthesis of silver and gold nanoparticles using peels of *Punica granatum*. *Adv. Mater. Lett.* **2012**, *3*, 376–380. [CrossRef]
21. Naik, S.K.; Chand, P.K. Silver nitrate and aminoethoxyvinylglycine promote in vitro adventitious shoot regeneration of pomegranate (*Punica granatum* L.). *J. Plant Physiol.* **2003**, *160*, 423–430. [CrossRef] [PubMed]
22. Ramirez-Tortoza, C.; Andersen, O.M.; Gardner, P.T.; Morrice, P.C.; Wood, S.G.; Duthie, S.J.; Collins, A.R.; Duthie, G.G. Anthocyanin-rich extract decreases indices of lipid peroxidation and DNA damage in vitamin E depleted rats. *Free Radic. Biol. Med.* **2001**, *46*, 1033–1037. [CrossRef]
23. Taguri, T.; Takashi, T.; Kouno, I. Antibacterial spectrum of plant polyphenols and extracts depending upon hydroxyphenyl structure. *Biol. Pharm. Bull.* **2006**, *29*, 2226–2235. [CrossRef] [PubMed]
24. Ahn, J.; Grun, I.U.; Mustapha, A. Effects of plant extracts on microbial growth, color change, and lipid oxidation in cooked beef. *Food Microbiol.* **2007**, *24*, 7–14. [CrossRef] [PubMed]
25. Elemike, E.E.; Onwudiwe, D.C.; Ekennia, A.C.; Ehiri, R.C.; Nnaji, N.J. Phytosynthesis of silver nanoparticles using aqueous leaf extracts of *Lippia citriodora*: Antimicrobial, larvicidal and photocatalytic evaluations. *Mater. Sci. Eng. C Mater. Biol. Appl.* **2017**, *75*, 980–989. [CrossRef] [PubMed]
26. Ovais, M.; Khalil, A.T.; Raza, A.; Khan, M.A.; Ahmad, I.; Islam, N.U.; Saravanan, M.; Ubaid, M.F.; Ali, M.; Shinwari, Z.K. Green-synthesis of silver nanoparticles via plant extracts: Beginning a new era in cancer theranostics. *Nanomedicine* **2016**, *11*, 3157–3177. [CrossRef] [PubMed]
27. Soman, S.; Ray, J.G. Silver nanoparticles synthesized using aqueous leaf extract of *Ziziphus oenoplia* (L.) Mill: Characterization and assessment of antibacterial activity. *J. Photochem. Photobiol. B Biol.* **2016**, *163*, 391–402. [CrossRef] [PubMed]
28. Kalaycioglu, Z.; Erim, F.B. Total phenolic contents, antioxidant activities, and bioactive ingredients of juices from pomegranate cultivars worldwide. *Food Chem.* **2017**, *221*, 496–507. [CrossRef] [PubMed]
29. Li, Y.; Yang, F.; Zheng, W.; Hu, M.; Wang, J.; Ma, S.; Deng, Y.; Luo, Y.; Ye, T.; Yin, W. *Punica granatum* (pomegranate) leaves extract induces apoptosis through mitochondrial intrinsic pathway and inhibits migration and invasion in non-small cell lung cancer in vitro. *Biomed. Pharmacother. Biomed. Pharmacother.* **2016**, *80*, 227–235. [CrossRef] [PubMed]
30. Martins, M.A.; Trindade, T. Os nanomateriais e a descoberta de novos mundos na bancada do químico. *Quím. Nova* **2012**, *35*, 1434–1446. [CrossRef]
31. Agnihotri, S.; Mukerji, S.; Mukerji, S. Size-controlled silver nanoparticles synthesized over the range 5–100 nm using the same protocol and their antibacterial efficacy. *RSC Adv.* **2014**, *4*, 3974–3983. [CrossRef]
32. Houk, L.R.; Challa, S.R.; Grayson, B.; Fanson, P.; Datye, A.K. The definition of "critical radius" for a collection of nanoparticles undergoing Ostwald ripening. *Langmuir* **2009**, *25*, 11225–11227. [CrossRef] [PubMed]
33. Prema, P.; Thangapandiyan, S.; Immanuel, G. CMC stabilized nano silver synthesis, characterization and its antibacterial and synergistic effect with broad spectrum antibiotics. *Carbohydr. Polym.* **2017**, *158*, 141–148. [CrossRef] [PubMed]
34. Gondikas, A.P.; Morris, A.; Reinsch, B.C.; Marinakos, S.M.; Lowry, G.V.; Hsu-Kim, H. Cysteine-induced modifications of zero-valent silver nanomaterials: Implications for particle surface chemistry, aggregation, dissolution, and silver speciation. *Environ. Sci. Technol.* **2012**, *46*, 7037–7045. [CrossRef] [PubMed]
35. Solomon, M.M.; Gerengi, H.; Umoren, S.A. Carboxymethyl Cellulose/Silver Nanoparticles Composite: Synthesis, characterization and Application as a Benign Corrosion Inhibitor for St37 Steel in 15% H_2SO_4 Medium. *ACS Appl. Mater. Interfaces* **2017**, *9*, 6376–6389. [CrossRef] [PubMed]
36. Liu, J.; Sonshine, D.A.; Shervani, S.; Hurt, R.H. Controlled release of biologically active silver from nanosilver surfaces. *ACS Nano* **2010**, *4*, 6903–6913. [CrossRef] [PubMed]
37. El Badawy, A.M.; Luxton, T.P.; Silva, R.G.; Scheckel, K.G.; Suidan, M.T.; Tolaymat, T.M. 2010. Impact of environmental conditions (pH, ionic strength, and electrolyte type) on the surface charge and aggregation of silver nanoparticles suspensions. *Environ. Sci. Technol.* **2010**, *44*, 1260–1266. [CrossRef] [PubMed]

38. Sadowski, Z.; Maliszewska, I.H.; Grochowalska, B.; Polowczyk, I.; Koźlecki, T. Synthesis of silver nanoparticles using microorganisms. *Mater. Sci.* **2008**, *26*, 419–424.

39. Salem, H.F.; Eid, K.A.M.; Saraf, M.A. Formulation and evaluation of silver nanoparticles as antibacterial and antifungal agents with a minimal cytotoxic effect. *Int. J. Drug Deliv.* **2011**, *3*, 293–304.

40. Leite, E.R.; Ribeiro, C. *Crystallization and Growth of Colloidal Nanocrystals*; Springer: New York, NY, USA, 2012; ISBN 978-1-4614-1308-0.

41. Malviya, S.; Jha, A.; Hettiarachchy, N. Antioxidant and antibacterial potential of pomegranate peel extracts. *J. Food Sci. Technol.* **2014**, *51*, 4132–4137. [CrossRef] [PubMed]

42. Bakkiyaraj, D.; Nandhini, J.R.; Malathy, B.; Pandian, S.K. The anti-biofilm potential of pomegranate (*Punica granatum* L.) extract against human bacterial and fungal pathogens. *Biofouling* **2013**, *29*, 929–937. [CrossRef] [PubMed]

43. Monteiro, D.R.; Gorup, L.F.; Silva, S.; Negri, M.; de Camargo, E.R.; Oliveira, R.; Barbosa, D.B.; Henriques, M. Silver colloidal nanoparticles: Antifungal effect against adhered cells and biofilms of *Candida albicans* and *Candida glabrata*. *Biofouling* **2011**, *27*, 711–719. [CrossRef] [PubMed]

44. Gogoi, N.; Babu, P.J.; Mahanta, C.; Bora, U. Green-synthesis and characterization of silver nanoparticles using alcoholic flower extract of *Nyctanthes arbortristis* and in vitro investigation of their antibacterial and cytotoxic activities. *Mater. Sci. Eng. C Mater. Biol. Appl.* **2015**, *46*, 463–469. [CrossRef] [PubMed]

45. Mohanta, Y.K.; Panda, S.K.; Bastia, A.K.; Mohanta, T.K. Biosynthesis of Silver Nanoparticles from *Protium serratum* and Investigation of their Potential Impacts on Food Safety and Control. *Front. Microbiol.* **2017**, *8*, 626. [CrossRef] [PubMed]

46. Asharani, P.V.; Lian Wu, Y.; Gong, Z.; Valiyaveettil, S. Toxicity of silver nanoparticles in zebrafish models. *Nanotechnology* **2008**, *19*, 255102. [CrossRef] [PubMed]

47. Ren, M.; Jin, Y.; Chen, W.; Huang, W. Rich capping ligand—Ag colloid interactions. *J. Phys. Chem. C* **2015**, *119*, 27588–27593. [CrossRef]

48. Gengan, R.M.; Anand, K.; Phulukdaree, A.; Chuturgoon, A. A549 lung cell line activity of biosynthesized silver nanoparticles using *Albizia adianthifolia* leaf. *Colloids Surf. B Biointerfaces* **2013**, *105*, 87–91. [CrossRef] [PubMed]

49. Delgado, N.T.; Rouver, W.D.; Freitas-Lima, L.C.; de Paula, T.D.; Duarte, A.; Silva, J.F.; Lemos, V.S.; Santos, A.M.; Mauad, H.; Santos, R.L.; et al. Pomegranate Extract Enhances Endothelium-Dependent Coronary Relaxation in Isolated Perfused Hearts from Spontaneously Hypertensive Ovariectomized Rats. *Front. Pharmacol.* **2016**, *7*, 522. [CrossRef] [PubMed]

50. Houston, D.M.; Bugert, J.; Denyer, S.P.; Heard, C.M. Anti-inflammatory activity of *Punica granatum* L. (Pomegranate) rind extracts applied topically to ex vivo skin. *Eur. J. Pharm. Biopharma.* **2017**, *112*, 30–37. [CrossRef] [PubMed]

51. Mgaya-Kilima, B.; Remberg, S.F.; Chove, B.E.; Wicklund, T. Physiochemical and antioxidant properties of roselle-mango juice blends; effects of packaging material, storage temperature and time. *Food Sci. Nutr.* **2015**, *3*, 100–109. [CrossRef] [PubMed]

52. Zhang, X.F.; Liu, Z.G.; Shen, W.; Gurunathan, S. Silver Nanoparticles: Synthesis, Characterization, Properties, Applications, and Therapeutic Approaches. *Int. J. Mol. Sci.* **2016**, *17*. [CrossRef] [PubMed]

53. De Oliveira, J.R.; de Castro, V.C.; das Gracas Figueiredo Vilela, P.; Camargo, S.E.; Carvalho, C.A.; Jorge, A.O.; de Oliveira, L.D. Cytotoxicity of Brazilian plant extracts against oral microorganisms of interest to dentistry. *BMC Complement. Altern. Med.* **2013**, *13*, 208. [CrossRef] [PubMed]

54. Waterman, P.G.; Mole, S. *Analysis of Phenolic Plant Metabolites*; Blackwell Scientific: Oxford, UK; Boston, MA, USA, 1994.

55. De Sousa, J.P.; Bueno, P.C.; Gregorio, L.E.; da Silva Filho, A.A.; Furtado, N.A.; de Sousa, M.L.; Bastos, J.K. A reliable quantitative method for the analysis of phenolic compounds in Brazilian propolis by reverse phase high performance liquid chromatography. *J. Sep. Sci.* **2007**, *30*, 2656–2665. [CrossRef] [PubMed]

56. Gorup, L.F.; Longo, E.; Leite, E.R.; Camargo, E.R. Moderating effect of ammonia on particle growth and stability of quasi-monodisperse silver nanoparticles synthesized by the Turkevich method. *J. Colloid Interface Sci.* **2011**, *360*, 355–358. [CrossRef] [PubMed]

57. Das, A.; Kumar, A.; Patil, N.B.; Viswanathan, C.; Ghosh, D. Preparation and characterization of silver nanoparticle loaded amorphous hydrogel of carboxymethylcellulose for infected wounds. *Carbohydr. Polym.* **2015**, *130*, 254–261. [CrossRef] [PubMed]

58. Gillum, A.M.; Tsay, E.Y.; Kirsch, D.R. Isolation of the Candida albicans gene for orotidine-5′-phosphate decarboxylase by complementation of *S. cerevisiae* ura3 and *E. coli* pyrF mutations. *Mol. Gen. Genet. MGG* **1984**, *198*, 179–182. [CrossRef] [PubMed]
59. Kuete, V.; Wiench, B.; Alsaid, M.S.; Alyahya, M.A.; Fankam, A.G.; Shahat, A.A.; Efferth, T. Cytotoxicity, mode of action and antibacterial activities of selected Saudi Arabian medicinal plants. *BMC Complement. Altern. Med.* **2013**, *13*, 354. [CrossRef] [PubMed]

Review

The Pros and Cons of the Use of Laser Ablation Synthesis for the Production of Silver Nano-Antimicrobials

Maria Chiara Sportelli [1,2]**, Margherita Izzi** [1]**, Annalisa Volpe** [2]**, Maurizio Clemente** [1]**,
Rosaria Anna Picca** [1]**, Antonio Ancona** [2,*]**, Pietro Mario Lugarà** [2]**, Gerardo Palazzo** [1] **and
Nicola Cioffi** [1,*]

[1] Dipartimento di Chimica, Università degli Studi di Bari "Aldo Moro", via E. Orabona 4, 70126 Bari, Italy;
   maria.sportelli@uniba.it (M.C.S.); m.izzi@studenti.uniba.it (M.I.); m.clemente8@studenti.uniba.it (M.C.);
   rosaria.picca@uniba.it (R.A.P.); gerardo.palazzo@uniba.it (G.P.)
[2] Institute of Photonics and nanotechnology-National Research Council (IFN-CNR), Physics Department
   "M. Merlin", Bari, Italy, via Amendola 173, 70126 Bari, Italy; annalisa.volpe@ifn.cnr.it (A.V.);
   pietromario.lugara@uniba.it (P.M.L.)
* Correspondence: antonio.ancona@uniba.it (A.A.); nicola.cioffi@uniba.it (N.C.); Tel.: +39-080-5442371 (A.A.);
   +39-080-544-2020 (N.C.); Fax: +39-080-5442219 (A.A.); +39-080-544-2026 (N.C.)

Received: 29 June 2018; Accepted: 27 July 2018; Published: 28 July 2018

Abstract: Silver nanoparticles (AgNPs) are well-known for their antimicrobial effects and several groups are proposing them as active agents to fight antimicrobial resistance. A wide variety of methods is available for nanoparticle synthesis, affording a broad spectrum of chemical and physical properties. In this work, we report on AgNPs produced by laser ablation synthesis in solution (LASiS), discussing the major features of this approach. Laser ablation synthesis is one of the best candidates, as compared to wet-chemical syntheses, for preparing Ag nano-antimicrobials. In fact, this method allows the preparation of stable Ag colloids in pure solvents without using either capping and stabilizing agents or reductants. LASiS produces AgNPs, which can be more suitable for medical and food-related applications where it is important to use non-toxic chemicals and materials for humans. In addition, laser ablation allows for achieving nanoparticles with different properties according to experimental laser parameters, thus influencing antibacterial mechanisms. However, the concentration obtained by laser-generated AgNP colloids is often low, and it is hard to implement them on an industrial scale. To obtain interesting concentrations for final applications, it is necessary to exploit high-energy lasers, which are quite expensive. In this review, we discuss the pros and cons of the use of laser ablation synthesis for the production of Ag antimicrobial colloids, taking into account applications in the food packaging field.

Keywords: silver nanoparticles; laser ablation synthesis in solution; nano-antimicrobials; food packaging

---

1. Introduction

Due to their unique properties [1], metal nanoparticles (NPs) have been used for applications in several fields, such as medicine and the biomedical sciences [2,3], cosmetics [4,5], food and agriculture [6–9], electronics [10], energy science [11], and catalysis [12], providing significant improvements in each area. This review focuses on laser ablation synthesis in solution (LASiS) nanotechnology and its specific potentialities in the food industry, with particular consideration to food packaging.

The growing demand for ready-to-eat food products, along with requirement of easy and safe transport, has led to the need to extend their shelf life, prevent foodborne diseases, minimize industrial processing, track them, and improve their preservation during storage. To this aim, the use of antimicrobial metal nanoparticles is continuously increasing. They are used in agriculture (e.g., pesticide and fertilizer delivery [13–16]), food processing (e.g., encapsulation of flavor or odor enhancers, food textural or quality improvement), food packaging (e.g., limitation of pathogen proliferation, gas sensors; UV protection, more impermeable polymer films), and nutrient supplements (e.g., nutraceuticals with higher stability and bioavailability) [17–19].

Inorganic or organic nanoparticles can either be placed on the surface of polymeric matrices used for food packaging or dispersed into their bulk [20]. The introduction of nanoparticles in a polymeric matrix aims to improve the properties of traditional packaging, e.g., containment and protection (ease of transportation and avoided leakage or break-up), foodstuffs preservation (protection against microbial contaminants, extended shelf-life), convenience (consumer-friendly products), and marketing and communication (real-time information about the quality of enclosed foodstuffs) [8]. In response to required features, food packaging employs innovative materials, which can be categorized as follows:

(1) Improved nanomaterials (the presence of nanoparticles in the polymeric matrix improves the mechanical and/or chemical properties of the packaging, but they are not in direct interaction with food);

(2) Active nanomaterials (dispersed nanoparticles into polymeric bulk enable the packaging to interact actively with environment and regulate the preservation of food);

(3) Intelligent nanomaterials (packaging is able to monitor and identify the state of the product, because of the integration of nanosensors and devices) [8,21].

Nanoparticles, and specifically silver nanoparticles (AgNPs), can be widely used for active packaging due to their antimicrobial properties [22]. Generally, organic antimicrobial materials are less stable at high temperatures compared to inorganic ones, whereas metal and metal oxide nanoparticles withstand harsher processing conditions [23].

The most common nanocomposites used as antimicrobial films for food packaging are based on AgNPs, which are well-known for their efficacy towards a wide range of microorganisms, with high temperature stability and low volatility [20].

Several reviews pick features and use common nanostructures employed in food packaging [8,20,21,24–31]. In each of them, AgNPs are taken into account, but few papers are exclusively focused on AgNPs in food packaging [23]. One of the earliest works on this specific topic was proposed by Rhim and coworkers; they produced chitosan–AgNP nanocomposite films and examined their bioactivity and mechanical properties [32].

In general, AgNP-based nanocomposites are stable and offer slow release of silver ions in the surrounding medium, resulting in a long-lasting antimicrobial activity [8]. The released amount of silver ions into the system is dependent not only on the properties of the nanoparticles themselves—for example NP size, shape, structure, composition, etc.—but also relies on external factors, including the properties of the surrounding medium: ionic strength, pH, composition, humidity, dissolved oxygen content, temperature, etc. [26].

This review is focused on AgNPs to be used in food packaging, and particularly on a peculiar NP synthesis approach. Special attention will be given to AgNPs synthetized by laser ablation, a method proposed by several groups as a green route to high-purity nanomaterials. The pros and cons of this technique for the production of Ag antimicrobial nanocolloids will be critically discussed.

2. Silver Nanoparticles

AgNPs are chemically stable [33,34] antimicrobial agents [35] providing strong activity towards a wide range of pathogenic microorganisms, including bacteria, yeasts, viruses, fungi, and parasites, even when low doses are used (full growth inhibition of bacteria can occur at a few mg/mL) [36]. Moreover, AgNPs are non-toxic to the human body at low concentrations [37]. The Occupational Safety and Health Administration (OSHA) and Mine Safety and Health Administration (MSHA) proposed that a permissible exposure limit (PEL) for metallic and most soluble Ag compounds should be 0.01 mg/m^3. Argentina, Bulgaria, Columbia, Jordan, Korea, New Zealand, Singapore, and Vietnam recognize the American Conference of Governmental Industrial Hygienists (ACGIH) threshold limit values (TLV) of 0.1 mg/m^3 for metallic Ag, while Austria, Denmark, Germany, the Netherlands, Norway, Switzerland, and Japan recognize 0.01 mg/m^3 as the occupational exposure limit for all forms [38]. According to the Registration, Evaluation and Authorization of Chemicals (REACH; Council of the European Union for chemicals and nanomaterials regulation), 0.01 ppb of Ag (for medical products) is not an environmental concern, even if this threshold cannot be interpreted as a safe concentration [39]. The European Food Safety Authority (EFSA) Panel on Food Additives and Nutrient Sources Added to Food did provide upper limits of Ag migration from packaging. Recommended values should not exceed 0.05 mg/L in water and 0.05 mg/kg in food. This implies that the evaluation of silver migration profiles is needed to assure antimicrobial effectiveness while complying with the current legislation, and that products for food packaging and food supplements containing AgNPs are not allowed in the EU, unless authorized [23]. Toxicity of AgNPs also depends on their size, as it generally increases upon decreasing size [40,41]. A smaller size results in the following characteristics: (1) greater tendency to enter into organisms; (2) a larger number of surface atoms available for diverse reactions; (3) more released Ag ions from the nanoparticles; and (4) more reactive oxygen species (ROS) production on the surface, which eventually results in an increased toxicity [40]. However, there are many papers that discuss in detail the properties and aspects connected to AgNP. toxicity, such as [38,42–49], to cite a few.

2.1. Synthesis Methods

The synthesis of supported (which is beyond the scope of this review) and colloidal AgNPs has been investigated extensively and, over the years, several techniques have been proposed for the synthesis of AgNPs. For a detailed view of such literature, we recommend review works and book chapters published in the last two years, such as [50–66], which deal with the synthesis of colloidal materials. Figure 1 summarizes the main approaches to synthetize AgNPs, including chemical reduction [67,68], photoreduction [69–71], microemulsion (reverse micelle) methods [72,73], electrochemical methods [74–76], evaporation-condensation processes [77], laser ablation [18], and biosynthesis [78,79].

Figure 1. Schematic representation of some of the main methods available for the synthesis of silver nanoparticles (AgNPs). Reprinted from [61], with permission from Elsevier.

Chemical reduction is the most common method for the preparation of AgNPs as stable colloidal dispersions. This method requires a reductant capable of transforming the silver salt into AgNPs, which is usually also used as a stabilizing or capping agent to ensure stability of colloids. Commonly used reducing agents are ascorbic acid, sodium borohydride, sodium citrate, ferulic acid, poly(ethylene glycol)-block copolymers, and hydrazine compounds [64,80–82]. Jokar et al. [83] prepared AgNPs for low-density polyethylene (LDPE) nanocomposites via chemical reduction, using polyethylene glycol (PEG) as a reducing agent, stabilizer, and solvent, with silver nitrate (AgNO$_3$) as metal precursor. To improve antibacterial activity of AgNPs, researchers usually use reductants and stabilizing agents which possess additional antibacterial properties. Cao and coworkers, for example, proposed the synthesis of AgNPs using AgNO$_3$ as a metal precursor, ascorbic acid as reducing agent, and chitosan as a stabilizer. Chitosan, which exhibits excellent biocompatibility, biodegradability, and antibacterial and antifungal activities, was used to prepare silver nanoparticles in many studies [84–86]. Sometimes, chemical reduction may be supported by microwave to achieve a more homogeneous heating process and speed up the reaction rate [87].

Biological synthesis of metal nanoparticles using biological agents such as bacteria, fungi, yeast, plant, and algal extracts is becoming more common due to the necessity to develop simple, cost-effective, and eco-friendly processes. The fundamental mechanism is an ordinary chemical reduction, with the difference that reductants are natural agents. Plants and their parts contain carbohydrates, fats, proteins, nucleic acids, pigments, and several types of secondary metabolites which can act as reducing agents to produce nanoparticles from metal salts without any toxic by-products [88]. Similarly, biomolecules such as enzymes, proteins and bio-surfactants present in microorganisms can serve as reducing agents [64]. The major phytochemicals responsible for reducing silver ions into AgNPs are terpenoids, glycosides, alkaloids, and phenolics (flavonoids, coumarins, ubiquinones, tannins, lignin, etc.) [89]. There are countless organisms which can be used for the synthesis of NPs. Bhoir and coworkers [89] used mint extract in the presence of silver nitrate as metal precursors and polyvinyl alcohol (PVA) as a capping material. Moreover, they demonstrated the use of these nanoparticles in food packaging: incorporating them in chitosan and gelatin blend they obtained improved mechanical and barrier properties for the chitosan–gelatin films, as well as antimicrobial activity for food packaging applications. Terenteva at al. [80] investigated the synthesis of AgNPs under the influence of flavonoids as reductants, and they found that quercetin, dihydroquercetin, rutin, and morin produced AgNPs better than chrysin, naringenin, and naringin.

Most of the synthetic techniques mentioned above may suffer from some drawbacks. With chemical synthesis, metal precursor, reductant, and stabilizing/capping agents, which are always needed to ensure stable chemical-synthetized colloids, are all present in the synthesis solution. However, they (or their by-products) can be toxic and unsafe for human health. For this reason, adverse products must be separated and removed from the final nanocolloids before their use in antibacterial, biomedical, or catalytic applications.

In general, LASiS is a low environmental impact technique which does not need metal precursors and reductants, and produces colloids of a relatively high purity as compared to chemical methods. In particular, the possibility of fragmenting a metal target without making use of capping and reducing agents intrinsically lowers the risk of contamination of the resulting colloid from unknown chemical agents and provides NPs with unique surface characteristics [90]. In the field of nano-antimicrobials, LASiS-generated NPs are expected to exhibit higher reactivity and antimicrobial effects in comparison to their chemically synthesized homologous NPs, due to the absence of ligands and/or stabilizers on the NP surface [91,92].

In addition, LASiS allows in situ conjugation of nanoparticles with biomolecules, which has sometimes proved to be more efficient than the ex situ conjugation required for chemically synthetized nanoparticles [93].

Hence, high-purity nanoparticles generated by laser ablation can be considered as a very promising agent for antimicrobial applications, especially in food packaging.

However, in terms of other preparation methods, laser ablation presents some drawbacks as well. Indeed, it incurs high investment costs because of the high price of laser system; to be economically convenient, a large number of colloids should be prepared, and fairly often. Moreover, lasers need a considerable amount of energy (although many other synthesis routes need a high energy consumption) [94], and the most diffused laser sources are not capable of producing nanomaterials on an industrial scale. A great amount of energy is necessary to deliver a good ablation efficiency. Ablation efficiency decreases with long ablation time because of significant number of NPs placed along the laser beam. This inconvenience can be solved through a careful choice of fluidics, e.g., by removing the as-prepared NPs from the optical path by using a flow-through system [95].

Barcikowski and Gökce et al. have recently demonstrated the production of nanocolloids at continuous multi-gram ablation rates (i.e., up to 4 g/h) for different metals and under tight composition control. They exploited a 500 W picosecond laser source working at 10 MHz repetition rate fully synchronized with a polygon scanner in order to reach a scanning speed up to 500 m/s [96,97]. Such a technical solution allows to spatially bypassing the laser-induced cavitation bubbles that prevent higher ablation rates at an MHz repetition rate due to the shielding effect. Therefore, we firmly believe LASiS represents a very interesting and versatile way to produce technologically relevant Ag nano-antimicrobials and in the following we will systematically discuss the aspects which make invaluable this technique.

2.2. Bioactivity of AgNPs

Many studies have been performed on the mechanism of action of AgNPs and the complete significance of AgNP bioactivity is still under investigation. A bird's-eye view of some relevant review works published in the last three years is provided in [36,52,66,98–113]. A variety of mechanisms may be involved in the antimicrobial activity of silver against a broad spectrum of organisms (Figure 2). Some of the commonly accepted mechanisms include silver–amino acid and silver–thiolate group interactions, silver–DNA interactions, generation of reactive oxidative species, and direct cell membrane damage [47,114].

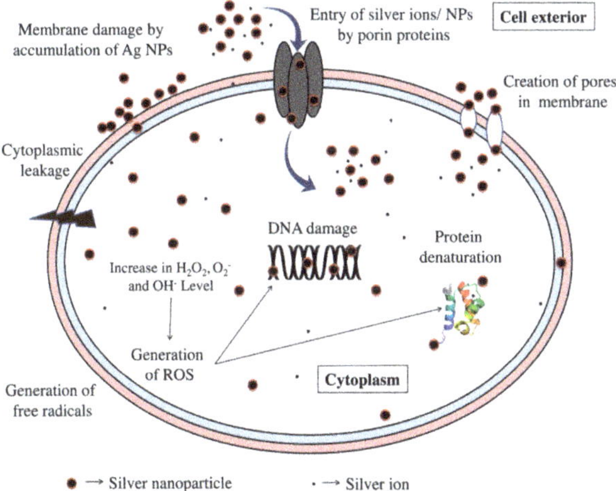

Figure 2. Schematic representation of the known mechanisms of antibacterial action of silver nanoparticles and silver ions. Reprinted from [108], with permission from Elsevier.

Zhang and his collaborators highlight that Ag^+ is expected to show high affinity to the soft base-like thiolate ligands, which are abundant in the bacterial membrane and subcellular structure (e.g., sulfur-containing proteins and enzymes), leading to the inhibition of crucial biological cellular functions [114]. It was found that intracellularly released Ag^+ ions interact with thiol groups of antioxidants such as glutathione (GSH), superoxide dismutase (SOD), and thioredoxin, leading to increased lipid peroxidation, oxidative stress, DNA damage, and subsequent apoptotic cell death [100].

Nomiya and coworkers found Ag$^+$ bonded to the amino acids, forming weak Ag–N bonds which replace biological bonds, resulting in the alteration of cell machinery [115]. AgNPs in the mitochondria could potentiate mitochondrial membrane potential collapse, disruption of the respiratory chain, oxidative stress, inhibition of ATP synthesis, and subsequent activation of the mitochondria-dependent intrinsic pathway of apoptosis. NPs are able to attach to the bacterial membrane by electrostatic interaction [100,114,116]. Lara et al. reported that the positive charge on the Ag$^+$ ion is crucial for its antimicrobial activity through the electrostatic attraction between the negatively-charged cell membrane of the microorganism and the positively-charged inorganic agent. According to these authors, one of the mechanisms of antibacterial action for AgNPs is the formation of pits in the cell wall due to AgNP accumulation in the bacterial membrane, which disturbs membrane permeability, resulting in membrane degradation and cell death [37]. Furthermore, AgNPs and Ag$^+$ ions can work as catalysts and increase generation of reactive oxygen species (ROS). ROS are usually present in cells in small amounts, but an excess can lead to oxidative stress [47,117]. Nanosilver is also known to interact with DNA and cause DNA damage [118]. DNA is responsible for the reproduction process. Any damage done to it will cause either mutation or death of the organism. It was found that AgNPs specifically interact with the exocyclic nitrogen present in the adenine, guanine, and cytosine bases, which leads to DNA changes [101]. Undoubtedly, both Ag$^+$ ions and AgNPs possess antimicrobial activity, but it is very hard to precisely discriminate between the effect of ions and those of nanosilver [44]. Li et al. [119] reported on a similar mode of action of Ag$^+$ ions compared to that of AgNPs, although with stronger antibacterial activity. Navarro et al. [120] suggested that AgNP toxicity could be explained by the release of Ag+ from the particles which damage cells.

AgNPs have been proved to be active against both Gram-positive and Gram-negative bacteria, as briefly listed in Table 1, where effect of both silver ions and AgNPs has been described on several microorganisms [103,121]. It is well accepted that AgNPs and Ag$^+$ give rise to different cellular uptake pathways (Figure 3) depending on whether Gram-positive or Gram-negative bacteria are considered, as also shown in some recent papers [103,122].

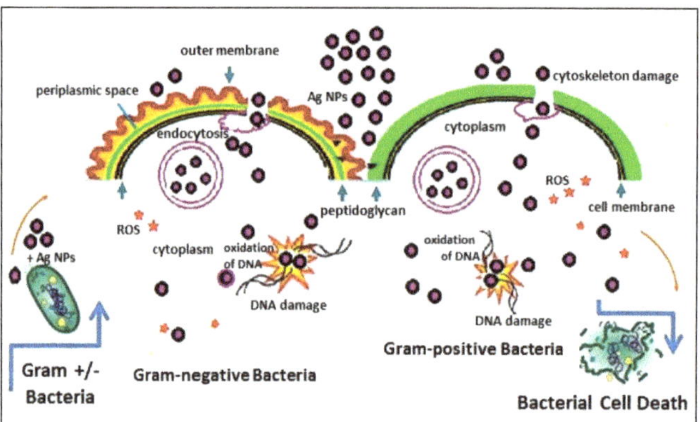

Figure 3. Schematic diagram of bactericidal activity of AgNPs on Gram-positive and Gram-negative bacteria. Reprinted from [105], an open access article distributed under the Creative Commons Attribution License. ROS: reactive oxygen species.

AgNPs have also been proved to be active against multi-resistant bacteria like methicillin-resistant *Staphylococcus aureus* (MRSA), as well as multidrug-resistant *Pseudomonas aeruginosa*, ampicillin-resistant *Escherichia coli O157:H7* and erythromycin-resistant *Streptococcus pyogenes* [37], or other pathogenic organisms such as *Bacillus subtilis*, *Vibrio cholera*, and *Syphilis typhus* [64]. Pazos-Ortiz and colleagues showed antibacterial activity of AgNPs dispersed in polycaprolactone

(PCL). They found that there is greater sensitivity towards *Escherichia coli*, *Klebsiella pneumoniae*, *Staphylococcus aureus*, and *Pseudomonas aeruginosa*, and poor results for *Bacillus subtilis* and *Streptococcus mutans* [123]. Gram-negative bacteria (e.g., *Escherichia coli*, *Pseudomonas aeruginosa*, and *Klebsiella pneumoniae*) were found to be, in general, more sensitive to AgNPs than Gram-positive ones (e.g., *Staphylococcus aureus*, *Streptococcus mutans*, and *Bacillus subtilis*), because of their negatively charged external membrane together with the thinner peptidoglycan layer, which allows adherence and the subsequent penetration of AgNPs [26,123].

Table 1. Activity of Ag$^+$ and/or AgNPs against selected bacterial strains. Adapted from [124], with permission from John Wiley and Sons.

Target Organism	Different Form of Silver	References
Staphylococcus aureus and *Escherichia coli*	Silver ions	[125]
Escherichiacoli	Silver nanoparticles	[126,127]
RNA viruses	Silver ions	[128]
Escherichia coli, *Vibrio cholerae*, *Pseudomonas aeruginosa*, and *Salmonella typhus*	Silver nanoparticles	[129]
Escherichia coli	Silver ions	[130]
Escherichia coli, *Salmonella typhi*, *Staphylococcus epidermidis*, *Streptococcus Aureus*	Silver nanoparticles	[131]
Phoma glomerata, *Phoma herbarum*, *Fusarium semitectum*, *Trichoderma specie* and *Candida albicans*	Silver nanoparticles	[132]
Escherichia coli, *Streptococcusaureus*, and *Pseudomonas aeruginosa*	Silver nanoparticles	[133]
P. aeruginosa, *S. aureus*, pathogenic fungi *Aspergillus flavus* and *Aspergillus niger*	Silver nanoparticles	[134]
S. aureus, *E. coli*, *Klebsiella pneumoniae*, *Bacillus subtilis*, *Enterococcus faecalis*, *P. aeruginosa*	Silver nanoparticles	[135]

The biological activity of AgNPs depends on factors including surface chemistry and morphology, size, shape, coating/capping agent, NP agglomeration and dissolution rate, particle reactivity in solution, and efficiency of ion release [136–140]. According to Riaz Ahmed and colleagues [100], smaller sized AgNPs (about <20 nm) with increased surface to volume ratio possess increased cell permeation capacity and a higher rate of Ag$^+$ ion release, thus increasing the potential for cytotoxicity and cell injury. They proposed an empiric scale of AgNP bioactivity depending on size (Figure 4) and they showed that cellular effects aggravate with size decrease. They also outlined how entity of DNA damage is not dependent on AgNP diameter.

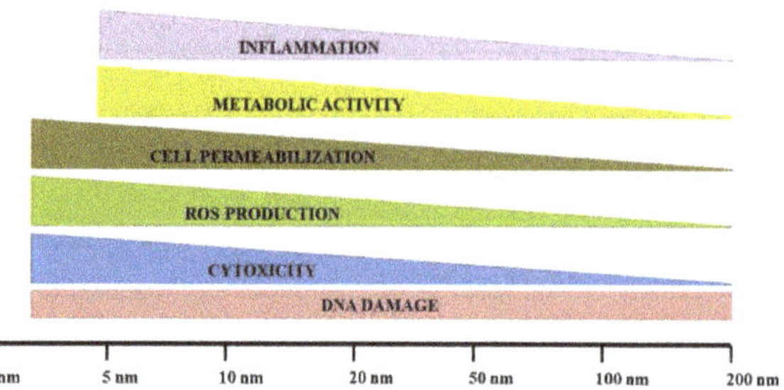

Figure 4. Size-dependent effects of AgNPs in vitro. In general, adverse cellular effects are associated with exposure to smaller AgNPs. One exception is DNA damage; the magnitude of response appears to not depend on AgNP size. Reprinted from [100], with permission from Elsevier.

It has been observed that the shape of AgNPs also causes a critical impact on their antimicrobial activity. Plate- and rod-shaped AgNPs showed higher antibacterial activity as compared to spherical AgNPs and thus they could be used in lower concentrations. In fact, it was observed that the bactericidal activity of plate- and rod-shaped AgNPs was favored by the presence of high atom density facets {111}, whereas, due to predominance of {100} facets on spherical AgNPs, the latter showed relatively lesser bactericidal activity [108,127]. The concentration of AgNPs is another important factor affecting toxicity. It is critical to determine the minimum concentration level of NPs that induces toxicity and its variation in different subjects. Table 2 summarizes a few works that discussed a concentration-dependent AgNP bioactivity.

Table 2. Effects of Ag-NPs at different ranges of concentration on different cell lines. Adapted from [52], with permission from Elsevier.

Concentration Range	Effects of AgNPs	References
25–75 µg/mL	In rat alveolar macrophage cell line, cytotoxicity increases in a concentration-dependent manner	[141]
5, 15, 40, 125 µg/mL	Cytotoxicity occurred through mitochondrial depolarization	[142]
20–250 µg/mL	Apoptosis and necrosis induced in an Hematopoietic stem cell (HSC) cell line	[143]
1, 2, 4 µg/mL	Cell viability decreased in a concentration-dependent manner	[144]
0.4 and 0.8 µg/mL	Arrest G1 phase in cell cycle in a Murine Macrophages cell line (#RAW 264.7)	[145]

AgNPs produced by laser ablation ensure surface cleanliness and the absence of capping agents, which could induce a potential shielding effect on the antimicrobial activity. Hence, we would expect that the antimicrobial activity of laser-produced nanoparticles would be higher compared with the bioactivity of colloids fabricated with other methods which result in NPs with a core-shell structure. Furthermore, reaction by-products of AgNPs synthesis may have potential toxicity. This poses serious issues for the authorization of use of these nanomaterials. However, there are only a few studies that analyze the bactericidal properties of AgNPs produced by laser ablation, and a systematic assessment of these aspects is still missing.

Perito and coworkers [146] tested the antimicrobial activity of AgNPs prepared either by nanosecond (ns) or picosecond (ps) laser ablation using a 1064-nm ablation wavelength, in pure water and in LiCl solution against two bacteria: *E. coli*, as a model for Gram-negative bacteria, and *B. subtilis*, as a model for Gram-positive bacteria. They found that silver colloids ablated in chloride solution exhibited higher antimicrobial activity compared to colloids ablated in pure water. They suggested that AgNPs are coated by a thin oxide layer which "activates" AgNP surface by the addition of a small quantity of LiCl, increasing metal surface reactivity due to the presence of positively-charged active sites. They also stated that bacterial growth inhibition is more effective with AgNPs having an average diameter lower than 10 nm (i.e., prepared with ns pulses), and this is in agreement with the findings of several other works [129,147–150]. As a general statement supported by different groups, it can be inferred that the antibacterial activity of AgNPs decreases with increasing particle size [88].

The effect of the AgNP concentration on the final biocidal properties, while intuitive and reasonable, is not obvious in the case of use of silver nano-antimicrobials because of the strict solubility limits that Ag^+ ions show in vivo, primarily due to the precipitation of insoluble salts such as AgCl, which could lower bioactivity. Nevertheless, Korshed and coworkers studied laser-generated AgNPs [151] and showed that the NP antibacterial effects against both Gram-negative and Gram-positive bacteria displayed significant dose dependency on AgNP concentration when investigating a range from 10 µg/mL to 50 µg/mL. Similar trends were observed by Pandey and colleagues, investigating a concentration range from 40 µg/mL to 600 µg/mL [122].

Zafar et al. [152] reported a comparison between the bioactivity of AgNPs produced by laser ablation and AgNPs produced by chemical reduction. The size of chemically synthesized nanoparticles was in the range of 30–40 nm, while the size range of laser ablated nanoparticles was 20–30 nm. Experiments were carried out at the same AgNP dose and laser-ablated nanoparticles provided maximum inhibition against each pathogen (*S. aureus*, *E. coli*, *Salmonella*). The reduced bioactivity of chemically synthesized NPs, as compared to laser-ablated ones, was interpreted as due to the adsorption of chemical species on their surface, producing adverse effects on their antibacterial action. However, NP size plays a key role in the antibacterial action, so the different bioactivity of laser-ablated AgNPs and chemically synthetized-AgNPs shown in this paper could also be ascribed to different NP sizes.

Based on the literature cited above, it is evident how both the effect of NP size and concentration strictly depends on the single microorganism involved. As a general interpretation, it is possible to state that, even though <10 nm NPs can perform a stronger antimicrobial action, it is better to maintain a conservative approach towards potential nanotoxicological issues arising from the use of such small materials. It is known, in fact, that risks related to NP penetration through main entrance pathways in the human body reach a maximum for dimensions below a critical value of ~50 nm in diameter [153]. Hence, in a growing number of applications of NP-based materials in real-life products, the presence of a (polymeric) matrix that immobilizes NPs appears to be extremely important. In fact, it can prevent and/or limit human exposure to bare (and potentially dangerous) NPs [154].

3. Laser Ablation Synthesis in Solution

In the laser ablation process, an extremely high energy is concentrated at a specific point on a solid target, to remove the material from surface. When a laser pulse irradiates the surface of a bulk material, electromagnetic radiation is adsorbed by the target electrons and energy transfers to material vibrational lattice. As a result, material is expelled from the surface in the form of a plasma plume (in which nanoparticles are formed) [155]. The plasma plume is confined, due to the high pressure exerted by the surrounding liquid, and the considerable temperature gradient between the plume and the liquid. During the plasma decay, the energy is transferred to the surrounding liquid, producing a layer of vapor with a volume approximately equal to that of the plasma, and shaping up a cavitation bubble. Soon after, the cavitation bubble undergoes a periodic evolution of further expansion and shrinkage until its collapse, after which nanoparticles are released into the environmental liquid [156].

The ablation rate is generally determined by laser parameters such as: wavelength, fluence, pulse duration and repetition rate, light absorption efficiency of the target material, transmission, and chemical composition of the liquid. Consequently, NP features depend on laser parameters as well as on the liquid medium. Typical requirements for laser ablation are a wavelength from UV-Vis to near infrared (NIR-IR), a laser fluence approximately comprised between 0.1 and 100 J/cm^2, pulse durations from nanosecond (ns) to picosecond (ps) and femtosecond (fs) [92,95].

These laser parameters can be used to tune several NP features, such as size, shape, surface properties, aggregation state, solubility, structure, and chemical composition [156,157]. As we have previously discussed, these features may affect the NP antimicrobial activities: hence, it is important to know how NP characteristics depend on some laser parameters.

3.1. Ablation Medium

Distilled or deionized water is the most frequently employed liquid medium for the LASiS synthesis of metal nanoparticles, as shown by Mafuné et al. [19,158–161]. Using water as synthesis medium could generate several oxide or hydroxide species because of reactions occurring between the target material and dissolved oxygen, or oxidizing reactions caused by the plasma-induced decomposition of water. Species (such as hydroxyl groups) can be adsorbed on the NP surface, which can lead to highly-charged surfaces that contribute to the electrostatic stabilization of the synthetized nanoparticles [156]. Water is a favorable medium in most ablation processes because it is cheap, safe, exhibits a high heat capacity, and does not absorb laser light [162].

Organic solvents have been also investigated for laser ablation processes, and the most commonly used ones are methanol, ethanol, isopropanol, acetonitrile, and ethylene glycol. When using organic solvents, the higher dipole moment of the solvent has been reported to result in a higher ablation efficiency and in smaller particles. This effect was attributed to the increased electrostatic interactions resulting from the higher molecular dipolar moment of the solvent molecules, which generates a stronger electric double layer at the NPs surface and enhances the repulsive force between NPs [156]. Comparing organic solvents with different viscosity and different dipole moment, it was found that the smallest and most stable AgNPs, with the narrowest size distribution, were obtained in acetone and 2-propanol. In fact, the former has high dipolar moment but low viscosity, while the latter has high viscosity and low dipolar moment. Hence, factors like solvent dipolar moment and viscosity play a fundamental role in avoiding NP agglomeration [163]. Furthermore, when using homologous solvents, such as alcohols with different chain lengths, it has been shown that short-chain alcohols (e.g., methanol and ethanol) result in unstable particles, whereas alcohols with chain lengths from C-3 to C-5 give rise to more stable and smaller particles as compared to those produced in alcohols with chain lengths exceeding C-5 [164].

Moura et al. [165] showed that ethanol and acetone can be good stabilizing environments to keep NPs free from precipitation and oxidation; however, organic environments resulted in a low process yield and a larger mean NP size compared to water. Figure 5 displays TEM images of NPs obtained by laser ablation of Au, Ag, and Fe bulk targets in different solvents with 9-ns pulses at 1064 nm and 10 J cm^{-2} [166]. Moura and coworkers [165] hypothesized that, when acetone molecules are adsorbed around the metal NP, a protective surface dipole layer is developed in the most external plane, inducing a repulsive interaction between nanoparticles. NP aggregation in ethanol can be more intense, since ethanol is a low-polarity solvent compared to acetone. However, it was reported that ablation processes performed in ethanol environment had a low ablation efficiency [165]. This was attributed to the ethanol decomposition during ablation process, promoting the formation of permanent gas bubbles. The latter, in combination with the ablated plasma plume and the as-formed NPs, may act as obstacles within the laser path, thus reducing the energy reaching the target. Recently, Kalus et al. [167] studied the effect of persistent microbubbles on nanoparticle productivity in laser synthesis of colloids, finding that the highest productivity and monodisperse quality is achieved in liquids with the lowest viscosities.

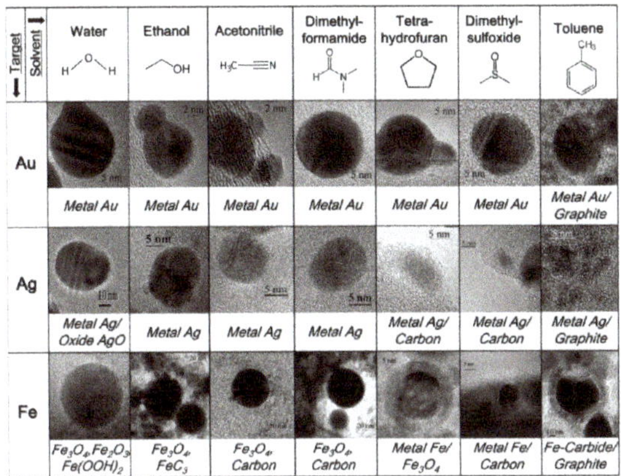

Figure 5. Summary of the NMs obtained by laser ablation of Au, Ag, and Fe bulk targets in different solvents with 9-ns pulses at 1064 nm and 10 J cm^{-2}. Reprinted from [166], an open access article distributed under the Creative Commons Attribution License.

Tajdidzadeh et al. [168] showed that NPs ablation efficiency in chitosan solution is higher than in ethylene glycol (EG), and that it is higher in EG than in deionized water due to plasma confinement on the Ag target (Figure 6). It is worth noting that the broad tail at high wavelength values in AgNP UV-Vis spectra reported in the following is known to be ascribable to NP agglomeration [169]. Those who are not familiar with the UV-vis spectra of nanoparticles should refer to [22] for fundamental information on these phenomena.

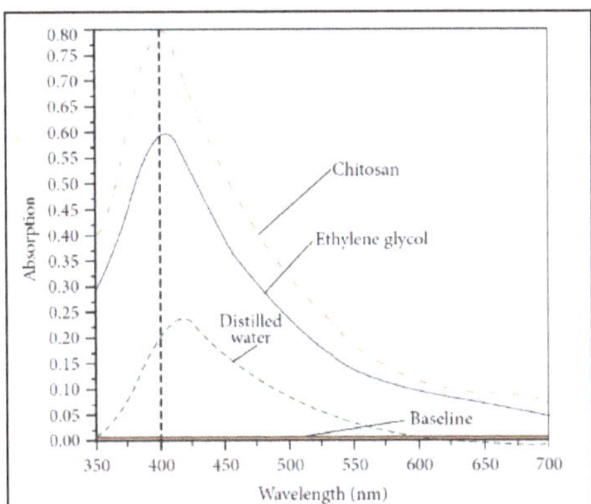

Figure 6. UV-visible absorption spectra of AgNPs prepared for 30 min ablation times in ethylene glycol, chitosan, and deionized water. Reprinted from [168], an open access article distributed under the Creative Commons Attribution License.

In the same paper it is shown by TEM how chitosan solution, which has higher density and viscosity than other liquids, produces a mean size decrement for AgNPs. The same authors also assumed that the plasma generated on the target surface is confined, generating local high pressures and thus etching the target surface. The process, called secondary ablation, can improve ablation efficiency. Additionally, the obtained chitosan-functionalized NPs were shown to be quite stable because the biopolymer acts as a capping agent.

Similar results was reported by Al-Azawi and coworkers [170], who synthetized AgNPs by laser ablation in three different solvents: water, ethanol, and polyvinylpyrrolidone (PVP). Indeed, the ablation efficiency for Ag colloids in ethanol was found to be the lowest, whereas in water it was higher than in PVP solution, corroborating the evidence that the ablation yield of AgNPs in organic solution is generally low. It was found that the efficiency of laser ablation increased, and the NP size decreased for the solvent with higher density and viscosity, which is in agreement with the findings of Moura and Tajdidzadeh.

Accordingly, also in Ganeev's work [171] colloidal silver surface plasmon resonance (SPR) absorbance in water and in ethanol noticeably decreased over time as compared with AgNPs in ethylene glycol, which resulted more stable because of high solvent viscosity. This may be ascribed to a higher solvent viscosity, which prevents NP flocculation.

Overall, we can state that the ablation efficiency is higher in an aqueous environment, but AgNPs are generally more stable in organic environment. This is correlated to solvent physicochemical properties, like the dipole moment and viscosity, which influence NP growth and stability. Higher solvent viscosity prevents NP flocculation and improves ablation efficiency and higher molecular dipolar moment of the solvent molecules generates a stronger electric double layer at the NP surface, which improves the repulsive forces between NPs, increasing their stability.

3.2. Pulse Duration

The effects of the pulse duration are dependent on the electron cooling time (electron-phonon coupling constant) of the material. For fs lasers, pulse duration is shorter than the electron cooling time; thus, the electron-lattice (phonon) coupling is negligible, and the ablation process can be considered as a solid–vapor transition. The ablation process associated with ns pulses is believed to be a thermal one, involving laser heating and melting [166,172]. For these reasons, during fs laser ablation, craters are clearer and more defined than during ns processes.

Tzuji et al. [173] reported a comparison between AgNPs ablated with ns and fs pulsed lasers. They showed that fs ablation yield was lower than nanosecond one. Sizes of ns-prepared particles were much dispersed, and they were irregularly shaped (Figure 7) as compared to fs-prepared particles.

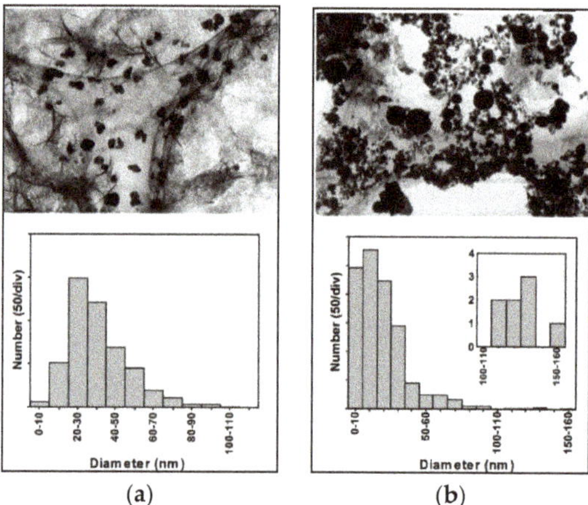

Figure 7. TEM images and size distribution of silver colloids prepared by (a) 120-fs and (b) 8-ns laser pulses. Reprinted from [173], with permission from Elsevier.

Barcikowski and colleagues [95] showed that fs pulses have an higher ablation rate than picosecond ones, but the reported process yield was about three times higher for ps laser ablation when compared to fs ablation. At the same time, it was shown that both ps and fs ultrashort pulsed lasers generated nanoparticles with comparable size distributions. A similar trend was found by Hamad [172], who reported a comparison between AgNPs produced by ns, ps, and fs pulses, and showed a higher ablation efficiency for fs pulses.

3.3. Laser Wavelength

According to the photon energy equation $E = hc/\lambda$, a shorter wavelength implies a greater energy. For instance, at a wavelength of 532 nm, green laser pulses have higher photon energy (2.33 eV) in comparison with those at a 1064-nm wavelength (1.16 eV). In general, the 532-nm wavelength is more effective at producing smaller AgNPs than the 1064-nm one. This is because the lower energy of 1064-nm photons results in less fragmentation, thus producing larger nanoparticles with a higher extinction coefficient in the near-infrared region. On the other hand, the fragmentation produced at 532 nm is higher not only because of the greater photon energy but also because this wavelength is in the range of the SPR peak of AgNPs, thus leading to a reduction of NP size in the colloidal solution [174].

The laser wavelength also determines the laser penetration into the metal target and, consequently, the ablation depth. This parameter decreases with the laser wavelength, thus indicating that the ablated mass per pulse may increase for longer wavelengths if reflectivity is the same [156].

However, it is important to highlight that the influence of the laser wavelength on NP properties still depends on all the other laser parameters, e.g., pulse energy and duration, liquid media, and radiation focus, above all. Solati and colleagues [175] investigated the effect of laser wavelength on the production of AgNPs in acetone (Figure 8); they used nanosecond pulses at 532 nm and 1064 nm and, for each wavelength, they worked at different laser fluences. Results showed that NP size was smaller for a 532 nm than 1064 nm wavelength. They also showed that a surface plasmon resonance (SPR) shift between colloids produced at different fluences is more evident at 1064 nm than at 532 nm.

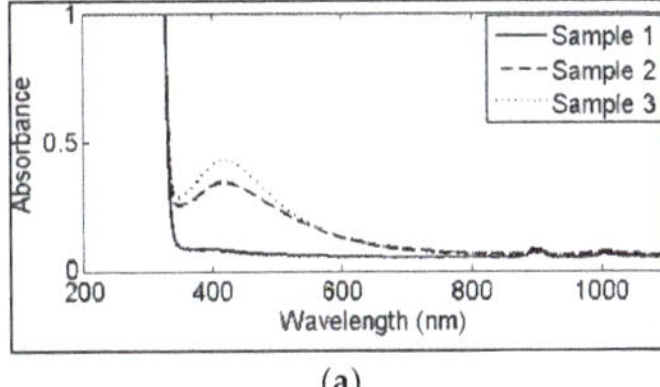
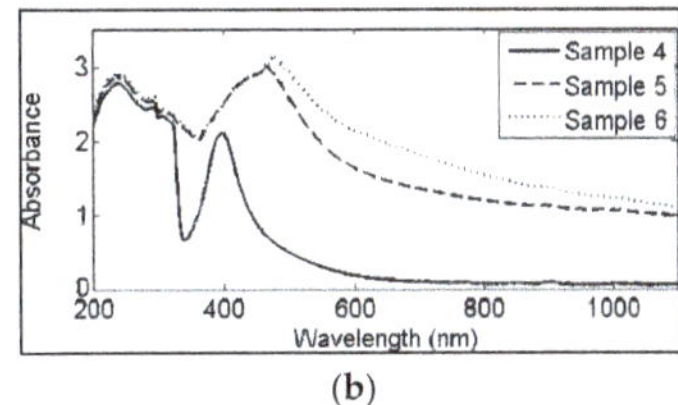

(a)   (b)

Figure 8. Absorption spectra of Ag nanoparticles in acetone prepared at (**a**) 532 nm wavelength; (**b**) 1064 nm wavelength at several fluences (samples 1 and 4: 14 J/cm^2, samples 2 and 5: 18 14 J/cm^2, samples 3 and 6: 22 J/cm^2). Reprinted from [175], with permission from Springer Nature.

The studies by Tsuji and coworkers [176,177] showed that ablation efficiency (evaluated by measuring interband absorption at 250 nm) increases for shorter wavelengths when radiation is unfocused with respect to the target, while it increases with the wavelength for tighter beam focusing conditions (Figure 9). Authors hypothesized that ablation efficiency depends on the laser fluence, which changed with beam focusing.

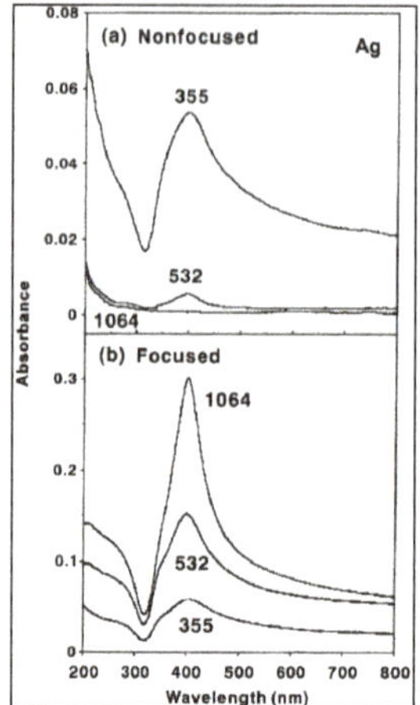

Figure 9. Absorption spectra of Ag colloidal solution prepared with various wavelength laser lights. (a) The laser beam was not focused and laser fluence was 900 mJ/cm^2; (b) The laser beam was focused and laser fluence was >12 J/cm^2. Reprinted from [176], with permission from Elsevier.

3.4. Laser Fluence and Energy Pulse

Laser fluence is a crucial laser parameter that determines the ablation efficiency. NP production yield is affected by the cavitation bubble. In general, the cavitation bubble lifetime increases with the laser fluence [178]. As a result, when the time interval (determined by the repetition rate of pulsed lasers) between two subsequent pulses spatially overlapping on the same spot onto the target is faster than the bubble lifetime, the latter substantially shields and reflects the incoming pulse, thus reducing the ablation rate [156].

Above the ablation threshold, increasing the laser fluence gradually increases the synthesis yield. In fact, Moura et al. [165] observed that the absorption intensity tended to be higher at increasing laser fluences, suggesting that AgNPs concentration increased. Dorranian and coworkers [179], for example, showed the increase of ablated mass upon increasing the fluence (Figure 10).

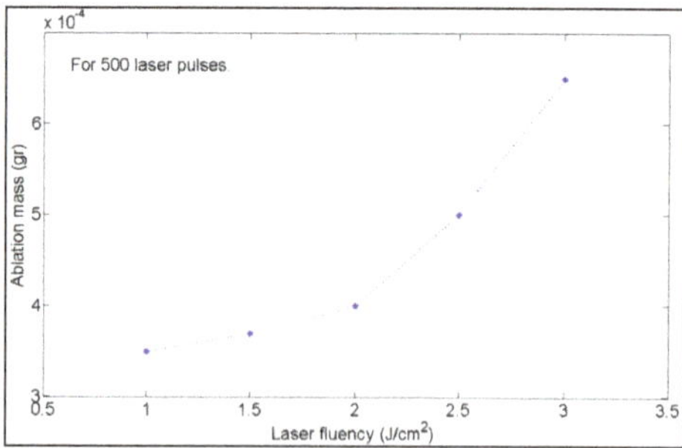

Figure 10. Ablated mass versus laser pulse fluency. Reprinted from [179], an open access article distributed under the Creative Commons Attribution License.

They also found that smaller average NP sizes were obtained by using higher laser fluence values, as shown in STEM (scanning transmission electron microscope) micrographs of Figure 11. This behavior was explained considering that, at high pulse energies, the ablation process is accompanied by melting of the target surface, with less evaporation and NP auto-absorption of laser light. This absorption leads to the formation of smaller particles as a result of fragmentation of larger ones.

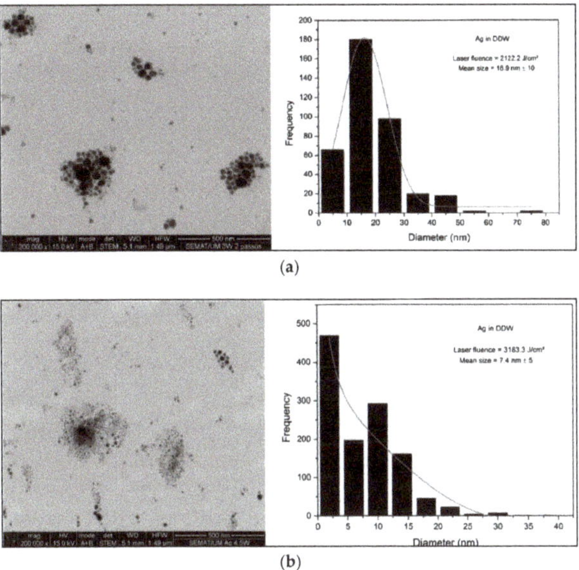

(a)

(b)

Figure 11. *Cont.*

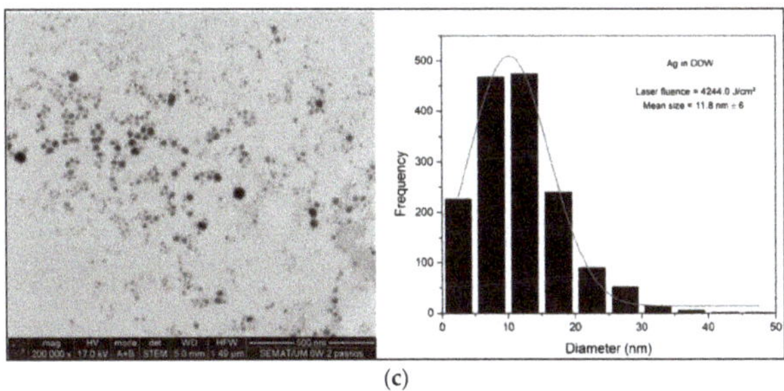

(c)

Figure 11. STEM images and corresponding histograms of silver colloidal nanoparticles prepared by laser ablation at different laser fluences (**a**) 2122.2 J/cm^2; (**b**) 3183.3 J/cm^2 and (**c**) 4244.0 J/cm^2. Reprinted from [165], with permission from Elsevier.

This trend was also found in other papers [175,179–181]. However, Nikolov et al. [182] demonstrated that the average particle size remained unchanged with the laser fluence at its fundamental wavelength (λ = 1064 nm), but it increased strongly with increasing laser fluence at the second harmonic wavelength (λ = 532 nm). On the contrary, Al-Azawi [170] found that the ablation efficiency increased (while the particle size decreased) when laser fluence reached its maximum; subsequently, the ablation efficiency rapidly decreased with increasing the laser fluence. This behavior was ascribed to the occurrence of auto-absorption processes leading to fragmentation of larger particles.

Ablation efficiency generally increases with increasing of pulse energy, as was shown by Valverde-Alva et al. [183] who found an increase of the SPR peak intensity when higher laser pulse energies were used. This result was attributed to the more concentrated AgNP colloids produced with higher pulse energies.

Syntheses of AgNPs using a 1064-nm wavelength Nd:YAG laser, with pulse frequency of 20 Hz and 4 ns of pulse duration, were also carried out by our group. We observed that, at a fixed ablation time of 20′, ablation rate increased with pulse energy. Moreover, AgNPs average size decreased with the energy. Moreover, we found that ablation rate grew with pulse energy for longer ablation times. This effect was attributed to the increasing concentration of AgNPs lying on the laser beam path, which caused an attenuation of the incident radiation owing to scattering phenomena.

3.5. Repetition Rate

The repetition rate (RR) is defined as the number of output laser pulses per unit time. Therefore, for a given laser power, reducing the repetition rate results in an increase of the pulse energy, thus yielding a higher ablation rate per pulse (and larger cavitation bubbles) because of the increased laser fluence [156]. In general, ablation efficiency and concentration of AgNPs increases with the repetition rate. To explain this phenomenon, Valverde-Alva measured the transmission of laser pulses through colloidal solutions and showed that it increased with the RR. Zamiri and coworkers [184] obtained a similar trend. They also investigated the variation of average AgNP size and showed that it increased with increasing RR, in contrast to the trends obtained by Menéndez-Manjón and Barcikowski [185]. However, in this last case, significantly higher repetition rates, spanning in a broad range from 100 to 5000 Hz, were used compared to the very limited range of RR from 10 to 40 Hz explored by Zamiri. For RR exceeding tens of kHz or even approaching the MHz regime, the pulse energy must be decreased to reduce the size and lifetime of the cavitation bubble, which otherwise would shield the laser radiation, thus reducing the ablation yield [97].

4. AgNPs in Food Packaging

AgNPs are a valid antimicrobial additive to control the microbial population in commercialized foodstuff. They have been proved to reduce, delay, or inhibit the growth of spoilage microorganisms, thus improving food shelf life.

Some recent reviews and book chapters discuss the use of AgNPs for this specific application [8,20,21,23–31,186,187]. These references highlight the antimicrobial activity of AgNPs, the mechanism of action against microorganisms, and the correlations existing between bioactivity and AgNP features such as size, shape, and concentration. In most of the literature in this field, AgNPs are embedded in packaging films and tested against foodborne microorganisms, or directly used to package fruit and vegetables, meat, and dairy products. For example, Costa and coworkers [188] prepared and used silver-montmorillonite (Ag-MMT) in packaging for freshly-sliced fruit salads, and found inhibition of microbial growth and increased shelf life. In another interesting work, Sivakumar and colleagues [189] showed a new method to produce nontoxic silver nanorods from dairy industry waste. They used them to control bacterial growth in milk during processing and storage, thus demonstrating extension of milk shelf-life. The most cited reviews and book chapters include many other examples of real-life use of AgNPs. Table 3 summarizes some of the aforementioned applications of AgNPs in food packaging.

Table 3. Applications of AgNPs in food industry. Adapted from [25], with permission from Elsevier.

Nanomaterial Product	Packaging Manufacturer	Country	NP Size	References
Nano-silver salad bowl	Changmin Chemicals	Korea	not reported	[190]
Nano silver baby mug cup and nurser	Baby Dream® Co., Ltd.	Korea	not reported	[191]
Fresh Box® food storage containers	BlueMoonGoods™	USA	not reported	[190]
FresherLonger™ containers and bags	SharperImage®	USA	25 nm and 1–100 nm	[192]
Nano-silver storage box	Quan Zhou Hu Zeng Nano Technology® Co., Ltd.	China	not reported	[190]
Plastic food containers and water bottle	A-Do Global	Korea	not reported	[191]
Fresh food containers	Oso Fresh	USA	40–60 nm	[193]
Smartwist food storage with nano-silver	Kinetic Go Green	USA	10–20 nm	[193]

The antimicrobial properties of packaging materials are based on the migration of antimicrobial substances from the packaging to the food, and/or to the headspace surrounding the food product. Thus, the migration of an active compound from the substrate is an intentional process, which is needed to exert the antimicrobial and protective action against undesirable food contaminants [194]. For these reasons, the efficiency of antimicrobial packaging is largely determined by a controlled release of the antimicrobial agent from the active material. A slow and gradual migration of these substances allows for maintaining an effective antimicrobial concentration on the product over time [194,195]. Nerin and colleagues [194] highlighted that antimicrobials needed to reach bacterial cells to exert their action, while other substances such as antioxidants can exert their action even without being in direct contact with foodstuffs and without releasing any agents. It is worth pointing out that other toxic substances can be released unintentionally into food. AgNP safety limits, both for the environment and human health, are different according to the various legislative authorities in Europe, the USA, Japan, and Australia [196]. For instance, in the United States and in Japan, only silver nitrate is regulated by law in the food and drinks sector, with a maximum allowed limit of 0.017 mg/kg for foodstuffs and 0.1 mg/kg for drinking water, respectively. As for nanosilver, colloidal solutions are accepted in the United States and commercialized as nutrition supplements (e.g., Mesosilver), with the claim of being highly beneficial for human health. In the medical field, different wound dressings containing nanocrystalline silver or silver ion-releasing systems are widely spread as well as indwelling devices, like prostheses or catheters [196]. To date, the European Union does not recommend silver for medicinal use, because of the lack of reliable information with respect to health-risk assessment. The European Food Safety Authority (EFSA) restricted food migration to a maximum of 0.05 mg/kg [196,197].

Besides silver ions, excessive release of entire silver nanoparticles from packaging can be dangerous to human health. Food and beverages produced with AgNPs added in their packaging are considered as the main source of exposure to nanoparticles through ingestion [198]. After ingestion, these nanoparticles undergo various chemical reactions, including agglomeration, adsorption, or binding with other food components, and reactions with acids and digestive enzymes. Internal systemic exposure to NPs can be hazardous since these particles are able to cross biological barriers and reach internal body tissues [154]. As a consequence, AgNPs may accumulate in tissues, resulting in changes in body nutrient profiles. In additions, nanoparticles may introduce toxic agents or viruses adsorbed on their surfaces, or induce production of oxyradicals at the cellular level [198,199]. Therefore, regulation is very important to minimize harmful consequences deriving from the use of nanoparticles, although there are still no internationally recognized research protocols or standards [198].

5. Conclusions

In this review, we focused on AgNPs, a very useful nanomaterial for active food packaging, paying particular attention to AgNPs synthetized by LASiS. After brief overviews of bioactivity pathways of AgNPs and of synthesis methods to produce silver nanocolloids, we focused on the LASiS method, describing in detail its working principles and the influence of the main laser parameters on the production yield and quality. A short overview of the use of AgNPs in food packaging was also proposed. Laser ablation is a green technique to produce stable Ag nanocolloids in a wide variety of dispersing media without using metal precursors and reductants. Highly pure colloids are produced with unique surface characteristics and without any by-products. These features, in principle, make AgNPs produced by LASiS some of the best candidates for antimicrobial applications. However, to date, the productivity is insufficient for direct use in the industrial sector. The current world record for LASiS NP productivity is 4 g/h; this value, although appealing, needs to be improved for LASiS to be used in the industrial sector. Methods to increase this productivity level are currently under development, exploiting high scanning speeds. This way, laser-induced cavitation bubbles are spatially bypassed at high repetition rates, and continuous multi-gram ablation rates have been already demonstrated for platinum, gold, silver, aluminum, copper, and titanium. LASiS requires a great amount of energy, and its scaling-up at the industrial production level is approaching the efficiency required for real-life applications, although there is still a need of for further technological improvement, mostly with regard to the technological laser solutions.

Author Contributions: M.C.S. and M.I. performed bibliographic research and wrote the first draft of the paper. A.V., M.C., R.A.P., A.A., P.M.L., G.P., and N.C. took part in scientific discussions and contributed to different sections of this review. A.A. and N.C. coordinated and revised the study.

Funding: This research was partially funded by: by Italian MIUR Project, grant number #PONa3 00369.

Acknowledgments: The authors gratefully acknowledge the Apulian Region and the Italian Ministry of Education, University and Research for having supported this research activity within the projects MICROTRONIC (Lab Network cod. 71) and "Reti di Laboratori Pubblici di Ricerca" (Lab Network cod. 56).

References

1. Lue, J.T. Physical Properties of Nanomaterials. In *Encyclopedia of Nanoscience and Nanotechnology;* Nalwa, H.S., Ed.; American Scientific Publishers: Valencia, CA, USA, 2007; Volume 10, pp. 1–46. ISBN 1-58883-058-6.

2. Daraee, H.; Eatemadi, A.; Abbasi, E.; Aval, S.F.; Kouhi, M.; Akbarzadeh, A. Application of gold nanoparticles in biomedical and drug delivery. *Artif. Cells Nanomed. Biotechnol.* **2016**, *44*, 410–422. [CrossRef] [PubMed]

3. Sun, H.; Jia, J.; Jiang, C.; Zhai, S. Gold Nanoparticle-Induced Cell Death and Potential Applications in Nanomedicine. *Int. J. Mol. Sci.* **2018**, *19*, 754. [CrossRef] [PubMed]

4. Yacoot, S.M.; Salem, N.F. A Sonochemical-assisted Simple and Green Synthesis of Silver Nanoparticles and its Use in Cosmetics. *Int. J. Pharmacol.* **2016**, *12*, 572–575. [CrossRef]

5. Jiménez-Pérez, Z.E.; Singh, P.; Kim, Y.-J.; Mathiyalagan, R.; Kim, D.-H.; Lee, M.H.; Yang, D.C. Applications of Panax ginseng leaves-mediated gold nanoparticles in cosmetics relation to antioxidant, moisture retention, and whitening effect on B16BL6 cells. *J. Ginseng Res.* **2018**, *42*, 327–333. [CrossRef] [PubMed]

6. Syed, B.; Tatiana, V.; Prudnikova, S.V.; Satish, S.; Prasad, N. Nanoagroparticles emerging trends and future prospect in modern agriculture system. *Environ. Toxicol. Pharmacol.* **2017**, *53*, 10–17. [CrossRef]

7. Kaphle, A.; Navya, P.N.; Umapathi, A.; Daima, H.K. Nanomaterials for agriculture, food and environment: Applications, toxicity and regulation. *Environ. Chem. Lett.* **2018**, *16*, 43–58. [CrossRef]

8. Sharma, C.; Dhiman, R.; Rokana, N.; Panwar, H. Nanotechnology: An Untapped Resource for Food Packaging. *Front. Microbiol.* **2017**, *8*, 1735. [CrossRef] [PubMed]

9. Srivastava, A.K.; Dev, A.; Karmakar, S. Nanosensors and nanobiosensors in food and agriculture. *Environ. Chem. Lett.* **2018**, *16*, 161–182. [CrossRef]

10. Ko, S.H. Low temperature thermal engineering of nanoparticle ink for flexible electronics applications. *Semicond. Sci. Technol.* **2016**, *31*, 073003. [CrossRef]

11. Liu, X.; Iocozzia, J.; Wang, Y.; Cui, X.; Chen, Y.; Zhao, S.; Li, Z.; Lin, Z. Noble metal-metal oxide nanohybrids with tailored nanostructures for efficient solar energy conversion, photocatalysis and environmental remediation. *Energy Environ. Sci.* **2017**, *10*, 402–434. [CrossRef]

12. Akbari, A.; Amini, M.; Tarassoli, A.; Eftekhari-Sis, B.; Ghasemian, N.; Jabbari, E. Transition metal oxide nanoparticles as efficient catalysts in oxidation reactions. *Nano-Struct. Nano-Objects* **2018**, *14*, 19–48. [CrossRef]

13. Baker, S.; Volova, T.; Prudnikova, S.V.; Satish, S.; Prasad, N. Nanoagroparticles emerging trends and future prospect in modern agriculture system. *Environ. Toxicol. Pharmacol.* **2017**, *53*, 10–17. [CrossRef] [PubMed]

14. Duhan, J.S.; Kumar, R.; Kumar, N.; Kaur, P.; Nehra, K.; Duhan, S. Nanotechnology: The new perspective in precision agriculture. *Biotechnol. Rep.* **2017**, *15*, 11–23. [CrossRef] [PubMed]

15. Ammar, A.S. Nanotechnologies associated to floral resources in agri-food sector. *Acta Agron.* **2018**, *67*, 146–159. [CrossRef]

16. Bhagat, Y.; Gangadhara, K.; Rabinal, C.; Chaudhari, G.; Ugale, P. Nanotechnology in agriculture: A review. *J. Pure Appl. Microbiol.* **2015**, *9*, 737–747.

17. Duncan, T.V. Applications of nanotechnology in food packaging and food safety: Barrier materials, antimicrobials and sensors. *J. Colloid Interface Sci.* **2011**, *363*, 1–24. [CrossRef] [PubMed]

18. Picca, R.A.; Di Maria, A.; Riháková, L.; Volpe, A.; Sportelli, M.C.; Lugarà, P.M.; Ancona, A.; Cioffi, N. Laser ablation synthesis of hybrid copper/silver Nanocolloids for prospective application as Nanoantimicrobial agents for food packaging. *MRS Adv.* **2016**, *1*, 3735–3740. [CrossRef]

19. Sportelli, M.C.; Volpe, A.; Picca, R.A.; Trapani, A.; Palazzo, C.; Ancona, A.; Lugarà, P.M.; Trapani, G.; Cioffi, N. Spectroscopic characterization of copper-chitosan Nanoantimicrobials prepared by laser ablation synthesis in aqueous solutions. *Nanomaterials* **2017**, *7*, 6. [CrossRef] [PubMed]

20. De Azeredo, H.M. Nanocomposites for food packaging applications. *Food Res. Int.* **2009**, *42*, 1240–1253. [CrossRef]

21. Ahmad, N.; Bhatnagar, S.; Dubey, S.D.; Saxena, R.; Sharma, S.; Dutta, R. Nanopackaging in Food and Electronics. In *Nanoscience in Food and Agriculture 4*; Sustainable Agriculture Reviews; Springer: Cham, Switzerland, 2017; pp. 45–97. ISBN 978-3-319-53111-3.

22. Sportelli, M.C.; Picca, R.A.; Cioffi, N. Nano-antimicrobials based on metals. In *Novel Antimicrobial Agents and Strategies*; Phoenix, D.A., Harris, F., Dennison, S.R., Eds.; Wiley-VCH Verlag GmbH & Co. KGaA: Weinheim, Germany, 2014; pp. 181–218. ISBN 978-3-527-67613-2.

23. Carbone, M.; Donia, D.T.; Sabbatella, G.; Antiochia, R. Silver nanoparticles in polymeric matrices for fresh food packaging. *J. King Saud Univ. Sci.* **2016**, *28*, 273–279. [CrossRef]

24. Farhoodi, M. Nanocomposite materials for food packaging applications: Characterization and safety evaluation. *Food Eng. Rev.* **2016**, *8*, 35–51. [CrossRef]

25. Hannon, J.C.; Kerry, J.; Cruz-Romero, M.; Morris, M.; Cummins, E. Advances and challenges for the use of engineered nanoparticles in food contact materials. *Trends Food Sci. Technol.* **2015**, *43*, 43–62. [CrossRef]

26. Hoseinnejad, M.; Jafari, S.M.; Katouzian, I. Inorganic and metal nanoparticles and their antimicrobial activity in food packaging applications. *Crit. Rev. Microbiol.* **2018**, *44*, 161–181. [CrossRef] [PubMed]

27. Kuswandi, B. Environmental friendly food nano-packaging. *Environ. Chem. Lett.* **2017**, *15*, 205–221. [CrossRef]

28. Llorens, A.; Lloret, E.; Picouet, P.A.; Trbojevich, R.; Fernandez, A. Metallic-based micro and nanocomposites in food contact materials and active food packaging. *Trends Food Sci. Technol.* **2012**, *24*, 19–29. [CrossRef]

29. Narayan, R.J.; Adiga, S.P.; Pellin, M.J.; Curtiss, L.A.; Stafslien, S.; Chisholm, B.; Monteiro-Riviere, N.A.; Brigmon, R.L.; Elam, J.W. Atomic layer deposition of nanoporous biomaterials. *Mater. Today* **2010**, *13*, 60–64. [CrossRef]

30. Piperigkou, Z.; Karamanou, K.; Engin, A.B.; Gialeli, C.; Docea, A.O.; Vynios, D.H.; Pavão, M.S.G.; Golokhvast, K.S.; Shtilman, M.I.; Argiris, A.; et al. Emerging aspects of nanotoxicology in health and disease: From agriculture and food sector to cancer therapeutics. *Food Chem. Toxicol.* **2016**, *91*, 42–57. [CrossRef] [PubMed]

31. De Azeredo, H.M. Antimicrobial nanostructures in food packaging. *Trends Food Sci. Technol.* **2013**, *30*, 56–69. [CrossRef]

32. Rhim, J.-W.; Hong, S.-I.; Park, H.-M.; Ng, P.K.W. Preparation and characterization of chitosan-based nanocomposite films with antimicrobial activity. *J. Agric. Food Chem.* **2006**, *54*, 5814–5822. [CrossRef] [PubMed]

33. Novikov, S.M.; Popok, V.N.; Evlyukhin, A.B.; Hanif, M.; Morgen, P.; Fiutowski, J.; Beermann, J.; Rubahn, H.-G.; Bozhevolnyi, S.I. Highly stable monocrystalline silver clusters for plasmonic applications. *Langmuir* **2017**, *33*, 6062–6070. [CrossRef] [PubMed]

34. Pugazhendhi, S.; Palanisamy, P.K.; Jayavel, R. Synthesis of highly stable silver nanoparticles through a novel green method using Mirabillis jalapa for antibacterial, nonlinear optical applications. *Opt. Mater.* **2018**, *79*, 457–463. [CrossRef]

35. Franci, G.; Falanga, A.; Galdiero, S.; Palomba, L.; Rai, M.; Morelli, G.; Galdiero, M. Silver nanoparticles as potential antibacterial agents. *Molecules* **2015**, *20*, 8856–8874. [CrossRef] [PubMed]

36. Le Ouay, B.; Stellacci, F. Antibacterial activity of silver nanoparticles: A surface science insight. *Nano Today* **2015**, *10*, 339–354. [CrossRef]

37. Lara, H.H.; Garza-Treviño, E.N.; Ixtepan-Turrent, L.; Singh, D.K. Silver nanoparticles are broad-spectrum bactericidal and virucidal compounds. *J. Nanobiotechnol.* **2011**, *9*, 30. [CrossRef] [PubMed]

38. Miyayama, T.; Arai, Y.; Hirano, S. Health Effects of Silver Nanoparticles and Silver Ions. In *Biological Effects of Fibrous and Particulate Substances*; Current Topics in Environmental Health and Preventive Medicine; Springer: Tokyo, Japan, 2016; pp. 137–147. ISBN 978-4-431-55731-9.

39. Hansen, S.F.; Baun, A. European regulation affecting nanomaterials—Review of limitations and future recommendations. *Dose-Response* **2011**, *10*, 364–383. [CrossRef] [PubMed]

40. Zhou, Q.; Liu, W.; Long, Y.; Sun, C.; Jiang, G. Toxicological effects and mechanisms of silver nanoparticles. In *Silver Nanoparticles in the Environment*; Springer: Berlin/Heidelberg, Germany, 2015; pp. 109–138. ISBN 978-3-662-46069-6.

41. Rai, M.; Yadav, A.; Cioffi, N. Silver Nanoparticles as Nano-antimicrobials: Bioactivity, benefits and bottlenecks. In *Nano-Antimicrobials*; Springer: Berlin/Heidelberg, Germany, 2012; pp. 211–224. ISBN 978-3-642-24427-8.

42. El-Ansary, A.; Al-Daihan, S. On the Toxicity of Therapeutically Used Nanoparticles: An Overview. Available online: https://www.hindawi.com/journals/jt/2009/754810/ (accessed on 23 June 2018).

43. Gaillet, S.; Rouanet, J.-M. Silver nanoparticles: Their potential toxic effects after oral exposure and underlying mechanisms—A review. *Food Chem. Toxicol.* **2015**, *77*, 58–63. [CrossRef] [PubMed]

44. Gupta, I.; Duran, N.; Rai, M. Nano-silver toxicity: Emerging concerns and consequences in human health. In *Nano-Antimicrobials*; Springer: Berlin/Heidelberg, Germany, 2012; pp. 525–548. ISBN 978-3-642-24427-8.

45. Jamuna, B.A.; Ravishankar, R.V. Environmental risk, human health, and toxic effects of nanoparticles. In *Nanomaterials for Environmental Protection*; Wiley-Blackwell: Hoboken, NJ, USA, 2014; pp. 523–535. ISBN 978-1-118-84553-0.

46. Levard, C.; Hotze, E.M.; Lowry, G.V.; Brown, G.E. Environmental transformations of silver nanoparticles: Impact on stability and toxicity. *Environ. Sci. Technol.* **2012**, *46*, 6900–6914. [CrossRef] [PubMed]

47. Marambio-Jones, C.; Hoek, E.M.V. A review of the antibacterial effects of silver nanomaterials and potential implications for human health and the environment. *J. Nanopart. Res.* **2010**, *12*, 1531–1551. [CrossRef]

48. Santos, C.A.D.; Seckler, M.M.; Ingle, A.P.; Gupta, I.; Galdiero, S.; Galdiero, M.; Gade, A.; Rai, M. Silver Nanoparticles: Therapeutical Uses, Toxicity, and Safety Issues. *J. Pharm. Sci.* **2014**, *103*, 1931–1944. [CrossRef] [PubMed]

49. Gonzalez, C.; Rosas-Hernandez, H.; Ramirez-Lee, M.A.; Salazar-García, S.; Ali, S.F. Role of silver nanoparticles (AgNPs) on the cardiovascular system. *Arch. Toxicol.* **2016**, *90*, 493–511. [CrossRef] [PubMed]

50. Abbasi, E.; Milani, M.; Aval, S.F.; Kouhi, M.; Akbarzadeh, A.; Nasrabadi, H.T.; Nikasa, P.; Joo, S.W.; Hanifehpour, Y.; Nejati-Koshki, K.; et al. Silver nanoparticles: Synthesis methods, bio-applications and properties. *Crit. Rev. Microbiol.* **2016**, *42*, 173–180. [CrossRef] [PubMed]

51. Abdelghany, T.M.; Al-Rajhi, A.M.H.; Abboud, M.A.A.; Alawlaqi, M.M.; Magdah, A.G.; Helmy, E.A.M.; Mabrouk, A.S. Recent advances in green synthesis of silver nanoparticles and their applications: About future directions. A review. *BioNanoScience* **2018**, *8*, 5–16. [CrossRef]

52. Akter, M.; Sikder, M.T.; Rahman, M.M.; Ullah, A.K.M.A.; Hossain, K.F.B.; Banik, S.; Hosokawa, T.; Saito, T.; Kurasaki, M. A systematic review on silver nanoparticles-induced cytotoxicity: Physicochemical properties and perspectives. *J. Adv. Res.* **2018**, *9*, 1–16. [CrossRef] [PubMed]

53. Beyene, H.D.; Werkneh, A.A.; Bezabh, H.K.; Ambaye, T.G. Synthesis paradigm and applications of silver nanoparticles (AgNPs), a review. *Sustain. Mater. Technol.* **2017**, *13*, 18–23. [CrossRef]

54. Calderón-Jiménez, B.; Johnson, M.E.; Montoro Bustos, A.R.; Murphy, K.E.; Winchester, M.R.; Vega Baudrit, J.R. Silver Nanoparticles: Technological advances, societal impacts, and metrological challenges. *Front. Chem.* **2017**, *5*, 6. [CrossRef] [PubMed]

55. De Matteis, V.; Cascione, M.; Toma, C.C.; Leporatti, S. Silver Nanoparticles: Synthetic Routes, In vitro toxicity and theranostic applications for cancer disease. *Nanomaterials* **2018**, *8*, 319. [CrossRef] [PubMed]

56. Javaid, A.; Oloketuyi, S.F.; Khan, M.M.; Khan, F. Diversity of Bacterial Synthesis of Silver Nanoparticles. *BioNanoScience* **2018**, *8*, 43–59. [CrossRef]

57. Khan, A.U.; Malik, N.; Khan, M.; Cho, M.H.; Khan, M.M. Fungi-assisted silver nanoparticle synthesis and their applications. *Bioprocess Biosyst. Eng.* **2018**, *41*, 1–20. [CrossRef] [PubMed]

58. Khatoon, U.T.; Rao, G.V.S.N.; Mantravadi, K.M.; Oztekin, Y. Strategies to synthesize various nanostructures of silver and their applications—A review. *RSC Adv.* **2018**, *8*, 19739–19753. [CrossRef]

59. Malik, B.; Pirzadah, T.B.; Kumar, M.; Rehman, R.U. Biosynthesis of nanoparticles and their application in pharmaceutical industry. In *Nanotechnology*; Springer: Singapore, 2017; pp. 235–252. ISBN 978-981-10-4677-3.

60. Pandiarajan, J.; Krishnan, M. Properties, synthesis and toxicity of silver nanoparticles. *Environ. Chem. Lett.* **2017**, *15*, 387–397. [CrossRef]

61. Pinto, R.J.B.; Nasirpour, M.; Carrola, J.; Oliveira, H.; Freire, C.S.R.; Duarte, I.F. Antimicrobial properties and therapeutic applications of silver nanoparticles and nanocomposites. In *Antimicrobial Nanoarchitectonics*; Grumezescu, A.M., Ed.; Elsevier: New York, NY, USA, 2017; Chapter 9; pp. 223–259. ISBN 978-0-323-52733-0.

62. Rafique, M.; Sadaf, I.; Rafique, M.S.; Tahir, M.B. A review on green synthesis of silver nanoparticles and their applications. *Artif. Cells Nanomed. Biotechnol.* **2017**, *45*, 1272–1291. [CrossRef] [PubMed]

63. Ramanathan, S.; Gopinath, S.C.B. Potentials in synthesizing nanostructured silver particles. *Microsyst. Technol.* **2017**, *23*, 4345–4357. [CrossRef]

64. Siddiqi, K.S.; Husen, A.; Rao, R.A.K. A review on biosynthesis of silver nanoparticles and their biocidal properties. *J. Nanobiotechnol.* **2018**, *16*, 14. [CrossRef] [PubMed]

65. Syafiuddin, A.; Salmiati; Salim, M.R.; Kueh, A.B.H.; Hadibarata, T.; Nur, H. A Review of silver nanoparticles: Research trends, global consumption, synthesis, properties, and future challenges. *J. Chin. Chem. Soc.* **2017**, *64*, 732–756. [CrossRef]

66. Khan, S.U.; Saleh, T.A.; Wahab, A.; Khan, M.H.U.; Khan, D.; Khan, W.U.; Rahim, A.; Kamal, S.; Khan, F.U.; Fahad, S. Nanosilver: New Ageless and Versatile Biomedical Therapeutic Scaffold. Available online: https://www.dovepress.com/nanosilver-new-ageless-and-versatile-biomedical-therapeutic-scaffold-peer-reviewed-fulltext-article-IJN (accessed on 23 June 2018).

67. Li, Z.; Wang, Y.; Yu, Q. Significant parameters in the optimization of synthesis of silver nanoparticles by chemical reduction method. *J. Mater. Eng. Perform.* **2010**, *19*, 252–256. [CrossRef]

68. Ajitha, B.; Divya, A.; Kumar, K.S.; Reddy, P.S. Synthesis of silver nanoparticles by soft chemical method: Effect of reducing agent concentration. In Proceedings of the International Conference on Advanced Nanomaterials Emerging Engineering Technologies, Chennai, India, 24–26 July 2013; pp. 7–10.

69. Xu, G.; Qiao, X.; Qiu, X.; Chen, J. Preparation and characterization of stable monodisperse silver nanoparticles via photoreduction. *Colloids Surf. A Physicochem. Eng. Asp.* **2008**, *320*, 222–226. [CrossRef]

70. Afify, T.A.; Saleh, H.H.; Ali, Z.I. Structural and morphological study of gamma-irradiation synthesized silver nanoparticles. *Polym. Compos.* **2017**, *38*, 2687–2694. [CrossRef]

71. Eid, M.; Araby, E. Bactericidal effect of poly(acrylamide/itaconic acid)–silver nanoparticles synthesized by gamma irradiation against pseudomonas aeruginosa. *Appl. Biochem. Biotechnol.* **2013**, *171*, 469–487. [CrossRef] [PubMed]

72. Fatema, U.K.; Rahman, M.M.; Islam, M.R.; Mollah, M.Y.A.; Susan, M.A.B.H. Silver/poly(vinyl alcohol) nanocomposite film prepared using water in oil microemulsion for antibacterial applications. *J. Colloid Interface Sci.* **2018**, *514*, 648–655. [CrossRef] [PubMed]

73. An, J.; Luo, Q.; Li, M.; Wang, D.; Li, X.; Yin, R. A facile synthesis of high antibacterial polymer nanocomposite containing uniformly dispersed silver nanoparticles. *Colloid Polym. Sci.* **2015**, *293*, 1997–2008. [CrossRef]

74. Thuc, D.T.; Huy, T.Q.; Hoang, L.H.; Tien, B.C.; Van Chung, P.; Thuy, N.T.; Le, A.-T. Green synthesis of colloidal silver nanoparticles through electrochemical method and their antibacterial activity. *Mater. Lett.* **2016**, *181*, 173–177. [CrossRef]

75. Yin, B.; Ma, H.; Wang, S.; Chen, S. Electrochemical Synthesis of silver nanoparticles under protection of poly(*N*-vinylpyrrolidone). *J. Phys. Chem. B* **2003**, *107*, 8898–8904. [CrossRef]

76. Cioffi, N.; Colaianni, L.; Pilolli, R.; Calvano, C.; Palmisano, F.; Zambonin, P. Silver nanofractals: Electrochemical synthesis, XPS characterization and application in LDI-MS. *Anal. Bioanal. Chem.* **2009**, *394*, 1375–1383. [CrossRef] [PubMed]

77. Baker, C.; Pradhan, A.; Pakstis, L.; Pochan, D.J.; Shah, S.I. Synthesis and antibacterial properties of silver nanoparticles. *J. Nanosci. Nanotechnol.* **2005**, *5*, 244–249. [CrossRef] [PubMed]

78. Velmurugan, P.; Iydroose, M.; Mohideen, M.H.A.K.; Mohan, T.S.; Cho, M.; Oh, B.-T. Biosynthesis of silver nanoparticles using *Bacillus subtilis* EWP-46 cell-free extract and evaluation of its antibacterial activity. *Bioprocess Biosyst. Eng.* **2014**, *37*, 1527–1534. [CrossRef] [PubMed]

79. Rajeshkumar, S.; Bharath, L.V. Mechanism of plant-mediated synthesis of silver nanoparticles—A review on biomolecules involved, characterisation and antibacterial activity. *Chemico-Biol. Interact.* **2017**, *273*, 219–227. [CrossRef] [PubMed]

80. Terenteva, E.A.; Apyari, V.V.; Dmitrienko, S.G.; Zolotov, Y.A. Formation of plasmonic silver nanoparticles by flavonoid reduction: A comparative study and application for determination of these substances. *Spectrochim. Acta Part A Mol. Biomol. Spectrosc.* **2015**, *151*, 89–95. [CrossRef] [PubMed]

81. Abou El-Nour, K.M.M.; Eftaiha, A.; Al-Warthan, A.; Ammar, R.A.A. Synthesis and applications of silver nanoparticles. *Arab. J. Chem.* **2010**, *3*, 135–140. [CrossRef]

82. Iravani, S.; Korbekandi, H.; Mirmohammadi, S.V.; Zolfaghari, B. Synthesis of silver nanoparticles: Chemical, physical and biological methods. *Res. Pharm. Sci.* **2014**, *9*, 385–406. [PubMed]

83. Jokar, M.; Rahman, R.A. Study of silver ion migration from melt-blended and layered-deposited silver polyethylene nanocomposite into food simulants and apple juice. *Food Addit. Contamin. Part A* **2014**, *31*, 734–742. [CrossRef] [PubMed]

84. Cao, X.L.; Cheng, C.; Ma, Y.L.; Zhao, C.S. Preparation of silver nanoparticles with antimicrobial activities and the researches of their biocompatibilities. *J. Mater. Sci.* **2010**, *21*, 2861–2868. [CrossRef] [PubMed]

85. Rehan, M.; El-Naggar, M.E.; Mashaly, H.M.; Wilken, R. Nanocomposites based on chitosan/silver/clay for durable multi-functional properties of cotton fabrics. *Carbohydr. Polym.* **2018**, *182*, 29–41. [CrossRef] [PubMed]

86. Regiel-Futyra, A.; Kus-Liśkiewicz, M.; Sebastian, V.; Irusta, S.; Arruebo, M.; Kyzioł, A.; Stochel, G. Development of noncytotoxic silver-chitosan nanocomposites for efficient control of biofilm forming microbes. *RSC Adv.* **2017**, *7*, 52398–52413. [CrossRef] [PubMed]

87. El-Naggar, M.E.; Shaheen, T.I.; Fouda, M.M.G.; Hebeish, A.A. Eco-friendly microwave-assisted green and rapid synthesis of well-stabilized gold and core-shell silver-gold nanoparticles. *Carbohydr. Polym.* **2016**, *136*, 1128–1136. [CrossRef] [PubMed]

88. Pollini, M.; Paladini, F.; Sannino, A.; Picca, R.A.; Sportelli, M.C.; Cioffi, N.; Nitti, M.A.; Valentini, M.; Valentini, A. Nonconventional routes to silver nanoantimicrobials: Technological issues, bioactivity, and applications. In *Nanotechnology in Diagnosis, Treatment and Prophylaxis of Infectious Diseases*; Kon, M.R., Ed.; Academic Press: Boston, FL, USA, 2015; Chapter 6; pp. 87–105. ISBN 978-0-12-801317-5.

89. Bhoir, S.A.; Chawla, S.P. Silver nanoparticles synthesized using mint extract and their application in chitosan/gelatin composite packaging film. *Int. J. Nanosci.* **2016**, *16*, 1650022. [CrossRef]

90. Amendola, V.; Meneghetti, M. Laser ablation synthesis in solution and size manipulation of noble metal nanoparticles. *Phys. Chem. Chem. Phys.* **2009**, *11*, 3805–3821. [CrossRef] [PubMed]

91. Szegedi, Á.; Popova, M.; Valyon, J.; Guarnaccio, A.; Stefanis, A.D.; Bonis, A.D.; Orlando, S.; Sansone, M.; Teghil, R.; Santagata, A. Comparison of silver nanoparticles confined in nanoporous silica prepared by chemical synthesis and by ultra-short pulsed laser ablation in liquid. *Appl. Phys. A* **2014**, *117*, 55–62. [CrossRef]

92. Zhang, J.; Chaker, M.; Ma, D. Pulsed laser ablation based synthesis of colloidal metal nanoparticles for catalytic applications. *J. Colloid Interface Sci.* **2017**, *489*, 138–149. [CrossRef] [PubMed]

93. Walter, J.G.; Petersen, S.; Stahl, F.; Scheper, T.; Barcikowski, S. Laser ablation-based one-step generation and bio-functionalization of gold nanoparticles conjugated with aptamers. *J. Nanobiotechnol.* **2010**, *8*, 21. [CrossRef] [PubMed]

94. Jendrzej, S.; Gökce, B.; Epple, M.; Barcikowski, S. How size determines the value of gold: Economic aspects of wet chemical and laser-based metal colloid synthesis. *ChemPhysChem* **2017**, *18*, 1012–1019. [CrossRef] [PubMed]

95. Barcikowski, S.; Menéndez-Manjón, A.; Chichkov, B.; Brikas, M.; Račiukaitis, G. Generation of nanoparticle colloids by picosecond and femtosecond laser ablations in liquid flow. *Appl. Phys. Lett.* **2007**, *91*, 083113. [CrossRef]

96. Sajti, C.L.; Sattari, R.; Chichkov, B.N.; Barcikowski, S. Gram scale synthesis of pure ceramic nanoparticles by laser ablation in liquid. *J. Phys. Chem. C* **2010**, *114*, 2421–2427. [CrossRef]

97. Streubel, R.; Barcikowski, S.; Gökce, B. Continuous multigram nanoparticle synthesis by high-power, high-repetition-rate ultrafast laser ablation in liquids. *Opt. Lett.* **2016**, *41*, 1486–1489. [CrossRef] [PubMed]

98. Abdelhamid, H.N.; Wu, H.-F. Proteomics analysis of the mode of antibacterial action of nanoparticles and their interactions with proteins. *Trends Anal. Chem.* **2015**, *65*, 30–46. [CrossRef]

99. Ahmad, V.; Jamal, Q.M.S.; Shukla, A.K.; Alam, J.; Imran, A.; Abaza, U.M. Bacilli as biological nano-factories intended for synthesis of silver nanoparticles and its application in human welfare. *J. Clust. Sci.* **2017**, *28*, 1775–1802. [CrossRef]

100. Riaz Ahmed, K.B.; Nagy, A.M.; Brown, R.P.; Zhang, Q.; Malghan, S.G.; Goering, P.L. Silver nanoparticles: Significance of physicochemical properties and assay interference on the interpretation of in vitro cytotoxicity studies. *Toxicology In Vitro* **2017**, *38*, 179–192. [CrossRef] [PubMed]

101. Durán, N.; Durán, M.; de Jesus, M.B.; Seabra, A.B.; Fávaro, W.J.; Nakazato, G. Silver nanoparticles: A new view on mechanistic aspects on antimicrobial activity. *Nanomedicine* **2016**, *12*, 789–799. [CrossRef] [PubMed]

102. Halbus, A.F.; Horozov, T.S.; Paunov, V.N. Colloid particle formulations for antimicrobial applications. *Adv. Colloid Interface Sci.* **2017**, *249*, 134–148. [CrossRef] [PubMed]

103. Kędziora, A.; Speruda, M.; Krzyżewska, E.; Rybka, J.; Łukowiak, A.; Bugla-Płoskońska, G. Similarities and differences between silver ions and silver in nanoforms as antibacterial agents. *Int. J. Mol. Sci.* **2018**, *19*, 444. [CrossRef] [PubMed]

104. Khalandi, B.; Asadi, N.; Milani, M.; Davaran, S.; Abadi, A.J.N.; Abasi, E.; Akbarzadeh, A. A review on potential role of silver nanoparticles and possible mechanisms of their actions on bacteria. *Drug Res.* **2017**, *11*, 70–76. [CrossRef] [PubMed]

105. Mosier-Boss, P.A. Review on SERS of bacteria. *Biosensors* **2017**, *7*, 51. [CrossRef] [PubMed]

106. Suresh, K.; Krishna, S.; Govender, P.; Adam, J.K. Nano silver particles in biomedical and clinical applications: Review. *J. Pure Appl. Microbiol.* **2015**, *9*, 103–112.

107. Natan, M.; Banin, E. From Nano to Micro: Using nanotechnology to combat microorganisms and their multidrug resistance. *FEMS Microbiol. Rev.* **2017**, *41*, 302–322. [CrossRef] [PubMed]

108. Pareek, V.; Gupta, R.; Panwar, J. Do physico-chemical properties of silver nanoparticles decide their interaction with biological media and bactericidal action? A review. *Mater. Sci. Eng. C* **2018**, *90*, 739–749. [CrossRef] [PubMed]

109. Rai, M.; Ingle, A.P.; Pandit, R.; Paralikar, P.; Gupta, I.; Chaud, M.V.; dos Santos, C.A. Broadening the spectrum of small-molecule antibacterials by metallic nanoparticles to overcome microbial resistance. *Int. J. Pharm.* **2017**, *532*, 139–148. [CrossRef] [PubMed]

110. Rai, M.; Deshmukh, S.D.; Ingle, A.P.; Gupta, I.R.; Galdiero, M.; Galdiero, S. Metal nanoparticles: The protective nanoshield against virus infection. *Crit. Rev. Microbiol.* **2016**, *42*, 46–56. [CrossRef] [PubMed]

111. Rudramurthy, G.R.; Swamy, M.K.; Sinniah, U.R.; Ghasemzadeh, A. Nanoparticles: Alternatives against drug-resistant pathogenic microbes. *Molecules* **2016**, *21*, 836. [CrossRef] [PubMed]

112. Tashi, T.; Gupta, N.V.; Mbuya, V.B. Silver nanoparticles: Synthesis, mechanism of antimicrobial action, characterization, medical applications, and toxicity effects. *J. Chem. Pharm. Res.* **2016**, *8*, 526–537.

113. Zhang, X.-F.; Shen, W.; Gurunathan, S. Silver nanoparticle-mediated cellular responses in various cell lines: An in vitro model. *Int. J. Mol. Sci.* **2016**, *17*, 1603. [CrossRef] [PubMed]

114. Zhang, H.; Wu, M.; Sen, A. Silver nanoparticle antimicrobials and related materials. In *Nano-Antimicrobials*; Springer: Berlin/Heidelberg, Germany, 2012; pp. 3–45. ISBN 978-3-642-24427-8.

115. Nomiya, K.; Takahashi, S.; Noguchi, R.; Nemoto, S.; Takayama, T.; Oda, M. Synthesis and characterization of water-soluble silver(I) complexes with l-Histidine (H2his) and (S)-(−)-2-Pyrrolidone-5-carboxylic Acid (H2pyrrld) showing a wide spectrum of effective antibacterial and antifungal activities. Crystal structures of chiral helical polymers [Ag(Hhis)]n and {[Ag(Hpyrrld)]2}n in the solid state. *Inorg. Chem.* **2000**, *39*, 3301–3311. [CrossRef] [PubMed]

116. AshaRani, P.V.; Low Kah Mun, G.; Hande, M.P.; Valiyaveettil, S. Cytotoxicity and genotoxicity of silver nanoparticles in human cells. *ACS Nano* **2009**, *3*, 279–290. [CrossRef] [PubMed]

117. Hajipour, M.J.; Fromm, K.M.; Akbar Ashkarran, A.; Jimenez de Aberasturi, D.; De Larramendi, I.R.; Rojo, T.; Serpooshan, V.; Parak, W.J.; Mahmoudi, M. Antibacterial properties of nanoparticles. *Trends Biotechnol.* **2012**, *30*, 499–511. [CrossRef] [PubMed]

118. McShan, D.; Ray, P.C.; Yu, H. Molecular toxicity mechanism of nanosilver. *J. Food Drug Anal.* **2014**, *22*, 116–127. [CrossRef] [PubMed]
119. Li, W.-R.; Sun, T.-L.; Zhou, S.-L.; Ma, Y.-K.; Shi, Q.-S.; Xie, X.-B.; Huang, X.-M. A comparative analysis of antibacterial activity, dynamics, and effects of silver ions and silver nanoparticles against four bacterial strains. *Int. Biodeterior. Biodegrad.* **2017**, *123*, 304–310. [CrossRef]
120. Navarro, E.; Piccapietra, F.; Wagner, B.; Marconi, F.; Kaegi, R.; Odzak, N.; Sigg, L.; Behra, R. Toxicity of silver nanoparticles to chlamydomonas reinhardtii. *Environ. Sci. Technol.* **2008**, *42*, 8959–8964. [CrossRef] [PubMed]
121. Gomaa, E.Z. Silver nanoparticles as an antimicrobial agent: A case study on *Staphylococcus aureus* and *Escherichia coli* as models for Gram-positive and Gram-negative bacteria. *J. Gen. Appl. Microbiol.* **2017**, *63*, 36–43. [CrossRef] [PubMed]
122. Pandey, J.K.; Swarnkar, R.K.; Soumya, K.K.; Dwivedi, P.; Singh, M.K.; Sundaram, S.; Gopal, R. Silver nanoparticles synthesized by pulsed laser ablation: As a potent antibacterial agent for human enteropathogenic gram-positive and gram-negative bacterial strains. *Appl. Biochem. Biotechnol.* **2014**, *174*, 1021–1031. [CrossRef] [PubMed]
123. Pazos-Ortiz, E.; Roque-Ruiz, J.H.; Hinojos-Márquez, E.A.; López-Esparza, J.; Donohué-Cornejo, A.; Cuevas-González, J.C.; Espinosa-Cristóbal, L.F.; Reyes-López, S.Y. Dose-dependent antimicrobial activity of silver nanoparticles on polycaprolactone fibers against gram-positive and gram-negative bacteria. *J. Nanomater.* **2017**, *2017*, 4752314. [CrossRef]
124. Rai, M.K.; Deshmukh, S.D.; Ingle, A.P.; Gade, A.K. Silver nanoparticles: The powerful nanoweapon against multidrug-resistant bacteria. *J. Appl. Microbiol.* **2012**, *112*, 841–852. [CrossRef] [PubMed]
125. Feng, Q.L.; Wu, J.; Chen, G.Q.; Cui, F.Z.; Kim, T.N.; Kim, J.O. A mechanistic study of the antibacterial effect of silver ions on *Escherichia coli* and *Staphylococcus aureus*. *J. Biomed. Mater. Res.* **2000**, *52*, 662–668. [CrossRef]
126. Sondi, I.; Salopek-Sondi, B. Silver nanoparticles as antimicrobial agent: A case study on *E. coli* as a model for Gram-negative bacteria. *J. Colloid Interface Sci.* **2004**, *275*, 177–182. [CrossRef] [PubMed]
127. Pal, S.; Tak, Y.K.; Song, J.M. Does the antibacterial activity of silver nanoparticles depend on the shape of the nanoparticle? a study of the gram-negative bacterium *Escherichia coli*. *Appl. Environ. Microbiol.* **2007**, *73*, 1712–1720. [CrossRef] [PubMed]
128. Butkus, M.A.; Labare, M.P.; Starke, J.A.; Moon, K.; Talbot, M. Use of aqueous silver to enhance inactivation of coliphage MS-2 by UV disinfection. *Appl. Environ. Microbiol.* **2004**, *70*, 2848–2853. [CrossRef] [PubMed]
129. Morones, J.R.; Elechiguerra, J.L.; Camacho, A.; Holt, K.; Kouri, J.B.; Ramírez, J.T.; Yacaman, M.J. The bactericidal effect of silver nanoparticles. *Nanotechnology* **2005**, *16*, 2346–2353. [CrossRef] [PubMed]
130. Yamanaka, M.; Hara, K.; Kudo, J. Bactericidal actions of a silver ion solution on *Escherichia coli*, studied by energy-filtering transmission electron microscopy and proteomic analysis. *Appl. Environ. Microbiol.* **2005**, *71*, 7589–7593. [CrossRef] [PubMed]
131. Ingle, A.; Gade, A.; Pierrat, S.; Sonnichsen, C.; Rai, M. Mycosynthesis of Silver Nanoparticles Using the Fungus *Fusarium acuminatum* and Its Activity Against Some Human Pathogenic Bacteria. Available online: http://www.eurekaselect.com/66921/article (accessed on 22 June 2018).
132. Gajbhiye, M.; Kesharwani, J.; Ingle, A.; Gade, A.; Rai, M. Fungus-mediated synthesis of silver nanoparticles and their activity against pathogenic fungi in combination with fluconazole. *Nanomedicine* **2009**, *5*, 382–386. [CrossRef] [PubMed]
133. Birla, S.S.; Tiwari, V.V.; Gade, A.K.; Ingle, A.P.; Yadav, A.P.; Rai, M.K. Fabrication of silver nanoparticles by Phoma glomerata and its combined effect against *Escherichia coli*, *Pseudomonas aeruginosa* and *Staphylococcus aureus*. *Letters Appl. Microbiol.* **2009**, *48*, 173–179. [CrossRef] [PubMed]
134. Govindaraju, K.; Tamilselvan, S.; Kiruthiga, V.; Singaravelu, G. Biogenic silver nanoparticles by *Solanum torvum* and their promising antimicrobial activity. *J. Biopestic.* **2010**, *3*, 394–399.
135. Namasivayam, S.K.R. Evaluation of anti bacterial activity of biocompatible polymer chitosan coated biogenic silver nanoparticles synthesized from *Klebsiella ornithinolytica*. *BioMedRx* **2013**, *1*, 459–563.
136. El Badawy, A.M.; Silva, R.G.; Morris, B.; Scheckel, K.G.; Suidan, M.T.; Tolaymat, T.M. Surface charge-dependent toxicity of silver nanoparticles. *Environ. Sci. Technol.* **2011**, *45*, 283–287. [CrossRef] [PubMed]
137. Jeong, Y.; Lim, D.W.; Choi, J. Assessment of size-dependent antimicrobial and cytotoxic properties of silver nanoparticles. *Adv. Mater. Sci. Eng.* **2014**, *2014*, 763807. [CrossRef]

138. Kumari, M.; Pandey, S.; Giri, V.P.; Bhattacharya, A.; Shukla, R.; Mishra, A.; Nautiyal, C.S. Tailoring shape and size of biogenic silver nanoparticles to enhance antimicrobial efficacy against MDR bacteria. *Microb. Pathog.* **2017**, *105*, 346–355. [CrossRef] [PubMed]

139. Zook, J.M.; Halter, M.D.; Cleveland, D.; Long, S.E. Disentangling the effects of polymer coatings on silver nanoparticle agglomeration, dissolution, and toxicity to determine mechanisms of nanotoxicity. *J. Nanopart. Res.* **2012**, *14*, 1165. [CrossRef]

140. Störmer, A.; Bott, J.; Kemmer, D.; Franz, R. Critical review of the migration potential of nanoparticles in food contact plastics. *Trends Food Sci. Technol.* **2017**, *63*, 39–50. [CrossRef]

141. Carlson, C.; Hussain, S.M.; Schrand, A.M.; Braydich-Stolle, L.; Hess, K.L.; Jones, R.L.; Schlager, J.J. Unique cellular interaction of silver nanoparticles: Size-dependent generation of reactive oxygen species. *J. Phys. Chem. B* **2008**, *112*, 13608–13619. [CrossRef] [PubMed]

142. Aueviriyavit, S.; Phummiratch, D.; Maniratanachote, R. Mechanistic study on the biological effects of silver and gold nanoparticles in Caco-2 cells—Induction of the Nrf2/HO-1 pathway by high concentrations of silver nanoparticles. *Toxicol. Lett.* **2014**, *224*, 73–83. [CrossRef] [PubMed]

143. Singh, R.P.; Bala, N. Comparative studies of cold and thermal sprayed hydroxyapatite coatings for biomedical applications—A review. In *Biomaterials Science: Processing, Properties and Applications II*; Wiley-Blackwell: Hoboken, NJ, USA, 2012; pp. 249–259. ISBN 978-1-118-51146-6.

144. Wang, J.; Rahman, M.F.; Duhart, H.M.; Newport, G.D.; Patterson, T.A.; Murdock, R.C.; Hussain, S.M.; Schlager, J.J.; Ali, S.F. Expression changes of dopaminergic system-related genes in PC12 cells induced by manganese, silver, or copper nanoparticles. *NeuroToxicology* **2009**, *30*, 926–933. [CrossRef] [PubMed]

145. Park, M.V.D.Z.; Neigh, A.M.; Vermeulen, J.P.; de la Fonteyne, L.J.J.; Verharen, H.W.; Briedé, J.J.; van Loveren, H.; de Jong, W.H. The effect of particle size on the cytotoxicity, inflammation, developmental toxicity and genotoxicity of silver nanoparticles. *Biomaterials* **2011**, *32*, 9810–9817. [CrossRef] [PubMed]

146. Perito, B.; Giorgetti, E.; Marsili, P.; Muniz-Miranda, M. Antibacterial activity of silver nanoparticles obtained by pulsed laser ablation in pure water and in chloride solution. *Beilstein J. Nanotechnol.* **2016**, *7*, 465–473. [CrossRef] [PubMed]

147. Agnihotri, S.; Mukherji, S.; Mukherji, S. Size-controlled silver nanoparticles synthesized over the range 5–100 nm using the same protocol and their antibacterial efficacy. *RSC Adv.* **2013**, *4*, 3974–3983. [CrossRef]

148. Lu, Z.; Rong, K.; Li, J.; Yang, H.; Chen, R. Size-dependent antibacterial activities of silver nanoparticles against oral anaerobic pathogenic bacteria. *J. Mater. Sci.* **2013**, *24*, 1465–1471. [CrossRef] [PubMed]

149. Martínez-Castañón, G.A.; Niño-Martínez, N.; Martínez-Gutierrez, F.; Martínez-Mendoza, J.R.; Ruiz, F. Synthesis and antibacterial activity of silver nanoparticles with different sizes. *J. Nanopart. Res.* **2008**, *10*, 1343–1348. [CrossRef]

150. Kőrösi, L.; Rodio, M.; Dömötör, D.; Kovács, T.; Papp, S.; Diaspro, A.; Intartaglia, R.; Beke, S. Ultrasmall, ligand-free Ag nanoparticles with high antibacterial activity prepared by pulsed laser ablation in liquid. *J. Chem.* **2016**, *2016*, 4143560. [CrossRef]

151. Korshed, P.; Li, L.; Liu, Z.; Wang, T. The molecular mechanisms of the antibacterial effect of picosecond laser generated silver nanoparticles and their toxicity to human cells. *PLoS ONE* **2016**, *11*, e0160078. [CrossRef] [PubMed]

152. Zafar, N.; Shamaila, S.; Nazir, J.; Sharif, R.; Shahid Rafique, M.; Ul-Hasan, J.; Ammara, S.; Khalid, H. Antibacterial action of chemically synthesized and laser generated silver nanoparticles against human pathogenic bacteria. *J. Mater. Sci. Technol.* **2016**, *32*, 721–728. [CrossRef]

153. Alam, M.N.; Roy, N.; Mandal, D.; Begum, N.A. Green chemistry for nanochemistry: Exploring medicinal plants for the biogenic synthesis of metal NPs with fine-tuned properties. *RSC Adv.* **2013**, *3*, 11935–11956. [CrossRef]

154. Sportelli, M.C.; Picca, R.A.; Cioffi, N. Recent advances in the synthesis and characterization of nano-antimicrobials. *Trends Anal. Chem.* **2016**, *84*, 131–138. [CrossRef]

155. Shih, C.-Y.; Streubel, R.; Heberle, J.; Letzel, A.; Shugaev, M.V.; Wu, C.; Schmidt, M.; Gökce, B.; Barcikowski, S.; Zhigilei, L.V. Two mechanisms of nanoparticle generation in picosecond laser ablation in liquids: The origin of the bimodal size distribution. *Nanoscale* **2018**, *10*, 6900–6910. [CrossRef] [PubMed]

156. Zhang, D.; Gökce, B.; Barcikowski, S. Laser synthesis and processing of colloids: Fundamentals and applications. *Chem. Rev.* **2017**, *117*, 3990–4103. [CrossRef] [PubMed]

157. Simakin, A.V.; Voronov, V.V.; Shafeev, G.A.; Brayner, R.; Bozon-Verduraz, F. Nanodisks of Au and Ag produced by laser ablation in liquid environment. *Chem. Phys. Lett.* **2001**, *348*, 182–186. [CrossRef]

158. Palazzo, G.; Valenza, G.; Dell'Aglio, M.; De Giacomo, A. On the stability of gold nanoparticles synthesized by laser ablation in liquids. *J. Colloid Interface Sci.* **2017**, *489*, 47–56. [CrossRef] [PubMed]

159. Mafuné, F.; Kohno, J.; Takeda, Y.; Kondow, T.; Sawabe, H. Structure and stability of silver nanoparticles in aqueous solution produced by laser ablation. *J. Phys. Chem. B* **2000**, *104*, 8333–8337. [CrossRef]

160. Mafuné, F.; Kondow, T. Selective laser fabrication of small nanoparticles and nano-networks in solution by irradiation of UV pulsed laser onto platinum nanoparticles. *Chem. Phys. Lett.* **2004**, *383*, 343–347. [CrossRef]

161. Mafuné, F.; Kohno, J.; Takeda, Y.; Kondow, T.; Sawabe, H. Formation and size control of silver nanoparticles by laser ablation in aqueous solution. *J. Phys. Chem. B* **2000**, *104*, 9111–9117. [CrossRef]

162. Kruusing, A. Underwater and water-assisted laser processing: Part 1—General features, steam cleaning and shock processing. *Opt. Lasers Eng.* **2004**, *41*, 307–327. [CrossRef]

163. Oseguera-Galindo, D.O.; Martínez-Benítez, A.; Chávez-Chávez, A.; Gómez-Rosas, G.; Pérez-Centeno, A.; Santana-Aranda, M.A. Effects of the confining solvent on the size distribution of silver NPs by laser ablation. *J. Nanopart. Res.* **2012**, *14*, 1133. [CrossRef]

164. Werner, D.; Hashimoto, S.; Tomita, T.; Matsuo, S.; Makita, Y. Examination of silver nanoparticle fabrication by pulsed-laser ablation of flakes in primary alcohols. *J. Phys. Chem. C* **2008**, *112*, 1321–1329. [CrossRef]

165. Moura, C.G.; Pereira, R.S.F.; Andritschky, M.; Lopes, A.L.B.; de Freitas Grilo, J.P.; do Nascimento, R.M.; Silva, F.S. Effects of laser fluence and liquid media on preparation of small Ag nanoparticles by laser ablation in liquid. *Opt. Laser Technol.* **2017**, *97*, 20–28. [CrossRef]

166. Amendola, V.; Meneghetti, M. What controls the composition and the structure of nanomaterials generated by laser ablation in liquid solution? *Phys. Chem. Chem. Phys.* **2013**, *15*, 3027–3046. [CrossRef] [PubMed]

167. Kalus, M.-R.; Bärsch, N.; Streubel, R.; Gökce, E.; Barcikowski, S.; Gökce, B. How persistent microbubbles shield nanoparticle productivity in laser synthesis of colloids—Quantification of their volume, dwell dynamics, and gas composition. *Phys. Chem. Chem. Phys.* **2017**, *19*, 7112–7123. [CrossRef] [PubMed]

168. Tajdidzadeh, M.; Azmi, B.Z.; Yunus, W.M.M.; Talib, Z.A.; Sadrolhosseini, A.R.; Karimzadeh, K.; Gene, S.A.; Dorraj, M. Synthesis of silver nanoparticles dispersed in various aqueous media using laser ablation. *Sci. World J.* **2014**, *2014*, 324921. [CrossRef] [PubMed]

169. Hao, E.; Schatz, G.C. Electromagnetic fields around silver nanoparticles and dimers. *J. Chem. Phys.* **2003**, *120*, 357–366. [CrossRef] [PubMed]

170. Al-Azawi, M.A.; Bidin, N.; Bououdina, M.; Abbas, K.N.; Al-Asedy, H.J.; Ahmed, O.H.; Thahe, A.A. The effects of the ambient liquid medium on the ablation efficiency, size and stability of silver nanoparticles prepared by pulse laser ablation in liquid technique. *J. Teknol.* **2016**, *78*, 7–11. [CrossRef]

171. Ganeev, R.A.; Baba, M.; Ryasnyanskii, A.I.; Suzuki, M.; Kuroda, H. Laser ablation of silver in different liquids: Optical and nonlinear optical properties of silver nanoparticles. *Opt. Spectrosc.* **2005**, *99*, 668–676. [CrossRef]

172. Hamad, A.; Li, L.; Liu, Z. A comparison of the characteristics of nanosecond, picosecond and femtosecond lasers generated Ag, TiO$_2$ and Au nanoparticles in deionised water. *Appl. Phys. A* **2015**, *120*, 1247–1260. [CrossRef]

173. Tsuji, T.; Kakita, T.; Tsuji, M. Preparation of nano-size particles of silver with femtosecond laser ablation in water. *Appl. Surface Sci.* **2003**, *206*, 314–320. [CrossRef]

174. Hamad, A.; Li, L.; Liu, Z. Comparison of characteristics of selected metallic and metal oxide nanoparticles produced by picosecond laser ablation at 532 and 1064 nm wavelengths. *Appl. Phys. A* **2016**, *122*, 904. [CrossRef]

175. Solati, E.; Mashayekh, M.; Dorranian, D. Effects of laser pulse wavelength and laser fluence on the characteristics of silver nanoparticle generated by laser ablation. *Appl. Phys. A* **2013**, *112*, 689–694. [CrossRef]

176. Tsuji, T.; Iryo, K.; Nishimura, Y.; Tsuji, M. Preparation of metal colloids by a laser ablation technique in solution: Influence of laser wavelength on the ablation efficiency (II). *J. Photochem. Photobiol. A Chem.* **2001**, *145*, 201–207. [CrossRef]

177. Tsuji, T.; Iryo, K.; Watanabe, N.; Tsuji, M. Preparation of silver nanoparticles by laser ablation in solution: Influence of laser wavelength on particle size. *Appl. Surf. Sci.* **2002**, *202*, 80–85. [CrossRef]

178. Reich, S.; Schönfeld, P.; Letzel, A.; Kohsakowski, S.; Olbinado, M.; Gökce, B.; Barcikowski, S.; Plech, A. Fluence threshold behaviour on ablation and bubble formation in pulsed laser ablation in liquids. *ChemPhysChem* **2017**, *18*, 1084–1090. [CrossRef] [PubMed]

179. Dorranian, D.; Tajmir, S.; Khazanehfar, F. Effect of laser fluence on the characteristics of Ag nanoparticles produced by laser ablation. *Soft Nanosci. Lett.* **2013**, *3*, 93–100. [CrossRef]
180. Mahdieh, M.H.; Fattahi, B. Size properties of colloidal nanoparticles produced by nanosecond pulsed laser ablation and studying the effects of liquid medium and laser fluence. *Appl. Surf. Sci.* **2015**, *329*, 47–57. [CrossRef]
181. Nikolov, A.S.; Nedyalkov, N.N.; Nikov, R.G.; Atanasov, P.A.; Alexandrov, M.T.; Karashanova, D.B. Investigation of Ag nanoparticles produced by nanosecond pulsed laser ablation in water. *Appl. Phys. A* **2012**, *109*, 315–322. [CrossRef]
182. Nikolov, A.S.; Nedyalkov, N.N.; Nikov, R.G.; Atanasov, P.A.; Alexandrov, M.T. Characterization of Ag and Au nanoparticles created by nanosecond pulsed laser ablation in double distilled water. *Appl. Surf. Sci.* **2011**, *257*, 5278–5282. [CrossRef]
183. Valverde-Alva, M.A.; García-Fernández, T.; Villagrán-Muniz, M.; Sánchez-Aké, C.; Castañeda-Guzmán, R.; Esparza-Alegría, E.; Sánchez-Valdés, C.F.; Llamazares, J.L.S.; Herrera, C.E.M. Synthesis of silver nanoparticles by laser ablation in ethanol: A pulsed photoacoustic study. *Appl. Surf. Sci.* **2015**, *355*, 341–349. [CrossRef]
184. Zamiri, R.; Zakaria, A.; Ahangar, H.A.; Darroudi, M.; Zamiri, G.; Rizwan, Z.; Drummen, G.P.C. The effect of laser repetition rate on the LASiS synthesis of biocompatible silver nanoparticles in aqueous starch solution. *Int. J. Nanomed.* **2013**, *8*, 233–244. [CrossRef]
185. Menéndez-Manjón, A.; Barcikowski, S. Hydrodynamic size distribution of gold nanoparticles controlled by repetition rate during pulsed laser ablation in water. *Appl. Surf. Sci.* **2011**, *257*, 4285–4290. [CrossRef]
186. Hosseini, H.; Shojaee-Aliabadi, S.; Hosseini, S.M.; Mirmoghtadaie, L. Nanoantimicrobials in Food Industry. In *Nanotechnology Applications in Food*; Oprea, A.E., Grumezescu, A.M., Eds.; Academic Press: Cambridge, MA, USA, 2017; Chapter 11; pp. 223–243. ISBN 978-0-12-811942-6.
187. Costa, C.; Conte, A.; Alessandro, M.; Nobile, D. Use of Metal Nanoparticles for Active Packaging Applications. In *Antimicrobial Food Packaging*; Barros-Velázquez, J., Ed.; Academic Press: San Diego, CA, USA, 2016; Chapter 31; pp. 399–406. ISBN 978-0-12-800723-5.
188. Costa, C.; Conte, A.; Buonocore, G.G.; Del Nobile, M.A. Antimicrobial silver-montmorillonite nanoparticles to prolong the shelf life of fresh fruit salad. *Int. J. Food Microbiol.* **2011**, *148*, 164–167. [CrossRef] [PubMed]
189. Sivakumar, P.; Sivakumar, P.; Anbarasu, K.; Pandian, K.; Renganathan, S. Synthesis of silver nanorods from food industrial waste and their application in improving the keeping quality of milk. *Ind. Eng. Chem. Res.* **2013**, *52*, 17676–17681. [CrossRef]
190. Maynard, A. *The Nanotechnology Consumer Products Inventory*; Woodrow Wilson International Center for Scholars: Washington, DC, USA, 2006; Volume 10.
191. Bouwmeester, H.; Dekkers, S.; Noordam, M.; Hagens, W.; Bulder, A.; De Heer, C.; ten Voorde, S.E.C.G.; Wijnhoven, S.; Sips, A. *Health Impact of Nanotechnologies in Food Production*; Netherlands National Institute for Public Health and the Environment: Bilthoven, The Netherlands, 2007.
192. Von Goetz, N.; Fabricius, L.; Glaus, R.; Weitbrecht, V.; Günther, D.; Hungerbühler, K. Migration of silver from commercial plastic food containers and implications for consumer exposure assessment. *Food Addit. Contamin. Part A* **2013**, *30*, 612–620. [CrossRef] [PubMed]
193. Echegoyen, Y.; Nerín, C. Nanoparticle release from nano-silver antimicrobial food containers. *Food Chem. Toxicol.* **2013**, *62*, 16–22. [CrossRef] [PubMed]
194. Nerin, C.; Silva, F.; Manso, S.; Becerril, R. The downside of antimicrobial packaging: Migration of packaging elements into food. In *Antimicrobial Food Packaging*; Barros-Velázquez, J., Ed.; Academic Press: San Diego, CA, USA, 2016; Chapter 6; pp. 81–93. ISBN 978-0-12-800723-5.
195. Biji, K.B.; Ravishankar, C.N.; Mohan, C.O.; Gopal, T.K.S. Smart packaging systems for food applications: A review. *J. Food Sci. Technol.* **2015**, *52*, 6125–6135. [CrossRef] [PubMed]
196. Castro-Mayorga, J.L.; Martínez-Abad, A.; Fabra, M.F.; Lagarón, J.M.; Ocio, M.J.; Sánchez, G. Silver-Based Antibacterial and Virucide Biopolymers: Usage and potential in antimicrobial packaging. In *Antimicrobial Food Packaging*; Barros-Velázquez, J., Ed.; Academic Press: San Diego, CA, USA, 2016; Chapter 32; pp. 407–416. ISBN 978-0-12-800723-5.
197. European Food Safety Authority (EFSA). Opinion of the Scientific Panel on food additives, flavourings, processing aids and materials in contact with food (AFC) on a request related to a 17th list of substances for food contact materials: Opinion of the scientific panel on food additives, flavourings, processing aids and materials in contact with food (AFC) on a request related to a 17t. *EFSA J.* **2008**, *6*, 601. [CrossRef]

198. Ranadheera, C.S.; Prasanna, P.H.P.; Vidanarachchi, J.K.; McConchie, R.; Naumovski, N.; Mellor, D. Nanotechnology in Microbial Food Safety. In *Nanotechnology Applications in Food*; Oprea, A.E., Grumezescu, A.M., Eds.; Academic Press: San Diego, CA, USA, 2017; Chapter 12; pp. 245–265. ISBN 978-0-12-811942-6.
199. Rossi, M.; Cubadda, F.; Dini, L.; Terranova, M.L.; Aureli, F.; Sorbo, A.; Passeri, D. Scientific basis of nanotechnology, implications for the food sector and future trends. *Trends Food Sci. Technol.* **2014**, *40*, 127–148. [CrossRef]

 *antibiotics*

Article

Tunable Silver-Functionalized Porous Frameworks for Antibacterial Applications

Mark A. Isaacs [1], Brunella Barbero [2], Lee J. Durndell [3], Anthony C. Hilton [4], Luca Olivi [5], Christopher M. A. Parlett [2], Karen Wilson [6,*] and Adam F. Lee [6,*]

1   Department of Chemistry, University College London, London WC1H 0AJ, UK; mark.isaacs@ucl.ac.uk
2   European Bioenergy Research Institute, Aston University, Birmingham B4 7ET, UK;
    brubarbero@gmail.com (B.B.); chrisparlett81@gmail.com (C.M.A.P.)
3   Inorganic Chemistry and Catalysis, Debye Institute for Nanomaterials Science, Utrect University,
    Universiteitsweg 99, 3584 CG Utrecht, The Netherlands; l.j.durndell@uu.nl
4   Life and Health Sciences, Aston University, Birmingham B4 7ET, UK; a.c.hilton@aston.ac.uk
5   Sincrotrone Trieste, 34149 Basovizza, Italy; luca.olivi@elettra.eu
6   School of Science, RMIT University, Melbourne, VIC 3001, Australia
*   Correspondence: karen.wilson2@rmit.edu.au (K.W.); adam.lee2@rmit.edu.au (A.F.L.);
    Tel.: +61-3-9925-2623 (A.F.L.)

Received: 22 June 2018; Accepted: 2 July 2018; Published: 3 July 2018

Abstract: Healthcare-associated infections and the rise of drug-resistant bacteria pose significant challenges to existing antibiotic therapies. Silver nanocomposites are a promising solution to the current crisis, however their therapeutic application requires improved understanding of underpinning structure-function relationships. A family of chemically and structurally modified mesoporous SBA-15 silicas were synthesized as porous host matrices to tune the physicochemical properties of silver nanoparticles. Physicochemical characterization by transmission electron microscopy (TEM), X-ray diffraction (XRD), X-ray photoelectron spectroscopy (XPS), X-ray absorption near-edge spectroscopy (XANES) and porosimetry demonstrate that functionalization by a titania monolayer and the incorporation of macroporosity both increase silver nanoparticle dispersion throughout the silica matrix, thereby promoting Ag_2CO_3 formation and the release of ionic silver in simulated tissue fluid. The Ag_2CO_3 concentration within functionalized porous architectures is a strong predictor for antibacterial efficacy against a broad spectrum of pathogens, including *C. difficile* and methicillin-resistant Staphylococcus aureus (MRSA).

Keywords: silver; antibacterial; titania; mesoporous; macroporous; surface functionalization

1. Introduction

Healthcare-associated infections (HCAIs) impact on millions of patients annually, and hundreds of thousands of deaths worldwide [1–4]. Current models predict that global mortality rates will approach 10 million annual deaths as a direct result of HCAIs by 2050 [5]. Approximately one-third of all such HCAIs are preventable [6], and while recent improvements in hygiene standards across the medical sector have for example significantly reduced the number of HCAIs per year in the UK, certain groups remain at very high risk of infection during hospital visits, and infection rates for some bacteria such as *Escherichia coli* have even increased [7–9].

The rise of drug-resistant bacteria which are disproportionately responsible for HCAIs, such as methicillin-resistant *Staphylococcus aureus* (MRSA) and extended-spectrum beta-lactamases (ESBL)-producing *Enterobacteriaceae* [2], has prompted a resurgence in the search for broad-spectrum antibiotics to tackle HCAIs. Silver is one such broad spectrum antibiotic, effective against Gram-positive (e.g., MRSA, *B. subtilis*) and Gram-negative bacteria (e.g., *E. coli*, *P. aeruginosa*) [10], and is consequently the subject of commercial [11] and academic interest for the treatment of existing HCAIs and drug-resistant organisms through its incorporation in surface coatings [12], nanotechnology [13–16], and pharmaceuticals [17–19].

A major advantage of nanoparticulate silver (and silver-based solid compounds) compared with salts or complexes is that oxidative dissolution is confined to exposed surfaces, offering a potential route to regulate the release of biologically active (commonly held to be Ag^+) [20] species permitting the design of long lasting antimicrobial therapies [15,21–23]. Control over silver dissolution kinetics in Ag-nanoparticle-(NP) based systems has been investigated through tuning particle size [23–26], the use of inert coatings to retard release [26], or control of silver speciation in nanocomposite to accelerate dissolution and hence antibacterial activity [13,14]. Silver has also shown recent promise as a promoter of conventional antibiotic therapies [27].

Here we explore the use of high surface area and porous metal oxide frameworks to control the dispersion and subsequent dissolution of surface silver. The impact of framework porosity and oxide termination on silver speciation, dissolution, and antibacterical efficacy, was studied using ordered mesoporous and hierarchical macroporous-mesoporous SBA-15 silicas [28,29], and titania functionalized analogues. Conformal titania surface coatings, and hierarchically porous architectures promote silver NP dissolution and activity against Gram-positive (*Staphylococcus aureus*) and Gram-negative (*Pseudomonas aeruginosa*) bacteria.

2. Results and Discussion

2.1. Materials Synthesis and Characterisation

Three families of ordered porous frameworks were prepared using soft or hard-soft dual templating approaches and subsequent titania functionalization (Figure 1). Mesoporous and macroporous-mesoporous (MM) SBA-15 silicas were prepared using a Pluronic 123 (P123) surfactant either alone or in conjunction with 400 nm colloidal polystyrene nanospheres respectively, according to literature methods. Titania modified variants were subsequently prepared adapting the procedure of Landau et al. [30] to preactivate surface silanols before the self-limited grafting of titanium isopropoxide under anhydrous conditions, and calcination. Repeated grafting cycles were used to create conformal TiO_2 monolayers encapsulating the silica supports. Silver nanoparticles were introduced to either the parent SBA-15, or TiO_2-grafted SBA-15 and TiO_2-grafted MM-SBA-15, frameworks by wet impregnation with varying silver nitrate concentrations.

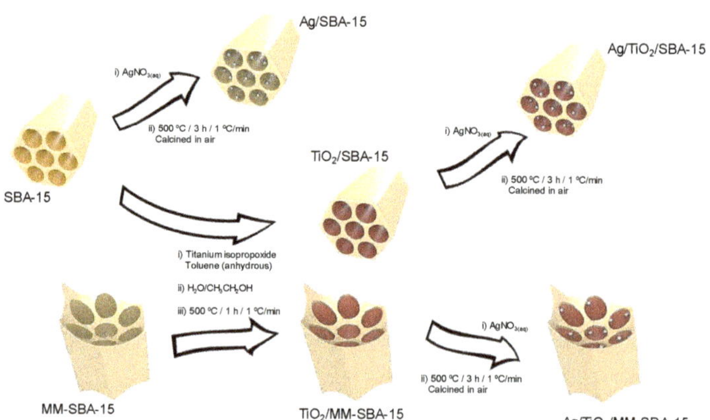

Figure 1. Synthesis of Ag-doped mesoporous silica and titania-functionalized mesoporous and mesoporous-macroporous silica materials.

Textural properties of the parent SBA-15 and MM-SBA-15 supports, before and after titania functionalization, were first characterized by transmission electronic microscopy (TEM), N_2 porosimetry, X-ray diffraction (XRD), and X-ray photoelectron spectroscopy (XPS). The expected hexagonally close-packed channels of the SBA-15 mesopore network were observed in all cases, with surface areas and mesopore diameters decreasing with successive TiO_2 grafting cycle (Figure 2 and Figure S1), reflecting pore narrowing and partial blockage of micropores in the silica framework [31]. Uniform mesopores of 6.3 nm diameter were in accordance with previous reports [29]. No titania crystallites were visible by TEM following grafting, indicating the formation of highly dispersed phase. Assuming a TiO_2 monolayer thickness of 0.355 nm [32], mesopore shrinkage across three grafting cycles from 6.3 to 5.3 nm for SBA-15 was consistent with deposition of a 1.4 monolayer (ML) titania coating, and from 6 to 5.2 nm for MM-SBA-15 consistent with 1.1 ML titania.

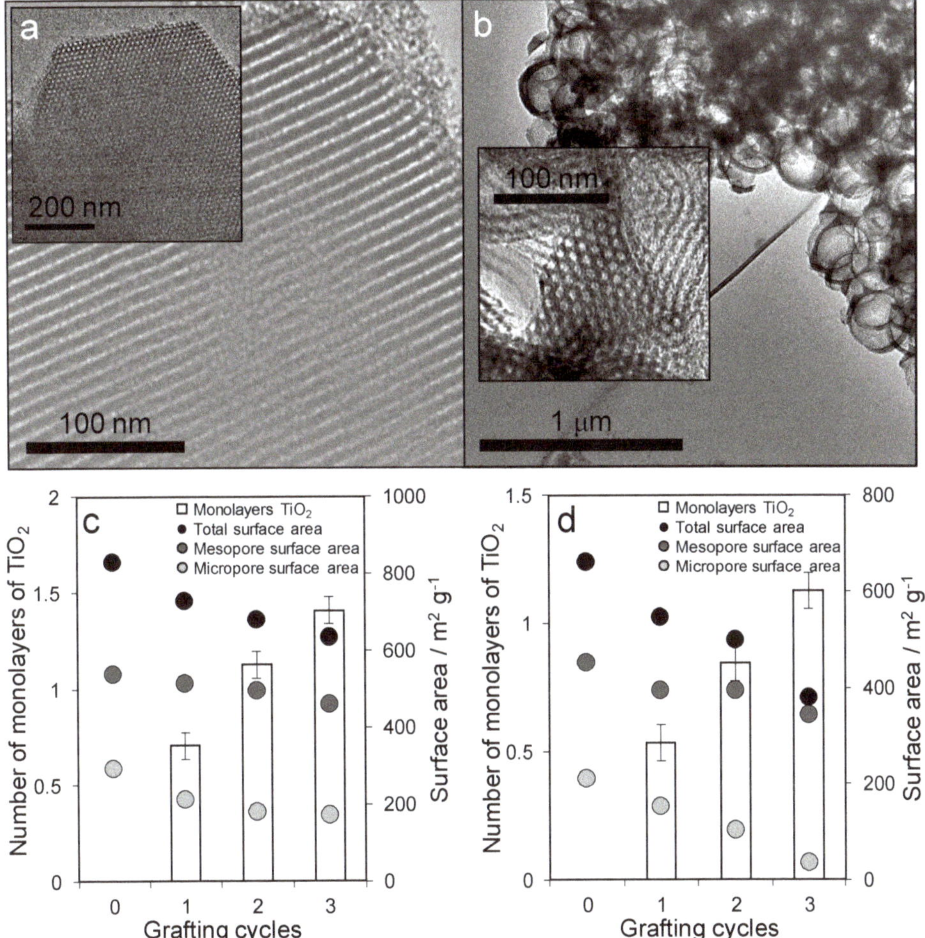

Figure 2. Bright-field transmission electron microscopy (TEM) images of (**a**) TiO_2/SBA-15 (3rd cycle grafting), and (**b**) TiO_2/MM-SBA-15 (3rd cycle grafting) highlighting ordered hexagonally close-packed mesopores and uniform macropores, and the evolution of textural properties and calculated TiO_2 monolayer thickness as a function of grafting cycle for (**c**) TiO_2/SBA-15 and (**d**) TiO_2/MM-SBA-15.

Low-angle X-ray diffraction (XRD) (Figure 3) confirmed that titania-functionalized frameworks retained the *P6mm* symmetry of the parent SBA-15 mesopore network. Wide angle XRD (Figure S2) showed no evidence for crystalline titania phases, consistent with TEM and the formation of conformal monolayers over the SBA-15 template. The surface chemical environment of Ti atoms within the monolayer coatings was examined by XPS, which reveals the presence of Ti-O-Si (associated with the silica-titania interface) and Ti-O-Ti species over both SBA-15 and MM-SBA-15 with Ti $2p_{5/2}$ binding energies of 459.9 eV and 458.5 eV respectively. The shift to higher binding energy for Ti coordinated to Si (through a bridging oxygen) reflects the higher electronegativity of the latter, and hence higher induced initial charge on the former. We attribute the small shift between the Ti-O-Ti environments in P25 and the titania monolayers to quantum size (initial and/or final state) effects.

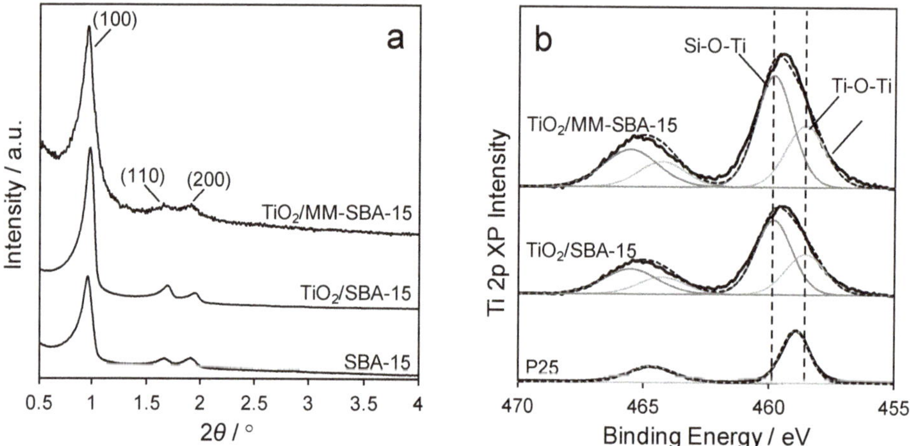

Figure 3. (**a**) Low angle X-ray diffraction (XRD), and (**b**) fitted Ti 2p XP spectra of TiO$_2$/SBA-15 and TiO$_2$/MM-SBA-15, alongside pure silica (SBA-15) and titania (P25) references.

Silver was subsequently deposited over the unfunctionalized SBA-15, and TiO$_2$/SBA-15 and TiO$_2$/MM-SBA-15 at three different nominal loadings spanning 0.3–3 wt% (see the Supplementary Materials (Table S1); elemental analysis revealed almost identical bulk loadings for the mesoporous materials, but systematically lower loadings for the hierarchical material. Surface and bulk silver loadings were generally in good agreement for all materials (see the Supplementary Materials Table S1), indicating a homogenous distribution throughout the porous frameworks. Physicochemical properties of the resulting silver species were determined by TEM, XPS, XRD, and X-ray absorption near edge spectroscopy (XANES). Only fcc metallic silver (JCPDS no. 04-0783) reflections were observed by XRD, which sharpened with increasing loading (see the Supplementary Materials Figure S3), indicative of crystallite growth. Volume averaged silver NP sizes were smaller for MM-SBA-15 (3.5–6.5 nm) than SBA-15 (4.5–8.5 nm) at comparable loadings (~0.3 wt%), evidencing higher metal dispersion over the hierarchically porous framework, and indeed fell below the threshold of detection for the lowest 0.25 wt% Ag/TiO$_2$/MM-SBA-15. Ag NPs were directly visualized by TEM (see the Supplementary Materials Figures S4–S6), which revealed a relatively broad size distribution spanning 1–20 nm over all three supports. The mean particle size increased (and size distribution narrowed) with Ag loading (see the Supplementary Materials Figure S7), with the smallest NPs observed over the hierarchical support in all cases, in accordance with XRD. Ag 3d XP spectra showed the existence of two distinct chemical environments with $3d_{5/2}$ spin-orbit split component binding energies of 368.0 eV and 370.5 eV consistent with Ag^0 and possibly a surface Ag$_2$CO$_3$ respectively (see the Supplementary Materials Figure S8) [33].

Note that discrimination of silver oxides and carbonate by XPS alone is complicated due to their similar core-level binding energies, and anomalous negative binding energy shift of the oxides relative to the metal [14,34,35] and therefore requires careful fitting to well-defined standards (difficult to guarantee due to the tendency of Ag_2O to oxidise, and of both oxides to adsorb atmospheric CO_2 and form surface carbonates [36]), or Auger parameter analysis employing the Ag $M_{4,5}NN$ Auger transition in conjunction with $3d_{5/2}$ core-level spectra [13]. Silver speciation as metal, oxide, or carbonate was therefore examined by linear combination fitting of Ag K-edge X-ray absorption near-edge spectroscopy (XANES, see the Supplementary Materials Figures S9–S11). In all cases, good spectral fits could only be obtained using Ag and Ag_2CO_3 components. Nanoparticle growth is accompanied by a continuous decrease in the proportion of surface and bulk silver carbonate for all frameworks (Figure 4). This switchover from electron deficient to metallic silver with particle size likely reflects the higher surface energy of the latter [37], which is thus favoured for larger particles. The bulk Ag_2CO_3 content determined by XANES was systematically higher than that of the surface determined by XPS, likely reflecting the distribution of particle sizes present in all materials; XPS is expected to more sensitive to larger (more metallic) particles preferentially located on the external surface. AgO and Ag_2O could not be detected by linear combination fitting of the XANES data.

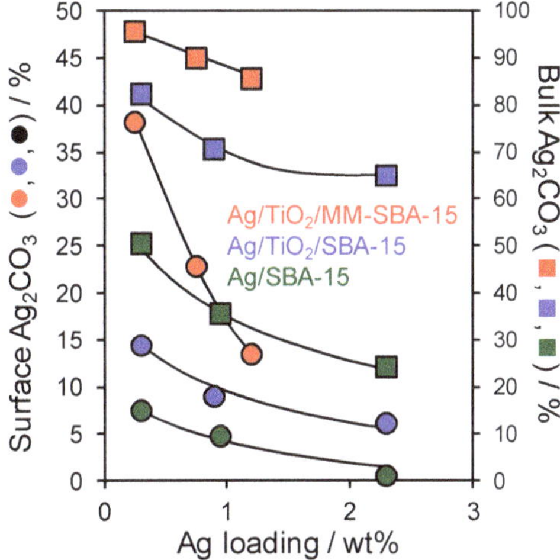

Figure 4. Surface (XPS) and bulk (XANES) silver speciation as Ag_2CO_3 of Ag/SBA-15, Ag/TiO_2/SBA-15, and Ag/TiO_2/MM-SBA-15 as a function of Ag loading.

2.2. Materials Performance Assaying

The release rate of ionic silver (Ag^+) from the preceding SBA-15 and MM-SBA-15 frameworks in a 0.5 M $NaNO_3$ solution at 37 °C was subsequently determined by ICP-MS analysis (Figure 5). Dissolution rates normalised to the mass of silver were inversely proportional to particle size (loading) for all frameworks, as reported for citrate stabilized silver metal nanoparticles [23] and core-shell silver-silica nanoparticles [26] indicating that the release of ionic silver occurred by a common mechanism, being solely dependent on the geometric surface area of the silver nanoparticles. Coating of the mesoporous SBA-15 by a conformal titania monolayer slightly decreased the average particle size over the bare silica framework, and hence increased the rate of silver dissolution, a phenomenon further enhanced by the introduction of macropores. The rate constant for Ag^+ dissolution was determined as 5244 $\mu mol \cdot h^{-1} \cdot nm^{-1} \cdot g^{1}_{Ag}$ by fitting the data (see the Supplementary Materials Figure S12) in accordance with the method of Zhang et al. [23] This is significantly higher than that found for Ag@SiO_2 core-shell nanoparticles (14.7 $\mu mol \cdot h^{-1} \cdot nm^{-1} \cdot g^{-1}_{Ag}$) [26], while the absolute release rate of the 0.3 wt% Ag/TiO$_2$/MM-SBA-15 is consistent with that previously observed for 0.05 wt% high surface area Ag-hydroxyapatite nanoparticles [14].

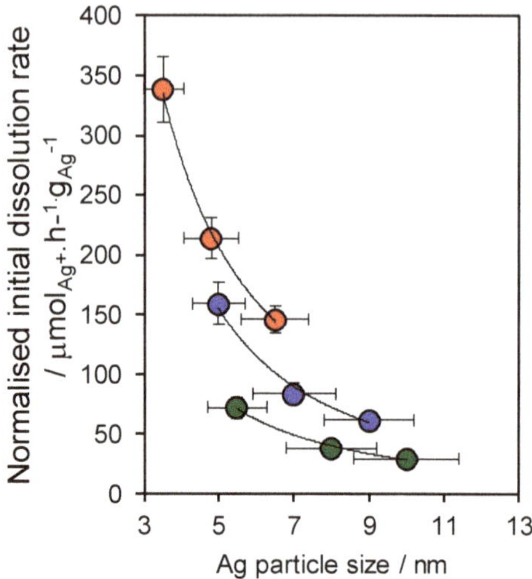

Figure 5. Ag$^+$ dissolution rates normalized to mass of Ag determined by XANES for Ag/SBA-15, Ag/TiO$_2$/SBA-15 and Ag/TiO$_2$/MM-SBA-15.

Antibacterial activity was subsequently assessed against a range of bacteria (*Pseudomonas aeruginosa, Staphylococcus aureus, Clostridium difficile, Escherichia coli, Bacillus subtilis* and MRSA) by zone plate inhibition (ZoI) (Figure 6) using simulated body fluid (SBF), in which the antimicrobial efficacy is proportional to the size of the colony-free zone surrounding the silver functionalized porous frameworks. None of the parent (silver-free) materials exhibited bactericidal properties. In all cases the ZoI values normalized to the mass of silver were inversely proportional to silver loading, mirroring the ionic silver release rates, and superior to those reported for hydroxyapatite (HA) supported Ag$_3$PO$_4$ composites or Ag@SiO$_2$ core-shell nanocomposites. For example, for *S. aureus* Ag/TiO$_2$/MM-SBA-15 achieved 70 mm·g$_{Ag}$$^{-1}$ versus 40–60 mm·g$_{Ag}$$^{-1}$ for a high area Ag-HA [14] or 23 mm·g$_{Ag}$$^{-1}$ for 4.5 nm Ag nanoparticles encapsulated by a 10.5 nm mesoporous silica shell [26]. The zone size (antibacterial performance) decreased in the sequence Ag/TiO$_2$/MM-SBA-15 > Ag/TiO$_2$/SBA-15 > Ag/SBA-15.

The logarithmic reduction method was also employed to obtain a more quantitative evaluation of antibacterial activity for representative Gram-positive and Gram-negative bacteria (*Staphylococcus aureus* and *Pseudomonas aeruginosa* respectively). Experiments were performed in the absence of light to eliminate any possible influence from photogenerated reactive oxygen species by the semiconducting titania coating. All three frameworks were inactive in the absence of silver (see the Supplementary Materials Figure S13). Corresponding decimal reduction times (D-values), i.e., the time to kill 90% of the bacteria, normalised to the mass of silver, also evidenced an inverse relationship with silver loading for all frameworks (Figure 7) consistent with their Ag$^+$ release rates. (Figures S14–S15). As for the ZoI assays, antibacterial activity followed the order Ag/TiO$_2$/MM-SBA-15 > Ag/TiO$_2$/SBA-15 > Ag/SBA-15, i.e., titania functionalization, and the introduction of macroporosity, both enhanced the potency of silver nanoparticles dispersed throughout an SBA-15 framework. Ag/MM-TiO$_2$/SBA-15 outperformed Ag/SBA-15 by ~105% against *Staphylococcus aureus* and ~103% against *Pseudomonas aeruginosa*. The present observations are consistent with previous reports of augmented antimicrobial efficacy arising from the preferential formation of highly soluble silver carbonate versus silver metal over γ-alumina supports [13].

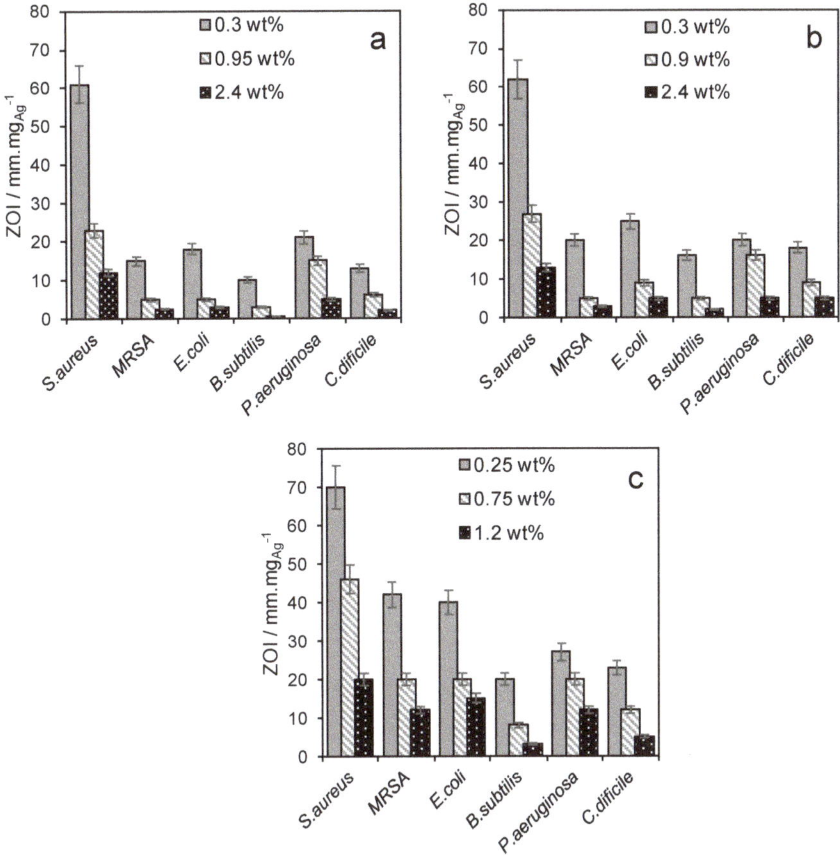

Figure 6. Zone of inhibition plots, normalized to mass of Ag, against a range of Gram-positive and Gram-negative bacteria for (**a**) Ag/SBA-15, (**b**) Ag/TiO$_2$/SBA-15 and (**c**) Ag/TiO$_2$/MM-SBA-15.

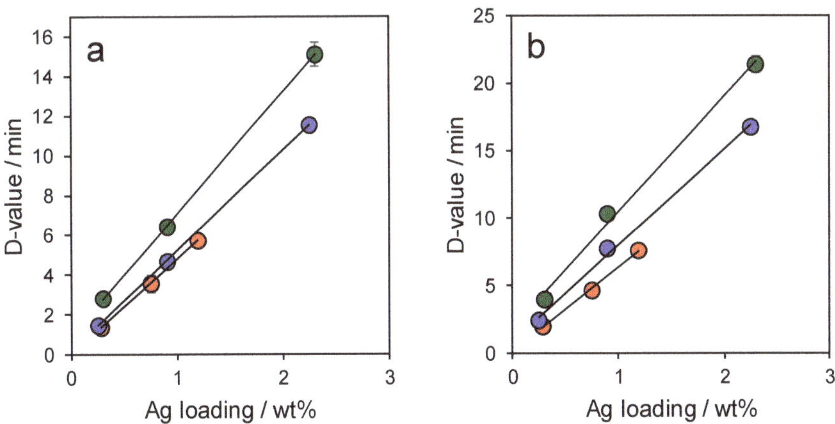

Figure 7. Decimal reduction times for Ag/TiO$_2$/MM-SBA-15, Ag/TiO$_2$/SBA-15 and Ag/SBA-15 against (**a**) *Staphylococcus aureus*, and (**b**) *Pseudomonas aeruginosa*.

3. Conclusions

Families of silver functionalized porous frameworks based around surfactant templated SBA-15 mesoporous silica were synthesized to investigate the impact of surface chemistry and pore architecture on antibacterial performance. Silver nanoparticles introduced by wet impregnation of $AgNO_3$ and subsequent thermal processing comprised mixed metal and carbonate phases, with the concentration of silver carbonate (and dissolution rate for ionic Ag^+) inversely proportional to particle size independent of the framework. The introduction of a conformal titania monolayer coating of SBA-15 improved the dispersion of silver nanoparticles, corresponding bulk and surface carbonate content, and hence ionic silver release kinetics. This observation likely reflects the ability of defective reducible metal oxides to act as nucleation centers for metals [38–40], thereby increasing the density of small nanoparticles, and amphoteric nature of titania and ability to capture atmospheric CO_2 [41] which can subsequently react with (weakly basic) silver oxides [42,43] at the nanoparticle interface. Additional rate enhancements for ionic silver were observed for hierarchical microporous-mesoporous SBA-15 frameworks, presumably due to their stabilization of even smaller silver nanoparticles and hence higher carbonate concentrations. Silver release rates from the porous frameworks directly correlated with broad spectrum antibacterial activity, with a 16–38% decrease in decimal reduction times for *Staphylococcus aureus* and *Pseudomonas aeruginosa* observed following titania functionalization, and a further 65–89% reduction following the introduction of macroporosity. The quantitative structure-function relationship identified between the concentration of Ag_2CO_3 and antibacterial efficacy will guide the development of future nanocomposite architectures, notably optimization of the reducible metal oxide coatings and pore networks to further promote ionic silver release.

4. Materials and Methods

Polystyrene bead templates were synthesized using a method developed by Vaudreuil et al. [44] in which styrene, divinyl benzene (comonomer, Sigma Aldrich, 80%) and potassium persulphate (initiator, Sigma Aldrich, >99%) were the reagents. The reaction was performed on a large scale in a 2-liter jacketed Radleys Reactor-Ready system at 90 °C. Deionized water (1.5 L) was introduced to the reactor, along with a Leibig condenser, thermocouple, and a nitrogen line at 1.5 bar pressure. The reactor was stirred at 300 rpm overnight to outgas the solution. Styrene (140 mL, Sigma Aldrich, >99%) and divinylbenzene (27 mL) were washed with NaOH (0.1 M) three times in separate separating funnels and added to the reaction vessel. Potassium persulfate (Sigma Aldrich, 0.35 g) was dissolved in deionized water (20 mL) at 80 °C. After 30 min of stirring (300 rpm) in the reactor at 90 °C, the potassium persulfate solution was added. After stirring for 3 h, the solid particles were recovered as a concentrated solution and stored in a freezer overnight, then the product was allowed to warm before being filtered, washed with ethanol and the beads dried at 80 °C overnight.

SBA-15 was prepared using a cooperative self-assembly method [29], in which a 2.6 wt% solution of Pluronic P123 triblock copolymer (poly(ethylene glycol)-poly(propylene glycol)-poly(ethylene glycol) (Sigma Aldrich), in 1.6-M HCl solution was stirred (500 rpm) at 35 °C. Tetraethyl orthoxysilicate (Sigma Aldrich, 98%) was then added to the mixture, at a molar ratio of 60:1 [TEOS]:[P-123]. The mixture was aged at 80 °C for 24 h without stirring in a sealed container in an oven. The resultant solid material was filtered, then washed with ethanol before drying in air at 100 °C overnight. Removal of the P123 framework was performed by calcination at 500 °C in a muffle furnace for 6 h with a ramp rate of 1 °C/min. Macropores were introduced by the addition of polystyrene beads to the SBA synthesis (2.3.1) at a weight ratio of 5.3:1 [PS beads]:[TEOS] in the initial mixture.

The grafting of titania onto the surface of the prepared silica materials was done using a modified procedure by Landau et al. [30] in which triethylamine is used to activate the surface silanols on the silica and allow the reaction to proceed at lower temperatures. To ensure a uniform coating of TiO_2, the reaction must be performed under completely dry conditions, due to the facile hydrolysis of the titania precursor, which will readily form large titania particles in the presence of water. The synthetic procedure involves mixing titanium isopropoxide (Sigma Aldrich,) in anhydrous toluene

(Aldrich, water content <0.002%), adding triethylamine (Sigma Aldrich, >99%) and MM-SBA-15 or SBA-15 material whilst stirring at 85 °C for 6 h under nitrogen flow. The concentration of titanium isopropoxide was 145 g/L, the molar ratio between titanium isopropoxide and SBA-15 was fixed at 3.5 and the triethylamine: SBA-15 weight ratio at 1.5 on a scale of 5 g of SBA-15/MM-SBA-15. After the reaction, the solid was separated by filtration, washed with toluene (300 mL) and inserted in a 0.5 wt% water-ethanol solution (500 mL) under stirring for 24 h. The resultant solid was washed with ethanol, dried in air in an oven at 90 °C for 24 h, then calcined for 1 h at 250 °C, 1 h 400 °C and finally for 4 h at 500 °C all at 1 °C/min.

Ag NPs were deposited via wet impregnation using a solution of aqueous silver nitrate (99.9%, Sigma-Aldrich). A slurry of the silver precursor and support (10 mL of 5–25 μM + 1 g support) was stirred for 18 h at room temperature before heating to 50 °C. After 5 h, agitation was ceased and the solid aged at 50 °C for a further 24 h to yield a dry powder. Dried samples were calcined at 500 °C (ramp rate 1 °C/min) in static air for 3 h.

XRD patterns were recorded on either a PANalytical X'pertPro diffractometer fitted with an X'celerator detector and Cu Kα (1.54 Å) source or a Bruker D8 Advance diffractometer fitted with a LynxEye high-speed strip detector and Cu Kα (1.54 Å) source. Both instruments were calibrated against either Si (PANalytical, Malvern, UK) or SiO_2 (Bruker, Billerica, MA, USA) standards. Low angle patterns were recorded for $2\theta = 0.3$–$8°$ with a step size of 0.01°. Wide angle patterns were recorded for $2\theta = 10$–$80°$ with a step size of 0.02°.

N_2 adsorption isotherms were recorded using a Nova 4000 porosimeter (Quantachrome, Boynton Beach, FL, USA), before which the samples were thoroughly degassed under vacuum at 120 °C for 2 h. T-plot analysis was used to calculate microporosity. Data was analyzed using NOVAWin version 11 (Quantachrome, Boynton Beach, FL, USA).

XPS analysis was recorded using a Kratos Axis Hsi X-ray photoelectron spectrometer (Kratos Analytical Ltd., Manchester, UK) using a monochromated Al K_α (1486.6 eV) anode. Data was charge corrected to adventitious carbon at 284.6 eV and analyzed using CASAXPS version 2.3.15. Ag 3D peaks were fitted using a Doniach-Sunjic modified Gaussian-Lorentz peak shape and a doublet separation of 6 eV.

TEM analysis was performed using a JEOL 2100 microscope (JEOL Ltd., Tokyo, Japan) with a LaB_6 source and HT of 180 kV. Particle sizing was performed using ImageJ 1.46r software (open source). Samples were prepared using a drop casting method in ethanol onto continuous carbon grids.

Dissolution rates were determined by stirring 10 mg of composite in 25 mL of 0.5 M $NaNO_3$ at 37 °C with periodic aliquots of the analyte solution measured for silver content using an Agilent 6130B single Quad (ESI) ICP-MS (Agilent Technologies, Santa Clara, CA, USA) calibrated against a range of silver concentrations made by serial dilution of a 1000 ppm Ag in 1% HNO3 standard from Sigma-Aldrich.

The antibacterial performance of all materials was evaluated against *Staphylococcus aureus* ATCC 6538, MRSA ATCC 33591, *Escherichia coli* NCTC 10418, *Bacillus subtilis* NCTC 8236, *Pseudomonas aeruginosa* ATCC 15442 and *Clostridium difficile* ATCC 9689, which are representative Gram-positive and Gram-negative problematic organisms found in hospital environments. Zone of inhibition (ZoI) tests were performed by inoculating the surface of a nutrient agar plate (Oxoid, Basingstoke, UK) with an excess volume (3 mL) of nutrient broth which had previously been inoculated and incubated to a cell density of ~108 colony-forming units (cfu)/mL as determined spectrophotometrically using a Perkin-Elmer Lambda 10 UV-Vis spectrophotometer. The liquid was manipulated by agitation to provide a confluent inoculum and the excess fluid removed to waste using a sterile pipette. Using a sterilized boring tool, 5 mm holes were then bored into the agar, and 100 μL of a solution of 10 mg of Ag nanocomposite in 5 mL of simulated body fluid (SBF, see the Supplementary Materials Table S2) dispensed into the borehole using a calibrated micropipette. SBF was prepared according to a method from Kokubo et al. [45] 750 mL of deionized water was stabilized at 37 °C with stirring, to this the following ions were added: NaCl (7.996 g, Sigma Aldrich >99%), $NaHCO_3$ (0.35 g, Sigma Aldrich

>99%), KCl (0.224 g, Sigma Aldrich >99%), $K_2HPO_4\cdot3H_2O$ (0.228 g, Sigma Aldrich >99%), $MgCl_2$ (0.305 g, Sigma Aldrich >99%), HCl (40 mL, 1 kmol/L, Fisher scientific 37%), $CaCl_2$ (0.278 g, Sigma Aldrich >99%), Na_2SO_4 (0.071 g, Sigma Aldrich >99%) and $(CH_2OH)3CNH_2$ (6.057 g, Sigma Aldrich 99%). Finally, the pH was adjusted to 7.35 using HCl solution (1 kmol/L, Fisher scientific 37%). Plates were then incubated at 37 °C overnight, photographed, and calibrated zone areas determined using ImageJ software (open source).

Quantitative antimicrobial activity was determined by logarithmic reduction [13,14]. Here, 5 mg of Ag nanocomposite material was added to an Eppendorf tube kept in dark conditions containing 1 mL of either *S. aureus* or *P. aeruginosa* in a nutrient broth at concentrations of 107 cfu·mL^{-1}. 100 μL aliquots of the resulting suspensions were subsequently removed at 0, 60, 240 min and 24 h, and added to a 1 mL solution of Tween 20 (Fisher, 1%), sodium dodecyl sulphate (Sigma-Aldrich, 0.4%) and sodium chloride (Sigma-Aldrich, 0.85%) in deionized water to neutralise any soluble silver species [13,14].

Each of the resulting neutralized solutions was serially diluted with phosphate buffered saline (PBS) prior to plating onto agar and incubation at 37 °C for 24 h. The experiments were all run with positive and negative controls of silver nitrate and without any nanocomposite respectively. After incubation, the number of colonies present on the agar was counted by visual inspection, and normalized relative to the initial colony count in the negative control at time t = 0 min to determine the logarithmic reduction of bacteria. All experiments were performed in triplicate, with mean values and standard deviations reported.

Supplementary Materials: The following are available online at http://www.mdpi.com/2079-6382/7/3/55/s1, Figure S1: (left) N_2 adsorption-desorption isotherms, and BJH pore size distributions for parent SBA-15, TiO_2/SBA-15 and TiO_2/MM-SBA-15. Type-IV isotherms with H1 hysteresis is characteristic of mesoporous SBA-15, Table S1: Elemental analysis of Ag/SBA-15, Ag/TiO_2/SBA-15 and Ag/TiO_2/MM-SBA-15, Figure S2: XRD patterns for the three prepared support materials, SBA-15, TiO_2/SBA-15 and TiO_2/MM-SBA-15, Figure S3: XRD patterns for (a) Ag/SBA-15; (b) Ag/TiO_2/SBA-15; and (c) Ag/TiO_2/MM-SBA-15 as a function of silver loading. All reflections associated with fcc silver metal., Figure S4: Particle size distributions from TEM for (a) 0.3 wt%; (b) 0.95 wt% and (c) 2.4 wt% Ag nanoparticles deposited on SBA-15, Figure S5: Particle size distributions from TEM (a) 0.3 wt%; (b) 0.9 wt% and (c) 2.4 wt% Ag nanoparticles deposited on TiO_2/SBA-15, Figure S6: Particle size distributions from TEM (a) 0.25 wt%; (b) 0.75 wt% and (c) 1.2 wt% Ag nanoparticles deposited on TiO_2/MM-SBA-15, Figure S6: Particle size distributions from TEM (a) 0.25 wt%, (b) 0.75 wt% and (c) 1.2 wt% Ag nanoparticles deposited on TiO2/MM-SBA-15, Figure S7: Mean particle size and standard deviation for Ag/SBA-15, Ag/TiO2/SBA-15, and Ag/TiO2/MM-SBA-15 as a function of bulk silver loading, Figure S8: Fitted Ag 3d XP spectra of (a) Ag/SBA-15; (b) Ag/TiO_2/SBA-15; and (c) Ag/TiO_2/MM-SBA-15 as a function of Ag loading. All spectra fitted to carbonate and metal spin-orbit doublets possessing common lineshapes and fixed binding energies, Figure S9: XANES profiles for (a) 0.3 wt%; (b) 0.95 wt% and (c) 2.4 wt% Ag/SBA-15, Figure S10: XANES profiles for (a) 0.3 wt%; (b) 0.9 wt% and (c) 2.4 wt% Ag/TiO_2/SBA-15, Figure S11: XANES profiles for (a) 0.25 wt%; (b) 0.75 wt% and (c) 1.2 wt% Ag/TiO_2/MM-SBA-15, Figure S12: Fitted dissolution kinetics for Ag/SBA-15, Ag/TiO_2/SBA-15, and Ag/TiO_2/MM-SBA-15 as a function of particle size, Figure S13: Zone of Inhibition assays for *Staphylococcus aureus*, MRSA, *Escherichia coli*, *Bacillus subtilis*, *Pseudomonas aeruginosa* and *Clostridium difficile* normalized to the mass of silver for (a) Ag/SBA-15; (b) Ag/TiO_2/SBA-15; and (c) Ag/TiO_2/MM-SBA-15, Figure S14: Logarithmic reductions for SBA-15, Ag/TiO_2/SBA-15, and Ag/TiO_2/MM-SBA-15 against (a) *Staphylococcus aureus* and (b) *Pseudomonas aeruginosa* after 24 h incubation, Figure S15: Logarithmic reductions for (a) Ag/SBA-15; (b) Ag/TiO_2/SBA-15; and (c) Ag/TiO_2/MM-SBA-15 against *Staphylococcus aureus* and *Pseudomonas aeruginosa*, Figure S16: Logarithmic colony forming units of (a) *S. aureus*; and (b) *P. aeruginosa* as a function of time for SBA-15, TiO_2/SBA-15, and TiO_2/MM-SBA-15, Table S1: Elemental analysis of Ag/SBA-15, Ag/TiO2/SBA-15 and Ag/TiO2/MM-SBA-15, Table S2: Ion concentrations in SBF solution.

Author Contributions: Formal analysis—M.A.I., A.C.H. and A.F.L.; Funding acquisition—K.W. and A.F.L.; Investigation—M.A.I.; Methodology—M.A.I., B.B., A.C.H. and L.O.; Supervision—K.W. and A.F.L.; Visualization—L.J.D. and C.M.A.P.; Writing—original draft, M.A.I.; Writing—review & editing, A.F.L.

Acknowledgments: We thank the Knowledge Economy Skills Scholarship (K.E.S.S.) program and Polymer Health Technology for providing funding for this work.

Conflicts of Interest: The authors declare no conflicts of interest.

References

1. Zarb, P.; Coignard, B.; Griskeviciene, J.; Muller, A.; Vankerckhoven, V.; Weist, K.; Goossens, M.; Vaerenberg, S.; Hopkins, S.; Catry, B. The European Centre for Disease Prevention and Control (ECDC) pilot point prevalence survey of healthcare-associated infections and antimicrobial use. *Eurosurveillance* **2012**, *17*, 20316. [CrossRef] [PubMed]

2. Hansen, S.; Schwab, F.; Zingg, W.; Gastmeier, P. Process and outcome indicators for infection control and prevention in European Acute Care Hospitals in 2011 to 2012-results of the prohibit study. *Eurosurveillance* **2018**, *23*, 1700513. [CrossRef] [PubMed]

3. Cassini, A.; Plachouras, D.; Eckmanns, T.; Sin, M.A.; Blank, H.-P.; Ducomble, T.; Haller, S.; Harder, T.; Klingeberg, A.; Sixtensson, M. Burden of six healthcare-associated infections on european population health: Estimating incidence-based disability-adjusted life years through a population prevalence-based modelling study. *PLoS Med.* **2016**, *13*, e1002150. [CrossRef] [PubMed]

4. Modic, M.; Nikiforov, A.; Leys, C.; Kuchakova, I.; De Vrieze, M.; Petrovska, M.; Zille, A.; Dinescu, G.; Mitu, B.; Cvelbar, U. *Atmospheric Pressure Plasma and Depositions of Antibacterial Coatings*; Meeting Abstracts; The Electrochemical Society: Pennington, NJ, USA, 2018; p. 1176.

5. O'Neill, J. Antimicrobial resistance: Tackling a crisis for the health and wealth of nations. *Rev. Antimicrob. Resist.* **2014**, 1–20. Available online: https://amr-review.org/sites/default/files/AMR%20Review% 20Paper%20-%20Tackling%20a%20crisis%20for%20the%20health%20and%20wealth%20of%20nations_1. pdf (accessed on 28 June 2018).

6. Peleg, A.Y.; Hooper, D.C. Hospital-acquired infections due to gram-negative bacteria. *N. Engl. J. Med.* **2010**, *362*, 1804–1813. [CrossRef] [PubMed]

7. Service, N.H. Clostridium Difficile, Staphylococcus Aureus Bacteraemia and Escherichia coli Bacteraemia, Surveillance Update. Available online: https://www.wales.nhs.uk/sites3/page.cfm?orgid=379&pid=67899 (accessed on 30 May 2018).

8. England, P.H. Annual Epidemiological Commentary Mandatory MRSA, MSSA and *E. coli* Bacteraemia and *C. difficile* Infection Data 2016/17. Available online: https://assets.publishing.service.gov.uk/government/ uploads/system/uploads/attachment_data/file/634675/Annual_epidemiological_commentary_2017. pdf (accessed on 2 July 2018).

9. Bhattacharya, A.; Nsonwu, O.; Johnson, A.; Hope, R. Estimating the incidence and 30-day all-cause mortality rate of *Escherichia coli* bacteraemia in England by 2020/21. *J. Hosp. Infect.* **2018**, *98*, 228–231. [CrossRef] [PubMed]

10. Percival, S.L.; Bowler, P.G.; Dolman, J. Antimicrobial activity of silver-containing dressings on wound microorganisms using an in vitro biofilm model. *Int. Wound J.* **2007**, *4*, 186–191. [CrossRef] [PubMed]

11. Furr, J.R.; Russell, A.D.; Turner, T.D.; Andrews, A. Antibacterial activity of actisorb plus, actisorb and silver nitrate. *J. Hosp. Infect.* **1994**, *27*, 201–208. [CrossRef]

12. Textor, T.; Fouda, M.M.; Mahltig, B. Deposition of durable thin silver layers onto polyamides employing a heterogeneous tollens' reaction. *Appl. Surf. Sci.* **2010**, *256*, 2337–2342. [CrossRef]

13. Buckley, J.J.; Gai, P.L.; Lee, A.F.; Olivi, L.; Wilson, K. Silver carbonate nanoparticles stabilised over alumina nanoneedles exhibiting potent antibacterial properties. *Chem. Commun.* **2008**, 4013–4015. [CrossRef] [PubMed]

14. Buckley, J.J.; Lee, A.F.; Olivi, L.; Wilson, K. Hydroxyapatite supported antibacterial Ag_3Po_4 nanoparticles. *J. Mater. Chem.* **2010**, *20*, 8056–8063. [CrossRef]

15. Lee, Y.-J.; Kim, J.; Oh, J.; Bae, S.; Lee, S.; Hong, I.S.; Kim, S.-H. Ion-release kinetics and ecotoxicity effects of silver nanoparticles. *Environ. Toxicol. Chem.* **2012**, *31*, 155–159. [CrossRef] [PubMed]

16. Luo, S.; Chen, J.; Chen, M.; Xu, W.; Zhang, X. Antibacterial activity of silver nanoparticles colloidal sol and its application in package film. *Adv. Mater. Res.* **2012**, *380*, 254–259. [CrossRef]

17. Ray, S.; Mohan, R.; Singh, J.K.; Samantaray, M.K.; Shaikh, M.M.; Panda, D.; Ghosh, P. Anticancer and antimicrobial metallopharmaceutical agents based on palladium, gold, and silver N-Heterocyclic carbene complexes. *J. Am. Chem. Soc.* **2007**, *129*, 15042–15053. [CrossRef] [PubMed]

18. Patil, S.; Deally, A.; Gleeson, B.; Müller-Bunz, H.; Paradisi, F.; Tacke, M. Novel benzyl-substituted N-Heterocyclic carbene–silver acetate complexes: Synthesis, cytotoxicity and antibacterial studies. *Metallomics* **2011**, *3*, 74–88. [CrossRef] [PubMed]

19. Almalioti, F.; MacDougall, J.; Hughes, S.; Hasson, M.M.; Jenkins, R.L.; Ward, B.D.; Tizzard, G.J.; Coles, S.J.; Williams, D.W.; Bamford, S. Convenient syntheses of cyanuric chloride-derived NHC ligands, their Ag (I) and Au (I) complexes and antimicrobial activity. *Dalton Trans.* **2013**, *42*, 12370–12380. [CrossRef] [PubMed]

20. Kumar, R.; Münstedt, H. Silver ion release from antimicrobial polyamide/silver composites. *Biomaterials* **2005**, *26*, 2081–2088. [CrossRef] [PubMed]

21. Liu, J.; Hurt, R.H. Ion release kinetics and particle persistence in aqueous nano-silver colloids. *Environ. Sci. Technol.* **2010**, *44*, 2169–2175. [CrossRef] [PubMed]

22. Zhao, X.; Xia, Y.; Li, Q.; Ma, X.; Quan, F.; Geng, C.; Han, Z. Microwave-assisted synthesis of silver nanoparticles using sodium alginate and their antibacterial activity. *Colloids Surf. A Physicochem. Eng. Asp.* **2014**, *444*, 180–188. [CrossRef]

23. Zhang, W.; Yao, Y.; Sullivan, N.; Chen, Y. Modeling the primary size effects of citrate-coated silver nanoparticles on their ion release kinetics. *Environ. Sci. Technol.* **2011**, *45*, 4422–4428. [CrossRef] [PubMed]

24. Peretyazhko, T.S.; Zhang, Q.; Colvin, V.L. Size-controlled dissolution of silver nanoparticles at neutral and acidic pH conditions: Kinetics and size changes. *Environ. Sci. Technol.* **2014**, *48*, 11954–11961. [CrossRef] [PubMed]

25. Ma, R.; Levard, C.; Marinakos, S.M.; Cheng, Y.; Liu, J.; Michel, F.M.; Brown, G.E., Jr.; Lowry, G.V. Size-controlled dissolution of organic-coated silver nanoparticles. *Environ. Sci. Technol.* **2011**, *46*, 752–759. [CrossRef] [PubMed]

26. Isaacs, M.A.; Durndell, L.J.; Hilton, A.C.; Olivi, L.; Parlett, C.M.; Wilson, K.; Lee, A.F. Tunable Ag@ Sio_2 core–shell nanocomposites for broad spectrum antibacterial applications. *RSC Adv.* **2017**, *7*, 23342–23347. [CrossRef]

27. Morones-Ramirez, J.R.; Winkler, J.A.; Spina, C.S.; Collins, J.J. Silver enhances antibiotic activity against gram-negative bacteria. *Sci. Transl. Med.* **2013**, *5*, 190ra81. [CrossRef] [PubMed]

28. Dhainaut, J.; Dacquin, J.-P.; Lee, A.F.; Wilson, K. Hierarchical macroporous-mesoporous SBA-15 sulfonic acid catalysts for biodiesel synthesis. *Green Chem.* **2010**, *12*, 296–303. [CrossRef]

29. Zhao, D.; Huo, Q.; Feng, J.; Chmelka, B.F.; Stucky, G.D. Nonionic triblock and star diblock copolymer and oligomeric surfactant syntheses of highly ordered, hydrothermally stable, mesoporous silica structures. *J. Am. Chem. Soc.* **1998**, *120*, 6024–6036. [CrossRef]

30. Landau, M.V.; Dafa, E.; Kaliya, M.L.; Sen, T.; Herskowitz, M. Mesoporous alumina catalytic material prepared by grafting wide-pore MCM-41 with an alumina multilayer. *Microporous Mesoporous Mater.* **2001**, *49*, 65–81. [CrossRef]

31. Parlett, C.M.A.; Durndell, L.J.; Machado, A.; Cibin, G.; Bruce, D.W.; Hondow, N.S.; Wilson, K.; Lee, A.F. Alumina-grafted SBA-15 as a high performance support for Pd-catalysed cinnamyl alcohol selective oxidation. *Catal. Today* **2014**, *229*, 46–55. [CrossRef]

32. Guo, X.-C.; Dong, P. Multistep coating of thick titania layers on monodisperse silica nanospheres. *Langmuir* **1999**, *15*, 5535–5540. [CrossRef]

33. NIST. Binding Energies of Ag 3d 5/2. Available online: https://srdata.nist.gov/xps/EngElmSrchQuery.aspx?EType=PE&CSOpt=Retri_ex_dat&Elm=Ag (accessed on 6 July 2017).

34. Gaarenstroom, S.W.; Winograd, N. Initial and final state effects in the esca spectra of cadmium and silver oxides. *J. Chem. Phys.* **1977**, *67*, 3500–3506. [CrossRef]

35. Weaver, J.F.; Hoflund, G.B. Surface characterization study of the thermal decomposition of ago. *J. Phys. Chem.* **1994**, *98*, 8519–8524. [CrossRef]

36. Waterhouse, G.I.N.; Bowmaker, G.A.; Metson, J.B. Oxidation of a polycrystalline silver foil by reaction with ozone. *Appl. Surf. Sci.* **2001**, *183*, 191–204. [CrossRef]

37. Nanda, K.; Maisels, A.; Kruis, F.; Fissan, H.; Stappert, S. Higher surface energy of free nanoparticles. *Phys. Rev. Lett.* **2003**, *91*, 106102. [CrossRef] [PubMed]

38. Jiawei, W.; Wenxing, C.; Chuanyi, J.; Lirong, Z.; Juncai, D.; Xusheng, Z.; Yu, W.; Wensheng, Y.; Chen, C.; Qing, P.; et al. Defect effects on TiO_2 nanosheets: Stabilizing single atomic site au and promoting catalytic properties. *Adv. Mater.* **2018**, *30*, 1705369.

39. Nolan, M.; Iwaszuk, A.; Lucid, A.K.; Carey, J.J.; Fronzi, M. Design of novel visible light active photocatalyst materials: Surface modified TiO_2. *Adv. Mater.* **2016**, *28*, 5425–5446. [CrossRef] [PubMed]

40. Yu, J.; Zhou, W.; Xiong, T.; Wang, A.; Chen, S.; Chu, B. Enhanced electrocatalytic activity of Co@N-doped carbon nanotubes by ultrasmall defect-rich TiO_2 nanoparticles for hydrogen evolution reaction. *Nano Res.* **2017**, *10*, 2599–2609. [CrossRef]

41. Bhattacharyya, K.; Danon, A.; Vijayan, B.K.; Gray, K.A.; Stair, P.C.; Weitz, E. Role of the surface lewis acid and base sites in the adsorption of CO_2 on titania nanotubes and platinized titania nanotubes: An in situ FT-IR study. *J. Phys. Chem. C* **2013**, *117*, 12661–12678. [CrossRef]

42. Xu, C.; Liu, Y.; Huang, B.; Li, H.; Qin, X.; Zhang, X.; Dai, Y. Preparation, characterization, and photocatalytic properties of silver carbonate. *Appl. Surf. Sci.* **2011**, *257*, 8732–8736. [CrossRef]

43. Slager, T.; Lindgren, B.; Mallmann, A.J.; Greenler, R.G. Infrared spectra of the oxides and carbonates of silver. *J. Phys. Chem.* **1972**, *76*, 940–943. [CrossRef]

44. Vaudreuil, S.; Bousmina, M.; Kaliaguine, S.; Bonneviot, L. Synthesis of macrostructured silica by sedimentation–aggregation. *Adv. Mater.* **2001**, *13*, 1310–1312. [CrossRef]

45. Kokubo, T.; Kushitani, H.; Sakka, S.; Kitsugi, T.; Yamamuro, T. Solutions able to reproduce in vivo surface-structure changes in bioactive glass-ceramic A-W[3]. *J. Biomed. Mater. Res.* **1990**, *24*, 721–734. [CrossRef] [PubMed]

Article

Plasmon Resonance of Silver Nanoparticles as a Method of Increasing Their Antibacterial Action

Alexander Yu. Vasil'kov [1], Ruslan I. Dovnar [2,*], Siarhei M. Smotryn [2], Nikolai N. Iaskevich [2] and Alexander V. Naumkin [1]

[1] Nesmeyanov Institute of Organoelement Compounds, Russian Academy of Sciences, 28 Vavilov st., Moscow 119991, Russia; alexandervasilkov@yandex.ru (A.Y.V.); naumkin@ineos.ac.ru (A.V.N.)

[2] Grodno State Medical University, 80 Gorky St., Grodno 230009, Belarus; s.smotrin@mail.ru (S.M.S.); inngrno@mail.ru (N.N.I.)

* Correspondence: dr_ruslan@mail.ru; Tel.: +375-297-868-643

Received: 14 July 2018; Accepted: 21 August 2018; Published: 22 August 2018

Abstract: In this article, a series of silver-containing dressings are prepared by metal-vapor synthesis (MVS), and their antibacterial properties are investigated. The antibacterial activity of the dressings containing silver nanoparticles (AgNPs) against some Gram-positive, and Gram-negative microorganisms (*Staphylococcus aureus, Staphylococcus haemolyticus, Pseudomonas aeruginosa, Klebsiella pneumoniae, Escherichia coli, Moraxella* spp.) has been determined. Based on the plasmon resonance frequency of these nanoparticles, the frequency of laser irradiation of the dressing was chosen. The gauze bandage examined showed pronounced antibacterial properties, especially to *Staphylococcus aureus* strain. When 470 nm laser radiation, with a power of 5 mW, was applied for 5 min, 4 h after inoculating the Petri dish, and placing a bandage containing silver nanoparticles on it, the antibacterial effect of the latter significantly increased—both against Gram-positive and Gram-negative microorganisms. The structure and chemical composition of the silver-containing nanocomposite were studied by transmission electron microscopy (TEM), X-ray photoelectron spectroscopy (XPS) and extended X-ray absorption fine structure (EXAFS). The synthesized AgNPs demonstrate narrow and monomodal particle size distribution with an average size of 1.75 nm. Atoms of metal in Ag/bandage system are mainly in Ag^0 state, and the oxidized atoms are in the form of Ag-Ag-O groups.

Keywords: antibacterial effect; laser irradiation; metal-vapour method; silver nanoparticles; TEM; XPS; EXAFS

1. Introduction

The widespread use of antibiotics for the treatment and prevention of bacterial diseases leads to a significant increase in the antibacterial resistance of microorganisms through its acquisition either through exogenous resistance genes or chromosomal mutations [1]. This stimulates not only the search for new antibacterial drugs, but also possible alternatives to the latter [2,3]. The scientists are tasked to find substances that effectively act simultaneously on Gram-positive, Gram-negative microorganisms and fungi, which are independent of the antibiotic resistance of the microorganism.

Silver and its compounds have been used in medicine since ancient times. Mass application of silver preparations began in the seventies of the 19th century [4,5]. Since then, numerous confirmations of antiviral, antibacterial, and immunomodulating activity of silver preparations have been received [6]. With the invention of antibiotics, which have more pronounced antibacterial properties, interest in the therapeutic properties of silver dramatically decreased. However, widespread use of antibiotics revealed by the end of the 20th century a number of their shortcomings and silver preparations were again actively studied and used [7].

The mechanism of antimicrobial effect of silver was carefully studied previously. It has now been established that silver ions are selectively toxic with respect to prokaryotic microorganisms with a weak effect on eukaryotic cells [6], including comparatively minimal toxicity to mammalian cells [8]. This is due to the fact that the concentration of silver ions or silver nanoparticles necessary for the death of prokaryotic cells (bacteria) is much lower than the concentration that causes the death of eukaryotic cells, including cells of the human body [9].

It should be noted that new prospects for the use of silver in medicine are opened in connection with the development of nanotechnology, an interdisciplinary field of science that deals with the creation, production and application of structures, devices and systems ranging in size from 1 to 1000 nm, although in practice they use the sizes ranging from 1 to 100 nm more often [10]. It is proved that the metal nanoparticles have unique properties, often differing from that of the solid metal [10]. This is due to the fact that the surface/bulk energy ratio of nanoparticles is much larger than that of compact metal [11].

As applied to medicine, this means that the nature of the interaction of a nanoparticle with a bacterium or fungus is significantly different from the impact of a compact metal on them and probably enhances their bactericidal or fungicidal activity [10].

Localized surface plasmon resonance is an optical phenomenon that is generated when light interacts with conductive nanoparticles that are smaller than the incident wavelength [12]. From the point of view of antibacterial properties of (silver nanoparticles) AgNPs, it is interesting to study how these properties change when plasmon resonance effect occurs.

One of the ecological and effective methods for producing mono- and bi-metallic nanoparticles and materials based on them is the metal-vapor synthesis (MVS). The method was used for preparation of silver-containing composite materials for medical applications [13]. It is assumed that MVS will be effective for the modification of wound dressings prepared from natural or synthetic materials with AgNPs.

In this regard, the wide introduction of dressings containing AgNPs correctly combined with the plasmon resonance effect can play a significant role in improving treatment of purulent wounds in the era of increasing antibiotic resistance of microorganisms.

The aim of this research is to study the antibacterial effect of a new dressing material, based on gauze bandage containing AgNPs prepared by MVS, and changing this effect under the influence of laser radiation.

2. Results

The structure and chemical composition of the silver-containing nanocomposite were studied by transmission electron microscopy (TEM), X-ray photoelectron spectroscopy (XPS) and extended X-ray absorption fine structure (EXAFS).

The TEM micrograph of the cotton fibre is shown in Figure 1. The photograph shows amorphous extended structure with a diameter in the range of 16–25 nm, and dark crystalline nanostructures, the density of which is higher than that of cotton. The enlarged image of the regions of ordered atoms on a surface marked with a square in Figure 1 is shown in Figure 2.

The EDS spectra obtained from dark nanostructures (not shown) contain the characteristic line AgLα = 2.98 keV, which allows attribute dark nanostructures to AgNPs. Most of AgNPs are in the form of chains and agglomerates, which is a characteristic feature of the systems prepared by MVS. The particle size distribution is narrow and monomodal. The average particle size is 1.75 $\pm$ 0.25 nm.

Simulation of the electron diffraction for the group of the atoms selected in Figure 2 is shown in Figure 3. It is seen from Figures 2 and 3 that groups of atoms form faces with interplanar distances d1 and d2. Calculating the ratio d1/d2 gives a value of 1.09. The angle between d1 and d2 is 52°. The values 1.09 and 52° indicate the presence of faces (111) and (200) of a face-centered cubic (fcc) structure in silver particles (for an ideal fcc structure: d(111)/d (200) = 1.15, the angle is 55°).

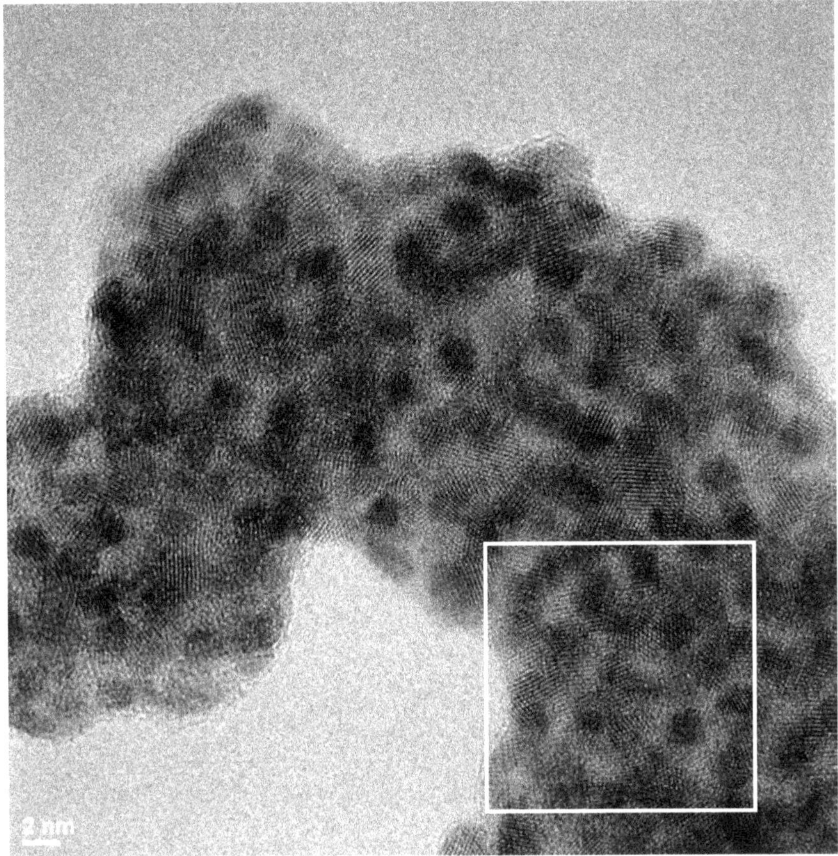

Figure 1. Transmission electron microscopy (TEM) micrograph of Ag-cotton system.

Based on the synthesis method and sample storage conditions, three models of the chemical composition of the particle surface can be proposed: Ag^0, Ag^+, and $Ag^0 + Ag^+$. With a high degree of probability, the mixed composition in crystalline form can be excluded from the consideration. Otherwise, "double reflexes" should be observed in Figure 3. The fcc lattice constants (**a**) for Ag^0 and Ag_2O are 4.08 Å and 4.76 Å, respectively. Calculation of **a** by the formula $a = d(hkl) \times (h^2 + k^2 + l^2)^{1/2}$, where h = 1, k = 1, l = 1; d (hkl) = 2.3 Å, gives a value of 3.98 Å. Consequently, there is reason to believe that the surface of the crystalline particles consists of metallic silver. However, presence of Ag^+ state in amorphous form cannot be excluded.

Figure 4 shows the survey spectrum of Ag black. Along with the peaks characteristic of silver atoms there are peaks of impurity carbon and oxygen atoms.

The determination of the chemical state of silver atoms in nanoparticles by the XPS method is a complex task. This is due to the fact that the spectral characteristics of the metal particles and the oxide particles are fairly close. According to NIST XPS Database [14] the binding energies of the Ag $3d_{5/2}$ peak for Ag, Ag_2O and AgO are in the ranges 367.9–368.4, 367.7–368.4, and 367.3–368.1 eV, respectively. One of the solutions to this problem is the use of the Auger parameter. However, as a rule, the concentration of silver nanoparticles in materials is low, whereas the registration of the Auger spectrum requires a significantly longer time than the recording of the photoelectron spectrum, which can lead to the reduction of silver from oxides under the action of X-ray irradiation [15].

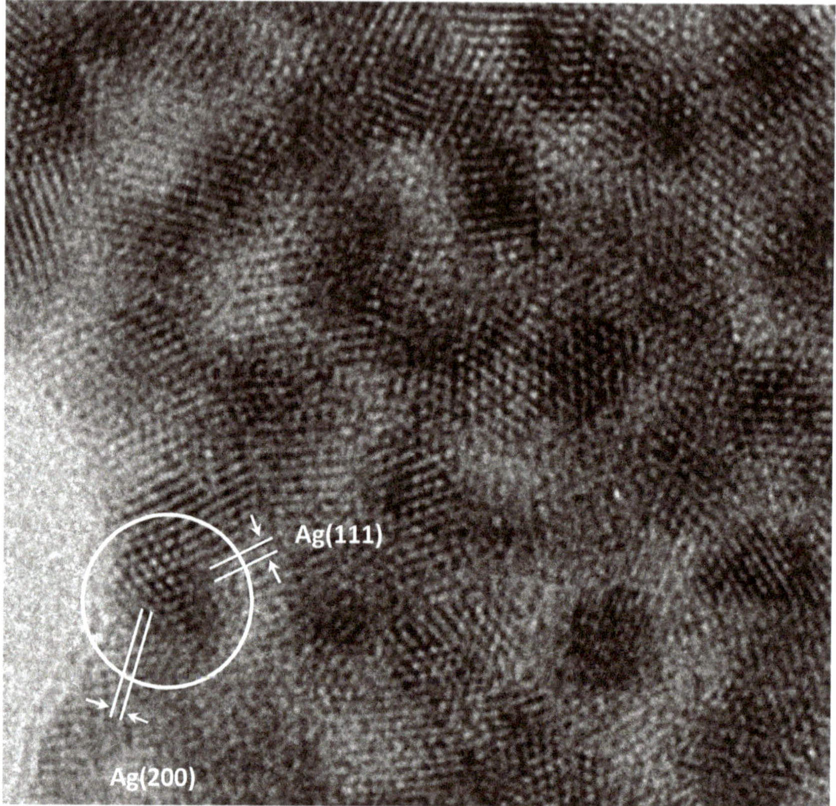

Figure 2. An enlarged image of the regions of ordered atoms on a surface, marked with a square in Figure 1.

Another approach may be based on controlled differential charging, in which separation of signals from regions with good and poor conductivity is possible. If the region has a metallic conductivity and is in good electrical contact with the sample holder, then it does not accumulate a charge due to the emission of secondary electrons. In contrast, in a region with poor conductivity, a positive charge is usually accumulated, the value of which depends on the secondary emission coefficient and conductivity, and it leads to the displacement of photoelectron peaks in the region of high binding energies. When a positive bias voltage U_b is applied, the stray electrons flow to the sample surface is increased and contributes to the compensation of the surface charge. In this case, narrowing of the photoelectron peaks and their shift toward higher binding energies are observed. Photoelectron peaks corresponding to regions with good conductivity should be shifted by U_b the amount of applied bias voltage, while from areas with a worse conductivity by a smaller amount. With a negative bias voltage, the flux stray electrons decreases, it leads to increase in charging on non-conducting regions and an increase in the interval between peaks corresponding to signals from the conducting and non-conducting regions. In this case, the signal from the conductive region must also be shifted by U_b.

Taking into account the difference in the electrical conductivity of metallic silver and its oxides, we attempted to separate signals from regions containing silver atoms in a metal state and other chemical states by controlled differential charging, changing the potential on the sample holder. This approach is widely used to determine the presence of different phases in a sample [16–21].

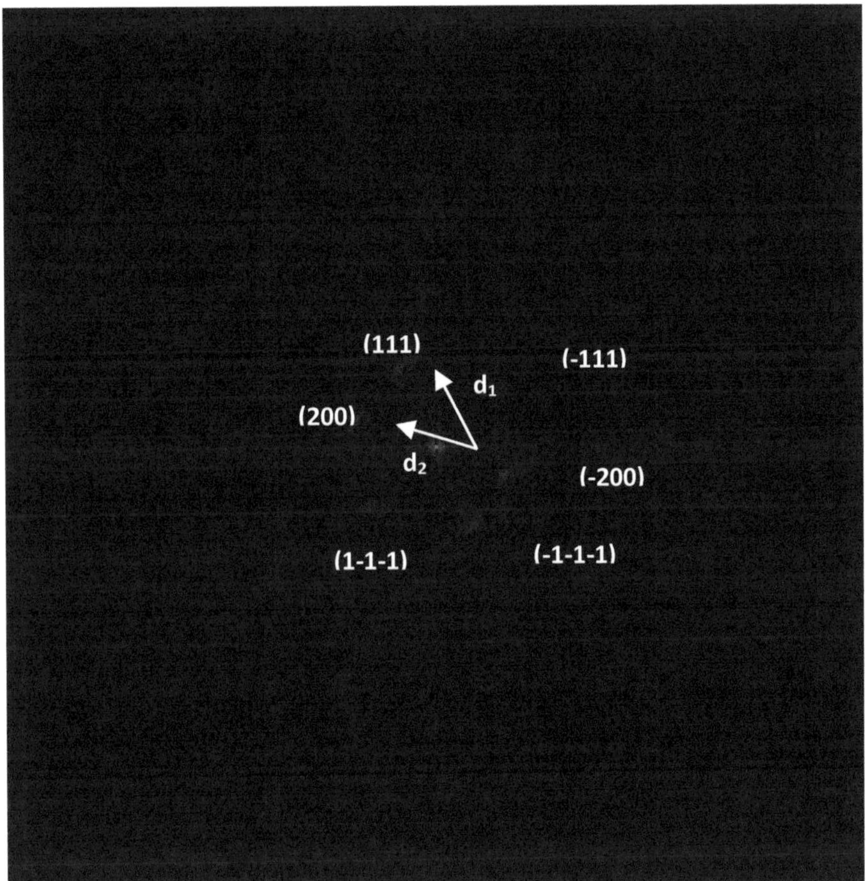

Figure 3. Simulation of the electron diffraction pattern for the group of atoms circled in Figure 2 and assignment for face-centered cubic (fcc) structure.

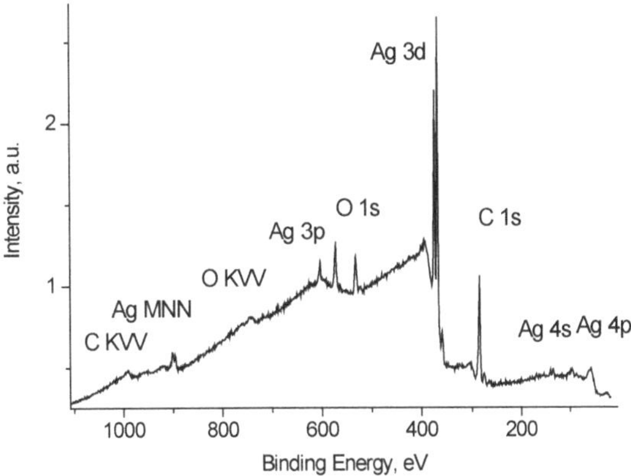

Figure 4. X-ray photoelectron spectroscopy (XPS) survey spectrum of Ag black.

To discriminate regions with different conductivities, or other words with different chemical states, a bias voltage $U_b = \pm 7$ V was supplied to the sample holder. Figure 5 shows the Ag 3d spectra of the Ag black measured at different bias voltage U_b applied to the sample holder. It is seen that binding energy of the Ag $3d_{5/2}$ and Ag $3d_{3/2}$ peaks is slightly depends on U_b. Table 1 presents the characterization of the Ag 3d spectra. The full widths at high maximum (FWHM) are rather different as well. Both the binding energies of the Ag 3d peaks and their FWHM values indicate that the spectra contain some states with different conductivities. Considering that the recording the Ag_2O and AgO spectra is accompanied by the surface charging [22,23], one can assume the presence of the Ag^+ and/or Ag^{2+} state in the Ag black. To determine the characteristics of the $Ag^{\delta+}$ state a subtraction of the spectrum measured at $U_b = 7$ V from the spectrum measured at $U_b = -7$ V was performed under the condition of their best coincidence in the high-energy region. The difference spectrum is presented in Figure 5. The binding energies of the Ag $3d_{5/2}$ and Ag $3d_{3/2}$ peaks 367.73 and 373.71 eV correspond to Ag_2O state [22]. The relative intensity of this state is no less than 0.27. This estimate is based on the fact that the spectrum measured at $U_b = 7$ V may contain Ag_2O state.

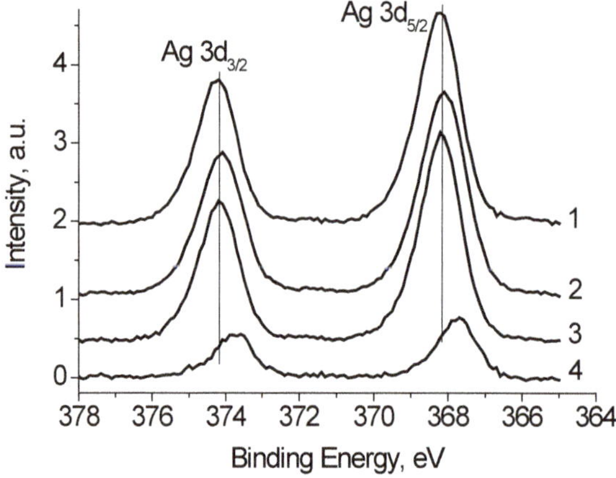

Figure 5. The Ag 3d photoelectron spectra of Ag black measured at different $U_b = +7$ (1), -7 (2), 0 V (3), ant difference spectrum (2)–(1). The spectra are corrected for U_b.

Table 1. Binding energies and full widths at high maximum (FWHMs) measured by XPS.

U_b	Ag $3d_{5/2}$, eV	Ag $3d_{3/2}$, eV	Ag $3d_{5/2}$ FWHM, eV
0 V	368.17	374.17	1.3
-7 V	368.09	374.11	1.4
$+7$ V	368.21	374.21	1.5
(-7 V)–($+7$ V)	367.73	373.71	1.3

The C 1s and O 1s spectra of Ag black show a strong dependence on U_b (Figure 6) which indicate that a large part of carbon and oxygen is has low conductivity. It should be noted that binding energy of the main C 1s peak measured at $U_b = +7$ V is 284.77 eV, and corresponds to that used for charge reference [14].

It follows that some of the carbon atoms have good conductivity, or in other words, is in close contact with silver atoms. To estimate this value, we use the fitting the C 1s spectrum measured at $U_b = -7$ V, when the best separation of the electron emission from regions with good and poor conductivity is realized. When the spectrum was fitted with some components, two restrictions were imposed: The width of the low-energy peak should describe the low-energy side by the best way, and the energy interval between the low-energy peak and the next should not be less than that at U_b = +7 V. The second restriction is imposed because of the possible manifestation of differential charging for regions containing C-O groups. The fraction of such carbon atoms is 0.47. A similar value of 0.48 was obtained for the spectrum measured at U_b = 0 V, whereas for the spectrum measured at U_b = +7 V it is 0.77. This is due to conductivity induced with the stray electrons, which make conducting regions that are not in direct contact with silver nanoparticles. Figure 7 shows fitting the corresponding C 1s spectra, and Table 2 contains corresponding data.

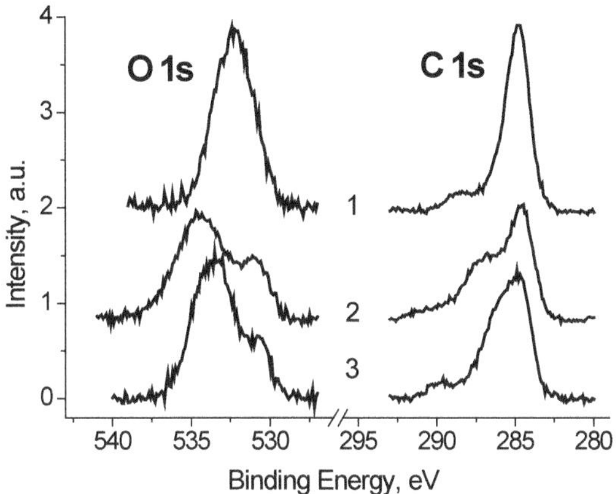

Figure 6. The C 1s and O 1s photoelectron spectra of Ag black measured at different U_b = +7 (1), −7 (2), 0 V (3).

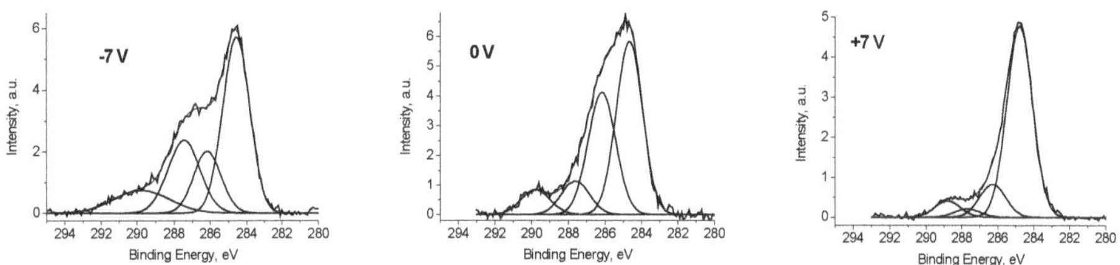

Figure 7. Fitting the C 1s photoelectron spectra of Ag black measured at U_b = −7, 0 and 7 V.

Table 2. Binding energies (E_b), peak widths (W) and relative intensities of the peaks deconvoluted in the C 1s and O 1s spectra of Ag black.

Ub		C-C/ C-H	C-OH/ C-O-C	C=O	C(O)O	Ag-O	C=O/ C(O*)O	C-OH/ C-O-C	C(O*)O	C(O)O*
		C 1s					**O 1s**			
0 V	E_b	284.7	286.2	287.6	289.8	530.6	532.5	533.5		534.7
	W	1.45	1.50	1.51	1.58	1.42	1.65	1.80		1.85
	I_{rel}	0.48	0.35	0.10	0.07	0.18	0.21	0.35		0.26
−7 V	E_b	284.6	286.2	287.4	289.8	530.9	533.2	534.9		536.7
	W	1.48	1.50	1.76	3.05	1.76	1.90	1.90		1.90
	I_{rel}	0.47	0.17	0.23	0.13	0.24	0.27	0.38	0.00	0.10
+7 V	E_b	284.8	286.3	287.7	288.8	530.7	532.4	532.5	531.8	533.8
	W	1.43	1.43	1.43	1.43	1.42	1.42	1.55	1.42	1.55
	I_{rel}	0.77	0.13	0.03	0.07	0.18	0.09	0.37	0.18	0.18

It should be noted that the fitting the C 1s spectra with some components measured at $U_b = -7$ and 0 V, presented in Figure 7 is largely conditional and do not reflect the real relative concentrations of CO_x groups. At the same time, applying a positive bias voltage $U_b = +7$ V practically compensates the surface charging, and Figure 7 (+7 V) reflects the real relative concentrations of CO_x groups.

It was found that the peaks in the low-energy region of the O 1s spectra, measured at $U_b = -7$ and 0 V (Figure 8), have close binding energies and intensities. Based on the reference data [21,22], the binding energy of these peaks of about 530.8 eV cannot be attributed to C-O bonds. Therefore, they should be attributed to the Ag-O bonds. And from the weak dependence of E_b and I_{rel} on U_b, one can assign binding energies of 530.6 and 530.9 eV to Ag-Ag-O state. The I_{rel} of this state, determined from the fitting the O 1s spectra measured at $U_b = -7$, 0 and 7 V, are 0.24, 0.18 and 0.18, respectively.

It should be stressed that relative intensities of CO groups in the O 1s spectrum measured at $U_b = 7$ V correspond those obtained from the relative C 1s spectrum. It means that the bias voltage of $U_b = 7$ V neutralize the surface charging.

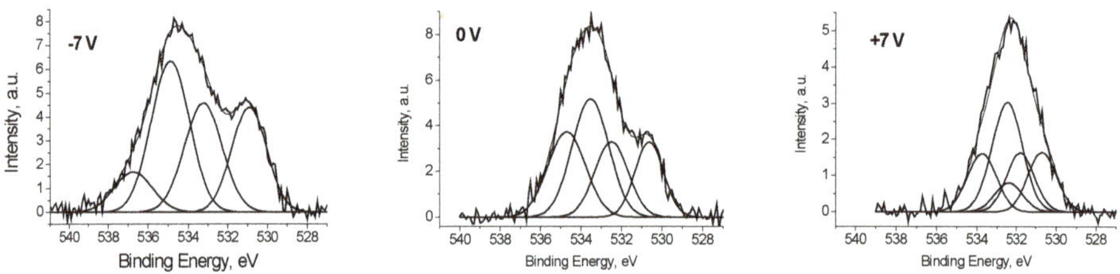

Figure 8. Fitting the O 1s photoelectron spectrum of Ag black measured at $U_b = -7$, 0 and 7 V.

Figure 9 shows the survey spectrum of the Ag/bandage system. Along with the peaks characteristic of silver, carbon and oxygen there are peaks of impurity silicon atoms.

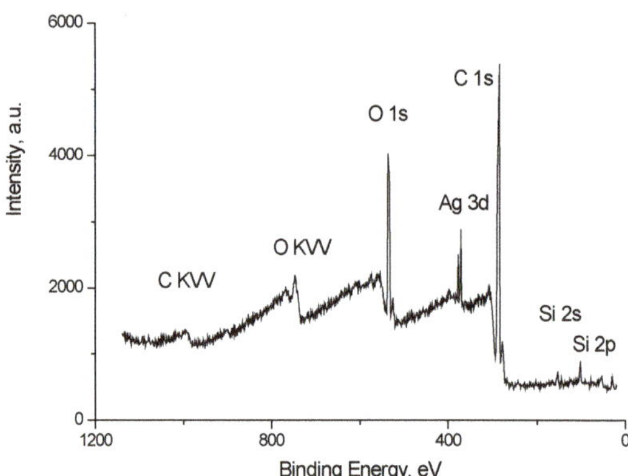

Figure 9. Survey spectrum of the Ag/bandage system.

In case of the Ag/bandage system the charge referencing was done using the C 1s spectrum of cellulose. The latter was simulated using the reference data [24] by considering the difference in peak resolution. The binding energy of 286.7 eV was assigned to C-OH state of cellulose. Figure 10 shows the C 1s spectrum of the Ag/bandage sample fitted with four Gaussian peaks at 286.7, 288.1, 284.8 and 283.13 eV. The first and the second peaks are assigned to cellulose [24]. The third peak is assigned to adventitious carbon, while the origin of the peak at 283.1 eV is not clear because it does not correspond to reference data for polymers [24]. However, it may be resulted either of differential charging or low-molecular weight species. Similar C 1s spectrum was recorded for Au/bandage system. It slightly differs in relative concentration of C-C/C-H peak and peak at 283.1 eV. The O 1s spectra of Ag/bandage and Au/bandage systems are practically indistinguishable. It means that the peak at 283.1 eV may be assigned to C-C/C-H state as well.

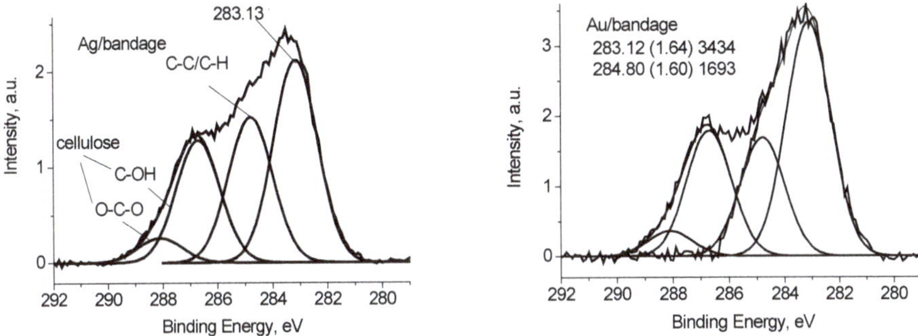

Figure 10. The C 1s photoelectron spectra of Ag/bandage and Au/bandage systems.

Figure 11 shows the C 1s spectra of Ag black and Ag/bandage system. It is clearly seen that the spectra are strongly different. The C 1s peak of the Ag/bandage system shifted to high binding energy region by 0.66 eV and its FWHM is 1 eV more than that of Ag black. These differences may be assigned to the size effect in photoelectron spectra which induces both energy shift to high binding energy and the peak broadening [25,26]. The transition from AgNPs in the Ag black, to their dispersion in the bandage followed with fairy large changes in size of AgNPs, and an increase of the proportion of the Ag^0 state was observed. Apparently, there was a partial reduction of silver and its stabilization by a modified layer of cellulose. However, as follows from a comparison of the O 1s spectra of the Ag black and Ag/bandage system (Figure 12), the low-energy side of the latter might be assigned to the Ag-Ag-O group.

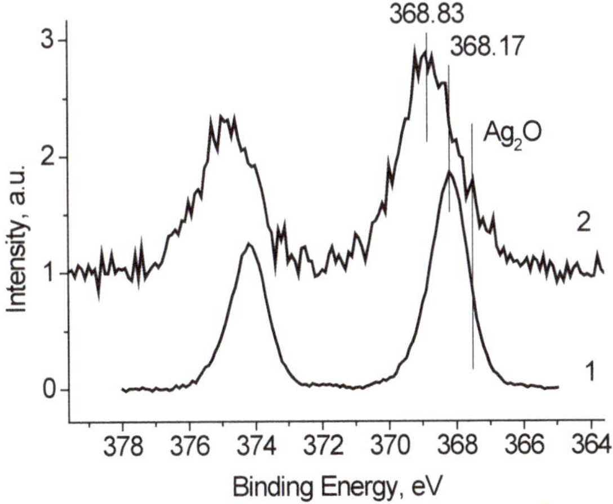

Figure 11. The C 1s photoelectron spectra of Ag black (1) and Ag/bandage system (2).

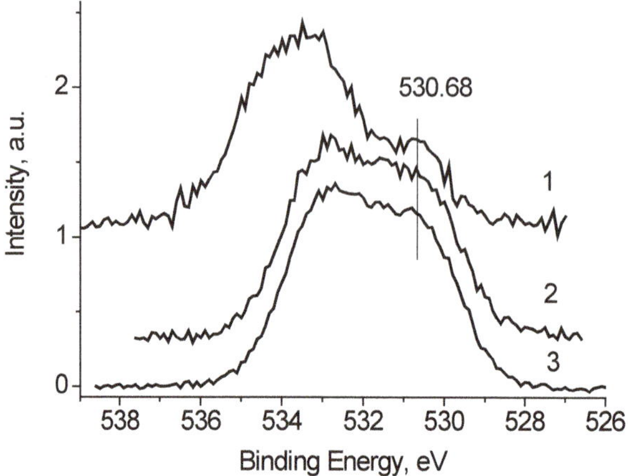

Figure 12. The O 1s photoelectron spectra of Ag black (1), Ag/bandage (2), and Au/bandage systems (3).

But, given the almost complete coincidence of the O 1s spectra of Ag/bandage and Au/bandage systems, this conclusion should be rejected. Thus, one can conclude that the basic state of Ag atoms in Ag/bandage system is Ag^0 state, whereas the oxidized silver is in the form of Ag-Ag-O groups, and, as follows from Figure 12 the proportion of oxidized state is small. This is in accordance with EXAFS and TEM data which indicate that silver atoms are mainly in Ag^0 state. It should be noted that EXAFS is not a surface-sensitive method as XPS and electron diffraction may be recorded only from the ordered regions.

Table 3 shows the number of colony-forming units (CFU) of the studied microbes along the perimeter of the bandage at a distance equal to the diameter of one colony on both sides of the edge in the form Me (Q_1; Q_3), together with the level of statistical significance, where Me is the median, Q_1—lower quartile, Q_3—upper quartile.

Table 3. The number of colony-forming units (CFU) of the studied microorganisms along the edge of the bandage at a distance to both sides of the edge equal to the diameter of one colony (Me (Q_1; Q_3)) and the level of statistical significance (p) between control groups and gauze with AgNPs.

Strain of Microorganism	Control (Normal Bandage)	AgNPs-Containing Bandage	p
Staphylococcus aureus	7.0 (6.0; 8.0)	0.0 (0.0; 1.0)	<0.001
Staphylococcus haemolyticus	11.0 (7.5; 14.5)	7.5 (6.0; 8.0)	0.049
Pseudomonas aeruginosa	5.0 (5.0; 6.0)	2.0 (1.0; 3.0)	<0.001
Klebsiella pneumoniae	6.5 (6.0; 7.0)	2.0 (1.0; 3.0)	<0.001
Escherichia coli	16.0 (13.5; 17.5)	6.0 (4.0; 7.0)	<0.001
Moraxella spp.	8.0 (5.0; 8.5)	3.0 (2.5; 4.0)	0.002

Due to the fact that the data of the control groups of different strains differ, in order to compare the antibacterial effect of the AgNPs-containing bandage with respect to different microorganisms, we calculated the percentage reduction factor. Table 4 presents the results of the study of the percentage reduction of CFU.

The results of the change in the antibacterial properties of the ordinary medical gauze bandage under the influence of laser irradiation are presented in Table 5, in the form Me (Q_1; Q_3), where Me is the median, Q_1 is the lower quartile, Q_3 is the upper quartile.

Table 4. Percentage reduction in the number of CFU towards control.

Microbial Strain	Percentage Reduction of CFU of the AgNPs-Containing Bandage
Staphylococcus aureus	95
Staphylococcus haemolyticus	34
Pseudomonas aeruginosa	58
Klebsiella pneumoniae	68
Escherichia coli	64
Moraxella spp.	55

Table 5. The number of CFU of the investigated microorganisms along the edge of the ordinary medical gauze bandage (Me (Q_1; Q_3)) and p—the level of statistical significance between groups.

Strain of Microorganism	Control (Normal Bandage)					
	Group 1	Group 2	$p1$	Group 3	$p2$	
	Without Laser Irradiation	Laser Irradiation after 2 h		Laser Irradiation after 4 h		
Staphylococcus aureus	6.0 (5.5; 7.0)	6.5 (5.0; 9.0)	0.682	6.5 (6.0; 8.0)	0.325
Staphylococcus haemolyticus	6.5 (5.5; 8.0)	7.0 (6.0; 8.0)	0.681	6.0 (4.0; 9.5)	0.620
Pseudomonas aeruginosa	14.0 (11.5; 15.5)	14.0 (13.0; 15.5)	0.861	14.5 (13.0; 16.5)	0.620
Klebsiella pneumoniae	5.0 (3.0; 5.0)	5.0 (4.0; 7.0)	0.223	4.0 (4.0; 5.0)	0.862
Escherichia coli	10.0 (9.0; 10.5)	10.0 (8.0; 12.0)	0.861	11.0 (8.5; 11.5)	0.380
Moraxella spp.	7.0 (4.0; 8.0)	6.5 (4.5; 8.0)	0.930	6.0 (5.0; 8.5)	0.838

$p1$ = the level of statistical significance between groups 1 and 2, and $p2$ = the level of statistical significance between groups 1 and 3.

Based on the data presented in Table 5, no statistically significant changes in the indices of the number of CFUs when using an ordinary medical gauze bandage without or in conjunction with laser irradiation have been identified. This indicates the absence of an antibacterial effect in laser radiation at a wavelength of 470 nm (blue region of the spectrum), a 5 mW power, and with exposure time of 5 min.

Table 6 shows the results of the change in the antimicrobial properties of the medical gauze bandage containing AgNPs, depending on the presence or absence of exposure to the laser and the time through which it was performed.

Table 6. The number of CFU of the investigated microorganisms along the edge of the medical gauze bandage containing AgNPs, (Me (Q_1; Q_3)) and p—the level of statistical significance between the groups.

Strain of Microorganism	AgNPs-Containing Bandage					
	Group 4	Group 5	$p1$	Group 6	$p2$	
	Without Laser Irradiation	Laser Irradiation after 2 h		Laser Irradiation after 4 h		
Staphylococcus aureus	5.0 (3.5; 5.5)	5.0 (4.0; 5.0)	0.904	4.0 (2.0; 5.0)	0.043
Staphylococcus haemolyticus	5.0 (4.0; 6.0)	5.5 (2.5; 6.5)	0.861	4.0 (2.5; 4.5)	0.011
Pseudomonas aeruginosa	7.0 (5.5; 8.0)	6.0 (5.0; 7.5)	0.250	4.0 (2.5; 5.0)	<0.001
Klebsiella pneumoniae	2.0 (1.5; 3.0)	2.0 (1.0; 3.0)	0.903	1.0 (1.0; 1.5)	0.016
Escherichia coli	6.5 (6.0; 7.0)	7.0 (5.5; 8.0)	0.727	5.0 (4.5; 6.0)	0.024
Moraxella spp.	3.5 (2.0; 4.0)	3.0 (2.0; 4.0)	0.723	2.0 (1.0; 3.0)	0.039

$p1$ = the level of statistical significance between groups 4 and 5, and $p2$ = the level of statistical significance between groups 4 and 6.

According to the data presented in Table 6, laser radiation in the blue region of the spectrum did not have a statistically significant increase in the antibacterial effect of AgNPs when applied two hours after inoculating the Petri dish and placing a bandage on it. However, when the laser treatment was used four hours after the seeding an increase in the antibacterial effect of this bandage was observed with respect to all studied microorganisms. The phenomenon is statistically significant in all groups of microorganisms. The difference can be observed in Table 7.

According to the data, presented in Table 7, there is no significant diversity between difference in percentage reduction of CFU of Gram-positive and Gram-negative microorganisms. Considering the fact that they differ in the structure of the cell wall, it can be concluded that the mechanism for increasing the antibacterial effect of the AgNPs-containing bandage cannot be explained solely by the effect on it. As follows from Table 7, laser radiation, when exposed four hours after inoculating the Petri dish and the application of a AgNPs-containing bandage on it allows to increase the antibacterial properties of the bandage by 15–24%, depending on the strain of the microorganism.

Table 7. Difference in percentage reduction of CFU between the groups of the AgNPs-containing bandage without laser irradiation and with laser irradiation four hours after inoculating the Petri dish and placing the bandage.

Strain of Microorganism	Difference in Percentage Reduction of CFU, %
Staphylococcus aureus	20
Staphylococcus haemolyticus	24
Pseudomonas aeruginosa	22
Klebsiella pneumoniae	22
Escherichia coli	15
Moraxella spp.	19

3. Discussion

It is believed that the antibacterial effect of silver ions is due primarily to the high affinity of the latter to sulfur or phosphorus [27]. Namely, it depends on the silver cations (Ag^+), which bind firmly to groups of electron donors in biological molecules containing silver, oxygen or nitrogen. Due to the large number of sulfur-containing proteins on the surface of the bacterial cell, AgNPs can interact with sulfur-containing proteins inside or outside the cell membrane, which affect the viability of the bacterial cell [27]. Silver ions act by replacing other necessary metal ions, such as Ca^{2+} or Zn^+ [28].

Simultaneously, it was suggested that silver ions (especially Ag^+) liberated from AgNPs can interact with phosphorus moieties in DNA, leading to inactivation of DNA replication [27] or react with sulfhydryl groups of metabolic enzymes of the chain of bacterial transport of electrons, causing their inactivation [29].

The release of silver ions is only one of the mechanisms of AgNPs action. AgNPs themselves have different physical, chemical and other properties than the solid silver from which they were obtained.

A number of authors suggest that the antibacterial effect of silver nanoparticles is due to the electrostatic interaction between negatively charged bacterial cells and positively charged nanoparticles [30].

Analysis of the electronic state of AgNPs showed that the core of the particles is zero-valent, while the shell is oxidized by interaction with the functional groups of the bandage and the surrounding medium. Earlier, analysis of the O 1s and valence band spectra showed the existence of four states of oxygen atoms on the Ag surface, which reflect the different states of silver atoms as well [31]. The discrimination of these in the Ag 3d spectrum is practically impossible because of insufficient overall energy resolution of electron energy analyzers. The surface atoms of nanoparticles, which, as a rule, are positively charged, should play the main role in interaction with the bacterial membranes. The Ag 3d spectra shoe that a large fraction (0.27) of the Ag atoms is in the Ag^+ state, which determines the antimicrobial effect. It is possible that laser radiation in the surface plasmon resonance mode changes the electronic state of AgNPs, which increases the antibacterial activity of the system as a whole.

The wavelength of laser radiation used in the study was chosen by considering the plasmon resonance effect. When the frequency of the external field coincides with the frequency of the localized surface plasmon, a resonance arises, leading to a sharp increase in the field on the surface of the particle and an increase in the absorption cross section [32].

Apparently, the statistically significant effect of enhancing the antibacterial properties of the medical gauze bandage containing AgNPs observed in the present study is caused by laser photothermolysis of bacteria when exposed to laser radiation.

Probably, AgNPs, when irradiated with a laser, absorb energy, which is transformed into heat. The resulting hyperthermia of AgNPs can lead to both local damage of the bacterial cell and intensification of metabolic processes around the heated AgNPs with the subsequent acceleration of silver ionization processes, which in turn leads to the death of the microbial cell.

4. Materials and Methods

The ordinary medical gauze bandage was used in the study, produced in the Republic of Belarus (State Standard 1172–93) both as a control and for the production of medical gauze bandages containing AgNPs.

The metal vapor technique was used for the preparation of Ag-bearing bandage (Scheme 1).

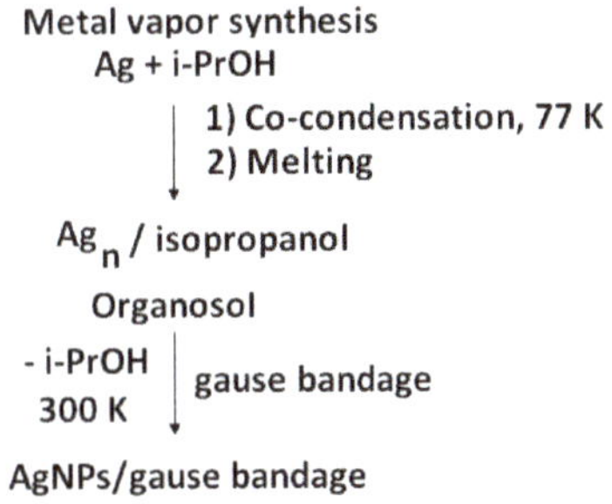

Metal vapor synthesis
Ag + i-PrOH

1) Co-condensation, 77 K
2) Melting

Ag_n / isopropanol

Organosol

- i-PrOH
300 K   gause bandage

AgNPs/gause bandage

Scheme 1. Synthesis of silver-containing nanocomposite.

The silver nanoparticles were obtained by the MVS from pieces of silver (99.99%) using an apparatus described elsewhere [33–36]. In the preparation of silver organosol, isopropanol was used as the organic dispersion medium, which was degassed in the vacuum prior to synthesis by alternating freeze-thaw cycles. Isopropanol (Fluka, 99.8%) was dried over zeolites 4 Å, and distilled under argon.

Silver was evaporated by resistive heating from a tantalum boat. During the synthesis, a vacuum of no more than 10^{-2} Pa was maintained in a 5-L quartz glass reactor with using a high-vacuum post. In a typical experiment, 120–150 mL of organic reagent was used in the synthesis and 0.1–0.12 grams of metal were evaporated. The supply of the organic reagent was adjusted with a fine adjustment valve. Before the synthesis, the glass reactor flask was evacuated, immersed in a vessel with liquid nitrogen, and then an organic reagent isopropanol was fed, which was condensed on the cooled walls of the reactor together with the metal vapors for about 1.5 h.

After the synthesis was completed, the cooling was stopped; the reactor was cut off from the vacuum post with a slide gate. Argon was fed to the reactor; the co-condensate of metal and organics was heated to the melting point. The obtained colloidal silver solution in isopropanol was impregnated with a bandage, which was before modification in a vacuum flask. The excess organosol was removed by drying in a vacuum of 1 Pa at a temperature of 80 °C.

Four strains of Gram-negative microbes (*Pseudomonas aeruginosa, Klebsiella pneumoniae, Escherichia coli, Moraxella* spp.) and two strains of Gram-positive bacteria (*Staphylococcus aureus, Staphylococcus haemolyticus*) were used. The strains were sowed from purulent wounds of patients of surgical departments in Grodno (Belarus).

The sampling for microbiological testing was performed in patients with purulent wounds using standard disposable sterile tampons by Heinz Herenz company, within an hour the material was delivered to a microbiological laboratory where a pure microbial culture was isolated and identified with a BioMerieux Vitek apparatus, antibiotic susceptibility of each microorganism was defined.

The bacteria sensitivity to the six most commonly used antibiotics in surgical hospitals in Belarus (amoxicillin (ACC), cephalexin (CFL), gentamicin (GEN), ciprofloxacin (CIP), cefazolin (CZ), erythromycin (ERI)) is presented in Table 8. Definition of antibacterial sensitivity was performed by diffusion into agar using discs.

Table 8. Antibacterial sensitivity of the strains of microbes used in the study.

Microbial Strain	Antibiotics					
	ACC	CZ	CFL	CIP	GEN	ERI
Staphylococcus aureus	S	S	S	S	S	R
Staphylococcus haemolyticus	R	S	R	S	S	S
Pseudomonas aeruginosa	R	R	R	R	R	R
Klebsiella pneumoniae	R	R	R	R	R	R
Escherichia coli	R	S	S	S	S	R
Moraxella spp.	R	R	R	R	R	I

S = the microorganism is sensitive to this antibiotic, I = moderate resistance of the microorganism to the antibiotic presented, R = the microorganism is resistant to this antibiotic. ACC = amoxicillin, CFL = cephalexin, GEN = gentamicin, CIP = ciprofloxacin, CZ = cefazolin, ERI = erythromycin.

Then, the isolated culture of the microbe was inoculated on sloping beef-extract agar, after 24 h cultivation, a sterile 0.85% solution of NaCl (5 mL) was flushed and diluted to the desired concentration with the same solution by successive inoculation into Petri dishes with agar of different concentrations of the microorganism. The desired concentration corresponded to the formation, after seeding, with a pipette, of 0.1 mL of the microbe suspension and placing the Petri dish in a 24 h thermostat for about 100 CFU. The following concentrations were used in the study: 0.5×10^{-6} for *Staphylococcus aureus*, 0.5×10^{-5} for *Staphylococcus haemolyticus*, *Klebsiella pneumoniae* and *Escherichia coli*, for *Pseudomonas aeruginosa* 1×10^{-7} and for *Moraxella* spp. 1×10^{-6}.

The obtained suspension of microorganisms (0.1 mL) was seeded on a Petri dish with beef-extract agar. Then, two bands of medical gauze bandage were placed on each cup, measuring 1.5×4 cm. A standard medical gauze bandage was used as a control, medical gauze bandage containing AgNPs was used in the experimental groups. After that, all Petri dishes were placed in a thermostat at a temperature of 37.0 °C on 1 day for cultivation. After 24 h, CFU were counted on both sides of the edge of the bandage at a distance from the edge equal to the diameter of one colony ad oculus and using a binocular magnifier glass.

In addition, the percentage decrease in the number of CFU was calculated by the Formula (1):

$$\text{Percentage reduction of CFU (\%)} = 100 \times (A - B)/A \qquad (1)$$

where A is the average value of the number of CFU along the edge of the bandage in the control groups; and B is the average value of the number of CFU in the groups with medical gauze bandage containing AgNPs.

Microbial strains were cultivated on the Pronadisa beef-extract agar manufactured by Laboratorios Conda, S.A., which was prepared and sterilized according to the instructions of the manufacturer.

Sterilization of the experimental and control samples of the bandage was carried out by autoclaving at 121 °C during 16 min with a Cliniklav-25 vacuum autoclave.

Laser irradiation was performed by a Rodnik-1 therapeutic laser apparatus, a wavelength of 470 ± 30 nm (blue region of the spectrum), with a power of 5 mW, for 5 min. Laser irradiation was performed 2 and 4 h after inoculating the Petri dish and placing the dressing on it.

The schematic view of the irradiation experiment is shown on the Figure 12.

Figure 13 shows a Petri dish with agar, and two pieces of bandage placed on it irradiated with a laser.

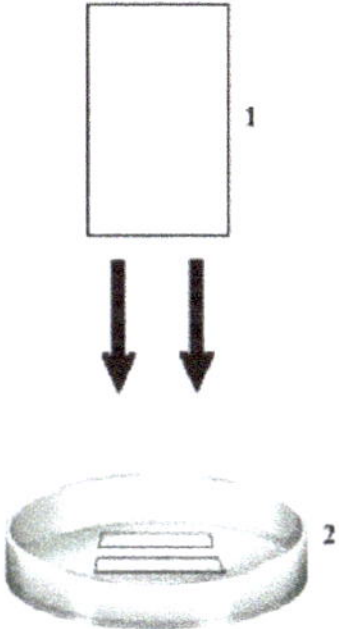

Figure 13. The scheme of the laser experiment; 1—laser and 2—two pieces of the bandage.

The frequency selected in the study is determined by the frequency of plasmon resonance of AgNPs, the power and time of the exposure are determined experimentally.

Statistical processing of the results was carried out using the program Statistica 10.0. The difference between the groups was estimated using the nonparametric Mann-Whitney U-test at a given 5% significance level.

To assess the degree of increase in the antibacterial effect, the difference between the percentage decrease in the number of CFUs in the group where a AgNPs-containing bandage without laser irradiation was used and the percentage reduction in the group with laser irradiation was calculated 4 h after seeding the Petri dish and placing the bandage on it using the Formula (2):

$$\text{Difference in the percentage reduction of CFU (\%)} = 100 \times (A - B)/A = 100 \times (A - C)/A \quad (2)$$

where A is the average value of the number of CFU along the edge of the bandage in the control group (ordinary medical gauze bandage without laser irradiation); B is the average value of the CFU in the band with medical gauze bandage containing AgNPs with laser irradiation four hours after seeding the Petri dish and placing the bandage on it; and C is the average value of the number of CFUs in the group with medical gauze bandage containing AgNPs without laser irradiation.

Micrographs of the Ag-cotton sample were obtained by transmission electron microscopy (TEM) using a JEOL JEM 2100F/UHR instrument with a resolution of 0.1 nm. Prior to the test, 0.1 g of the sample was placed in 30 mL of isopropanol and sonicated for 300 s. A drop of the resulting mixture was placed on a carbon-coated copper TEM grid and dried for 1 h. The size of the Ag-containing particles was determined as the maximum linear dimension. To construct a histogram of the particle size distribution, the TEM data on 192 particles were processed. Identification of the chemical composition of the particles and the surface of the samples was carried out using an energy dispersive analysis (EDA) on a JED-2300 instrument. The interplanar distances in structures visible in high resolution TEM photographs were calculated using the electron diffraction patterns obtained with the fast Fourier transform in the ImajeJ 1.4 code. To assign the electron diffraction patterns to the crystallographic faces of the silver compounds, a crystallographic JCPDS database was used.

The X-ray photoelectron spectra were recorded using a XSAM-800 spectrometer (Kratos, Manchester, UK) with Mg Kα radiation at an operating power of 90 W of the X-ray tube in the fixed analyzer transmission mode at ~20 °C and a pressure in the analytical chamber of ~10^{-8} Pa. Survey and high-resolution spectra of appropriate core levels were recorded with step sizes of 1 and 0.1 eV, respectively. The photoelectron spectra were approximated by Gauss function or the sum of Gauss functions, and the background caused by secondary electrons and photoelectrons that lost energy, was approximated by the straight line. The energy scale of spectrometer was calibrated according to the standard procedure, while considering the following binding energies: 932.7, 368.3, and 84.0 eV for $Cu2p_{3/2}$, $Ag3d_{5/2}$, and $Au4f_{7/2}$, respectively. Quantification was performed using atomic sensitivity factors included in the software of the spectrometer. The samples were fixed to the sample holder by double-sided conductive adhesive tape. Sample charging was corrected by referencing to the C-C/C-H state deconvoluted in the C ls spectrum (284.8 eV).

5. Conclusions

1. The Ag-cotton nanocomposites synthesized via MVS has a pronounced antimicrobial activity against some Gram-positive and Gram-negative bacteria.

2. Antibacterial properties of medical gauze bandage containing AgNPs significantly increase when exposed to laser radiation. The effect of enhancing the antibacterial properties of the nanocomposite apparently associated with the effect of localized plasmon resonance of AgNPs.

3. TEM data indicate that the particle size distribution is narrow and monomodal with average particle size of 1.75 nm. EDA shows that interplanar distances correspond to metallic nanoparticles. XPS and EXAFS demonstrate that Ag atoms in Ag/bandage system is Ag^0, the oxidized silver is in the form of Ag-Ag-O groups.

Author Contributions: S.M.S. and N.N.I. designed the study; R.I.D., A.Y.V. and A.V.N. performed experiments; A.Y.V. provided reagents; A.Y.V., A.V.N. and S.M.S. provided technical advice; R.I.D. and A.V.N. drafted the manuscript; A.Y.V., A.V.N., S.M.S., and N.N.I. edited the manuscript.

Funding: This investigation was carried out with the State Assignment of Fundamental Research to the A. N. Nesmeyanov Institute of Organoelement Compounds of the RAS. This research received no external funding.

Acknowledgments: The authors thank CCU MSU and S.V. Savilov for TEM measurements and Yan V. Zubavichus for measurements and analysis EXAFS data.

Conflicts of Interest: The authors declare no conflict of interest.

References

1. Qiao, M.; Ying, G.G.; Singer, A.C.; Zhu, Y.G. Review of antibiotic resistance in China and its environment. *Environ. Int.* **2018**, *110*, 160–172. [CrossRef] [PubMed]

2. MacGowan, A.; Macnaughton, E. Antibiotic resistance. *Medicine* **2017**, *45*, 622–628. [CrossRef]

3. Chaudhary, A.S. A review of global initiatives to fight antibiotic resistance and recent antibiotics' discovery. *Acta Pharm. Sin. B* **2016**, *6*, 552–556. [CrossRef] [PubMed]

4. Marx, D.E.; Barillo, D.J. Silver in medicine: The basic science. *Burns* **2014**, *40*, S9–S18. [CrossRef] [PubMed]

5. Barillo, D.J.; Marx, D.E. Silver in medicine: A brief history BC 335 to present. *Burns* **2014**, *40*, S3–S8. [CrossRef] [PubMed]

6. Kodurua, J.R.; Kailasa, S.K.; Bhamore, J.R.; Kim, K.; Dutta, T.; Vellingiri, K. Phytochemical-assisted synthetic approaches for silver nanoparticles antimicrobial applications: A review. *Adv. Colloid Interface Sci.* **2018**, *256*, 326–339. [CrossRef] [PubMed]

7. Beyene, H.D.; Werkneh, A.A.; Kassa, H.; Ambaye, T.G. Synthesys paradigm and applications of silver nanoparticles (AgNPs), a review. *Sustain. Mater. Technol.* **2017**, *13*, 18–23.

8. Vazquez-Muñoz, R.; Borrego, B.; Juárez-Moreno, K.; García-García, M.; Mota Morales, J.D.; Bogdanchikova, N.; Huerta-Saquero, A. Toxicity of silver nanoparticles in biological systems: Does the complexity of biological systems matter? *Toxicol. Lett.* **2017**, *276*, 11–20.

9. Kvitek, L.; Panacek, A.; Prucek, R.; Soukupova, J.; Vanickova, M.; Kolar, M.; Zboril, R. Antibacterial activity and toxicity of silver—Nanosilver versus ionic silver. *J. Phys. Conf. Ser.* **2011**, *304*, 1–8. [CrossRef]

10. Kargozar, S.; Mozafari, M. Nanotechnology and nanomedicine: Start small, think big. *Mater. Today* **2018**, *5*, 15492–15500. [CrossRef]

11. Merisko-Liversidge, E.M.; Liversidge, G.G. Drug nanoparticles: Formulating poorly water-soluble compounds. *Toxicol. Pathol.* **2008**, *36*, 43–48. [CrossRef] [PubMed]

12. Petryayeva, E.; Krull, U.J. Localized surface plasmon resonance: Nanostructures, bioassays and biosensing—A review. *Anal. Chim. Acta* **2011**, *706*, 8–24. [CrossRef] [PubMed]

13. Vasil'kov, A.; Naumkin, A.; Nikitin, L.; Volkov, I.; Podshibikhin, V.; Lisichkin, G. Ultrahigh molecular weight polyethylene modified with silver nanoparticles prepared by metal-vapour synthesis. *AIP Conf. Proc.* **2008**, *1042*, 255–257.

14. NIST X-ray Photoelectron Spectroscopy Database. Available online: http://srdata.nist.gov/xps/ (accessed on 4 April 2018).

15. Lützenkirchen-Hecht, D.; Strehblow, H.H. Anodic silver (II) oxides investigated by combined electrochemistry, ex situ XPS and in situ X-ray absorption spectroscopy. *Surf. Interface Anal.* **2009**, *41*, 820–829. [CrossRef]

16. Bukhtiyarov, V.I.; Prosvirin, I.P.; Kvon, R.I. Application of differential charging for analysis of electronic properties of supported silver. *J. Electron Spectrosc. Relat. Phenom.* **1996**, *77*, 7–14. [CrossRef]

17. Suzer, S. Differential charging in X-ray photoelectron spectroscopy: A nuisance or a useful tool? *Anal. Chem.* **2003**, *75*, 7026–7029. [CrossRef] [PubMed]

18. Sohn, Y. SiO_2 nanospheres modified by Ag nanoparticles: Surface charging and CO oxidation activity. *J. Mol. Catal. A Chem.* **2013**, *379*, 59–67. [CrossRef]

19. Cao, Y.; Dai, W.-L.; Deng, J.-F. The oxidative dehydrogenation of methanol over a novel Ag/SiO_2 catalyst. *Appl. Catal. A* **1997**, *158*. [CrossRef]

20. Tan, B.J.; Sherwood, P.M.; Klabunde, K.J. XPS studies of gold films prepared from nonaqueous gold colloids. *Langmuir* **1990**, *6*, 105–113. [CrossRef]

21. Dubey, M.; Gouzman, I.; Bernasek, S.L.; Schwartz, J. Characterization of self-assembled organic films using differential charging in X-ray photoelectron spectroscopy. *Langmuir* **2006**, *22*, 4649–4653. [CrossRef] [PubMed]

22. Hoflund, G.B.; Weaver, J.F.; Epling, W.S. Ag_2O XPS spectra. *Surf. Sci. Spectra* **1994**, *3*, 157–162. [CrossRef]

23. Hoflund, G.B.; Weaver, J.F.; Epling, W.S. AgO XPS spectra. *Surf. Sci. Spectra* **1998**, *3*, 163–168. [CrossRef]

24. Beamson, G.; Briggs, D. *High Resolution XPS of Organic Polymers: The Scienta ESCA300 Database*; Wiley: Chichester, UK, 1992.

25. Shin, H.S.; Choi, H.C.; Jung, Y.; Kim, S.B.; Song, H.J.; Shin, H.J. Chemical and size effects of nanocomposites of silver and polyvinyl pyrrolidone determined by X-ray photoemission spectroscopy. *Chem. Phys. Lett.* **2004**, *383*, 418–422. [CrossRef]

26. Schnippering, M.; Carrara, M.; Foelske, A.; Kötz, R.; Fermín, D.J. Electronic properties of Ag nanoparticle arrays. A Kelvin probe and high resolution XPS study. *Phys. Chem. Chem. Phys.* **2007**, *9*, 725–730. [CrossRef] [PubMed]

27. Feng, Q.L.; Wu, J.; Chen, G.Q.; Cui, F.Z.; Kim, T.N.; Kim, J.O. A mechanistic study of the antibacterial effect of silver ions on Escherichia coli and Staphylococcus aureus. *J. Biomed. Mater. Res.* **2000**, *52*, 662–668. [CrossRef]

28. Betts, A.J.; Dowling, D.P.; McConnell, M.L.; Pope, C. The influence of platinum on the performance of silver-platinum antibacterial coatings. *Mater. Des.* **2005**, *26*, 217–222. [CrossRef]

29. Gupta, A.; Maynes, M.; Silver, S. Effects of halides on plasmid-mediated silver resistance in *Escherichia coli*. *Appl. Environ. Microbiol.* **1998**, *64*, 5042–5045. [PubMed]

30. Stoimenov, P.K.; Klinger, R.L.; Marchin, G.L.; Klabunde, K.J. Metal oxide nanoparticles as bactericidal agents. *Langmuir* **2002**, *18*, 6679–6686. [CrossRef]

31. Boronin, A.I.; Koscheev, S.V.; Zhidomirov, G.M. XPS and UPS study of oxygen states on silver. *J. Electron Spectrosc. Relat. Phenom.* **1998**, *96*, 43–51. [CrossRef]

32. Pompa, P.P.; Martiradonna, L.; Torre, A.D.; Sala, F.D.; Manna, L.; Vittorio, M.D.; Calabi, F.; Cingolani, R.; Rinaldi, R. Metal-enhanced fluorescence of colloidal nanocrystals with nanoscale control. *Nat. Nanotechnol.* **2006**, *1*, 126–130. [CrossRef] [PubMed]

33. Blackborow, J.R.; Young, D. Reactivity and Structure Concepts in Organic Chemistry. In *Metal Vapour Synthesis in Organometallic Chemistry*; Hafner, K., Leah, J.-M., Ree, C.W., Schleyer, P.v.R., Trost, B.M., Zahraník, R., Eds.; Springer: Berlin/Heidelberg, Germany, 1979; Volume 9.

34. Klabunde, K.J. *Free Atoms, Clusters and Nanoscale Particles*; Academic Press: New York, NY, USA, 1994.
35. Vasil'kov, A.Y.; Naumkin, A.V.; Volkov, I.O.; Podshibikhin, V.L.; Lisichkin, G.V.; Khokhlov, A.R. XPS/TEM characterisation of Pt–Au/C cathode electrocatalysts prepared by metal vapour synthesis. *Surf. Interface Anal.* **2010**, *42*, 559–563. [CrossRef]
36. Abd-Elsalam, K.A.; Vasil'kov, A.Y.; Said-Galiev, E.E.; Rubina, M.S.; Khokhlov, A.R.; Naumkin, A.V.; Shtykova, E.V.; Alghuthaymi, M.A. Bimetallic blends and chitosan nanocomposites: Novel antifungal agents against cotton seedling damping-off. *Eur. J. Plant Pathol.* **2017**. [CrossRef]

Article

Nanosynthesis of Silver-Calcium Glycerophosphate: Promising Association against Oral Pathogens

Gabriela Lopes Fernandes [1], Alberto Carlos Botazzo Delbem [2], Jackeline Gallo do Amaral [2], Luiz Fernando Gorup [3,4], Renan Aparecido Fernandes [1,5], Francisco Nunes de Souza Neto [3], José Antonio Santos Souza [2], Douglas Roberto Monteiro [6], Alessandra Marçal Agostinho Hunt [7], Emerson Rodrigues Camargo [3] and Debora Barros Barbosa [1,*]

[1] Department of Dental Materials and Prosthodontics, School of Dentistry, Araçatuba, São Paulo State University (UNESP), Araçatuba 16015-050, São Paulo, Brazil; fernandesgabriela@hotmail.com (G.L.F.); renanfernandes@fai.com.br (R.A.F.)

[2] Department of Pediatric Dentistry and Public Health, School of Dentistry, Araçatuba, São Paulo State University (UNESP), Araçatuba 16015-050, São Paulo, Brazil; adelbem@foa.unesp.br (A.C.B.D.); jackelineamaral@gmail.com (J.G.d.A.); joseanonio_249@hotmail.com (J.A.S.S.)

[3] Department of Chemistry, Federal University of São Carlos, São Carlos 13565-905, São Paulo, Brazil; lfgorup@gmail.com (L.F.G.); francisco_nsn@yahoo.com.br (F.N.d.S.N.); camargo@ufscar.br (E.R.C.)

[4] FACET—Department of Chemistry, Federal University of Grande Dourados, Dourados 79804-970, Mato Grosso do Sul, Brazil

[5] Department of Dentistry, University Center of Adamantina (UNIFAI), Adamantina 17800-000, São Paulo, Brazil

[6] Graduate Program in Dentistry (GPD—Master's Degree), University of Western São Paulo (UNOESTE), Presidente Prudente 19050-920, São Paulo, Brazil; douglasmonteiro@hotmail.com

[7] Department of Microbiology and Molecular Genetics, Michigan State University, East Lansing, MI 48823, USA; alehunt@msu.edu

* Correspondence: debora@foa.unesp.br; Tel.: +55-18-3636-3284

Received: 27 April 2018; Accepted: 25 June 2018; Published: 27 June 2018

Abstract: Nanobiomaterials combining remineralization and antimicrobial abilities would bring important benefits to control dental caries. This study aimed to produce nanocompounds containing calcium glycerophosphate (CaGP) and silver nanoparticles (AgNP) by varying the reducing agent of silver nitrate (sodium borohydride (B) or sodium citrate (C)), the concentration of silver (1% or 10%), and the CaGP forms (nano or commercial), and analyze its characterization and antimicrobial activity against ATCC *Candida albicans* (10231) and *Streptococcus mutans* (25175) by the microdilution method. Controls of AgNP were produced and silver ions (Ag^+) were quantified in all of the samples. X-ray diffraction, UV-Vis, and scanning electron microscopy (SEM) analysis demonstrated AgNP associated with CaGP. Ag^+ ions were considerably higher in AgCaGP/C. *C. albicans* was susceptible to nanocompounds produced with both reducing agents, regardless of Ag concentration and CaGP form, being Ag10%CaGP-N/C the most effective compound (19.5–39.0 µg Ag mL^{-1}). While for *S. mutans*, the effectiveness was observed only for AgCaGP reduced by citrate, also presenting Ag10%CaGP-N the highest effectiveness (156.2–312.5 µg Ag mL^{-1}). Notably, CaGP enhanced the silver antimicrobial potential in about two- and eight-fold against *C. albicans* and *S. mutans* when compared with the AgNP controls (from 7.8 to 3.9 and from 250 to 31.2 µg Ag mL^{-1}, respectively). The synthesis that was used in this study promoted the formation of AgNP associated with CaGP, and although the use of sodium borohydride (B) resulted in a pronounced reduction of Ag^+, the composite AgCaGP/B was less effective against the microorganisms that were tested.

Keywords: silver; calcium glycerophosphate; nanoparticles; *Candida albicans*; *Streptococcus mutans*

---

1. Introduction

The synthesis and study of properties of new biomaterials has been emphasized lately with the improvement of nanotechnology. In this context, the development of nanomaterials has been the focus of many areas of chemistry, physics, and materials science because of the promising characteristics that these materials exhibit [1].

Nanotechnology aims to manipulate particles by creating new structures with favorable properties in many areas, such as medicine and dentistry [2], and new alternatives of treatment for oral pathologies are emerging. Metallic nanoparticles, in particular silver nanoparticles (AgNP), have been studied as an alternative antimicrobial agent against a broad spectrum of species in the control of oral biofilms [3–5]. Although there are several studies where AgNP are used as antimicrobial agents, their mechanism of action is not completely understood. Kim et al. [6] and Besinis et al. [4] related their antimicrobial action to the toxicity resulting from free metal ions dissolution from the surface of the AgNP. In addition, AgNP would lead to oxidative stress through the generation of reactive oxygen species (ROS), interacting with cytoplasmic and nucleic acid components by inhibiting enzymes of the respiratory chain and changing the permeability of the cytoplasmatic bacterial membrane [7–11].

Among oral pathologies, dental caries is one of the most common diseases in humans that relates to genetics, saliva, and diet of the host [9]. *Streptococcus mutans* is the main cariogenic microorganism owing to its ability to produce acids and glucans from sugar metabolism, which exceed the buffering capacity of saliva [9–11] and leads by a localized and irreversible destruction of the tooth structure [9,12]. However, recent evidence indicates the presence of *C. albicans* and *S. mutans* in oral biofilms, suggesting that the interaction between them can lead to the development of caries [9,13,14]. *C. albicans* colonization depends on the presence of the bacteria, which, besides promoting adhesion sites, act as a carbon source for yeast growth. On the other hand, yeasts reduce the levels of oxygen for streptococci [9]. Studies have shown the resistance of many microorganisms to antimicrobial agents currently used [15,16].

Studies since the 1930s [17] have reported the importance of using calcium phosphate derivatives for favouring the remineralization process in dental caries. Calcium glycerophosphate (CaGP) is an organic phosphate salt with anti-caries properties being demonstrated in studies carried out in monkeys [18] and in rats [19]. It is action in dental biofilms may be related to the increase of calcium and phosphate levels [20], buffering capacity [18], and reduction of the mass of the biofilms [21]. Becauses that it seems to interact with dental tissues [22], CaGP has been incorporated in dentifrices [23,24]. Do Amaral et al. [25] and Zaze et al. [26], when associating CaGP (0.25%) in toothpastes with fluoride at low concentrations, found the same efficacy against caries in enamel when compared to dentifrices that were supplemented with a higher concentration of fluoride demonstrating CaGP be an good option for oral products to both prevent caries and avoid fluorose in dental tissues.

The use of a biomaterial containing both an antimicrobial and a compound acting as a source of calcium phosphate for dental remineralization would have a great impact on the prevention and control of dental caries. Therefore, this study aimed to produce nanocompounds containing calcium glycerophosphate (CaGP) and silver nanoparticles (AgNP) by varying the reducing agent of silver nitrate (sodium borohydride or sodium citrate), the concentration of silver (1% or 10%), and the CaGP forms (nano or microparticulated), and analyze its characterization and antimicrobial activity against ATCC strains of *Candida albicans* and *Streptococcus mutans*.

2. Results

2.1. Synthesis and Characterization of Ag-CaGP Nanocomposites

UV-Vis absorption spectroscopy (UV-Vis) showed that Ag-CaGP nanocomposites presented silver in nanosized dimensions in all of the nanocomposites synthesized, regardless of the reducing agent used. It was demonstrated by the presence of an intense absorption peak, denominated plasmonic band, which occurred between 420 and 450 nm (Figure 1a, Figure S1). It characterizes noble metal nanoparticles, with strong absorption band being observed in the visible region [27]. The CaGP did not exhibit absorption peak in the visible region of the electromagnetic spectrum.

X-ray diffraction (XRD) pattern indicated that all of the Ag-CaGP nanocomposites were composed of AgNP and CaGP for confirming the presence of silver in Ag-CaGP nanocomposites through comparison of the nanoparticles and CaGP. The typical powder XRD pattern of the prepared CaGP showed diffraction peaks at 2θ = 6.30°, 12.3°, 26.4°, 41.1°, and 44.2° (Figure 1b, Figure S2), and the corresponding crystallographic form (PDF No. 1-17) [28]. The typical powder XRD pattern of the silver nanoparticles showed (Figure 1b) diffraction peaks at 2θ = 38.2°, 44.4°, 64.6°, 77.5°, and 81.7°, which can be indexed to (111), (200), (220), (311), and (222) planes of pure silver with face-centered cubic system (PDF No. 04-0783).

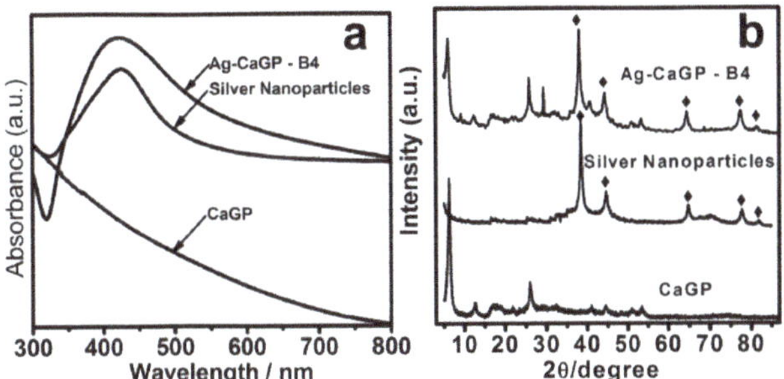

Figure 1. (**a**) UV-Vis (**b**) XRD pattern of Ag-CaGP (B4 group) nanocomposite, silver nanoparticles, and CaGP.

Nanostructured materials that exhibit a pattern of small nanoparticles scattered on a larger surface, similar to glass bead embellishments on a Christmas tree, are generally classified as a decorated material. The scanning electron microscopy (SEM) images of Figure 2 show this typical pattern, with spherical silver nanoparticles (indicated by arrows) decorating the surface of the CaGP microparticles in all synthesized nanocomposites containing 10% Ag (B4; B8; C4). In addition, transmission electron microscopy (TEM) was performed for the nanocomposite B4 (Figure S3).

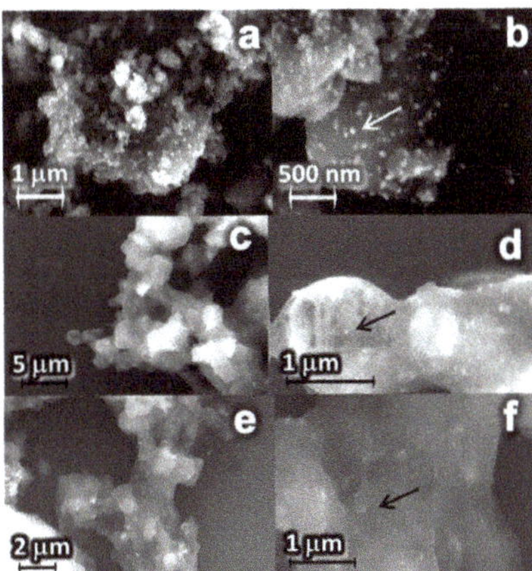

Figure 2. SEM images of the Ag-CaGP nanocomposites: B4 (**a,b**), C4 (**c,d**), and B8 (**e,f**) at different magnifications. The arrows indicate silver nanoparticles on the surface of CaGP in B4 and in the bulk of CaGP in C4 and B8.

The energy-dispersive X-ray sprectroscopy (EDS) clearly showed the outline of Ag-CaGP nanocomposites in all micrographs. Also, Figure 3 (B4), Figure 4 (C4) and Figures S4–S6, the two-dimensional (2D) images were constructed by analyzing the energy released from the issuance Si Kα, O Kα, P Kα, Ca Kα, and Ag Kα, indicating the distribution of these elements on the demarcated area in the micrograph.

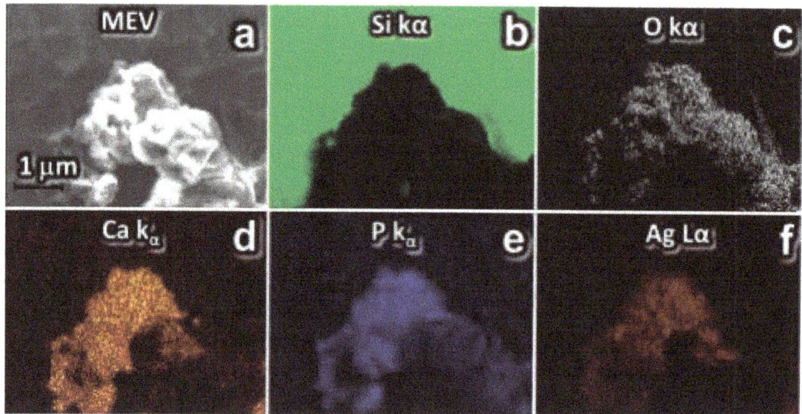

Figure 3. SEM and EDS mapped in 2D elements issuance Si Kα, O Kα, P Kα, Ca Kα, and Ag Kα false color. Analysis of the distribution of silver nanoparticles on the Ag-CaGP for sample B4: (**a**) SEM image; (**b**) chemical mapping of silicon element present in the substrate, where the electron beam was focused directly on the substrate and is showed in green color, and the dark regions the beam was focused in Ag-CaGP nanocomposite B4; (**c**–**f**) oxygen, calcium, phosphorus, and silver, respectively, demonstrating they are constituents of the Ag-CaGP.

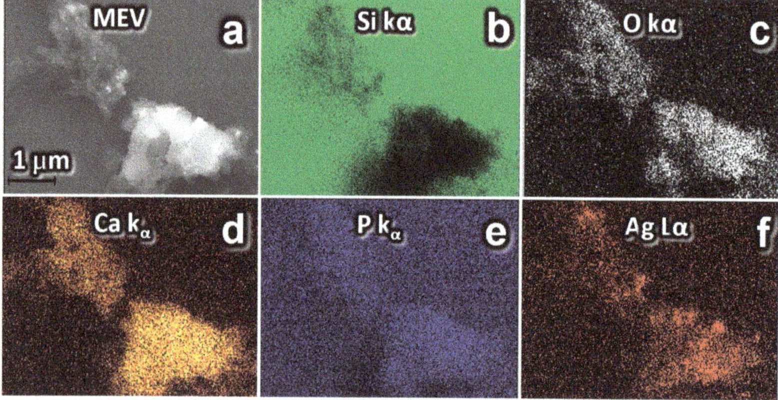

Figure 4. SEM and EDS mapped in 2D elements issuance Si Kα, O Kα, P Kα, Ca Kα, and Ag Kα false color. Analysis of the distribution of silver nanoparticles on the Ag-CaGP for sample C4: (**a**) SEM image; (**b**) chemical mapping of silicon element present in the substrate, where the electron beam was focused directly on the substrate and is showed in green color, and the dark regions the beam was focused in Ag-CaGP nanocomposite C4; and, (**c**–**f**) oxygen, calcium, phosphorus, and silver, respectively, demonstrating they are constituents of the Ag-CaGP.

2.2. Minimum Inhibitory Concentration

The results showed that the MIC values were related to the synthesis process and the Ag concentration used (Table 1). Nanocomposites that were obtained using $Na_3C_6H_5O_7$ as reducing agent showed the most effective antimicrobial activity against *C. albicans* and *S. mutans*. In these composites, the lowest MIC values were observed for those containing 10% of Ag (C3 and C4), being between 19.05 and 39.05 µg/mL for *C. albicans* and 156.2 and 625 µg/mL for *S. mutans*. The nanocomposites that were synthesized using $NaBH_4$ as reducing agent and isopropanol as solvent showed fungicidal effect varying between 100 and 1600 µg/mL, whilst no effect against *S. mutans* was observed. While the nanocomposites synthesized using the same reducing agent and deionized water as solvent did not show any effect against both microorganisms. In addition to the MICs found for the synthesized compounds, it was carried out the microdilution assay to find the MIC values for the solutions containing only AgNP or CaGP diluted in deionized water, besides the other compounds used in the synthesis reaction as reducing and surfactant agents. These data are showed in Table 2.

Table 1. Values of minimum inhibitory concentration (MIC) of the nanocompounds based on µg of AgCaGP mL^{-1} and on µg of Ag mL^{-1} in each ones, synthesized using sodium borohydride (Group B) and sodium citrate (Group C), and silver ions concentration (µg Ag^+/mL) in all nanocompounds tested.

GROUP	Ag %/CaGP Form	Solvent	MIC (µg AgCaGP mL^{-1}/µg Ag mL^{-1})		µg Ag^+/mL
			C. albicans	*S. mutans*	
B1	1/M	I	>1600/>16	>1600/>16	2.83
B2	1/N	I	400–1600/0.40–16	>1600/>16	4.46
B3	10/M	I	400–800/40–80	>1600/>16	10.81
B4	10/N	I	100–200/10–20	>1600/>16	63.34
B5	1/M	W	>1600/>16	>1600/>16	0.44
B6	1/N	W	>1600/>16	>1600/>16	2.76
B7	10/M	W	>1600/>16	>1600/>16	5.97
B8	10/N	W	>1600/>16	>1600/>16	15.63
C1	1/M	W	156.2–312.5/1.56–3.12	1250/12.5	305.43
C2	1/N	W	156.2–312.5/1.56–3.12	1250/12.5	168.14
C3	10/M	W	39.0/3.9	312.5–625/31.2–62.5	506.73
C4	10/N	W	19.5–39.0/1.9–3.9	156.2–312.5/15.6–31.2	487.95

Table 2. Values of minimum inhibitory concentrations (MIC) (µg/mL) and silver ions (Ag^+) (µg Ag^+/mL) of the control solutions: AgNP reduced by sodium citrate ($Na_3C_6H_5O_7$), and by sodium borohydride ($NaBH_4$); silver nitrate ($AgNO_3$); nanoparticulated CaGP (CaGP-nano), and CaGP in commercial form (CaGP-commercial); sodium citrate and surfactant ($Na_3C_6H_5O_7$+NH-PM), and sodium borohydride and surfactant ($NaBH_4$+NH-PM).

Controls	MIC		Ag^+
	C. albicans	*S. mutans*	
AgNP ($Na_3C_6H_5O_7$)	7.8	250	107.2
AgNP ($NaBH_4$)	62.5	125	576.2
$AgNO_3$	5.3	21.2	-
CaGP-nano	>5000	>5000	-
CaGP-commercial	>5000	>5000	-
$Na_3C_6H_5O_7$+NH-PM	>400	>400	-
$NaBH_4$+NH-PM	>1500	>1500	-

2.3. Determination of Ag^+ Concentration

The Ag^+ concentration of all the nanocomposites containing Ag (AgNP and Ag-CaGP) is showed in Table 1. For samples that were obtained through $NaBH_4$ route (B1–B8), a reduction of ionic silver higher than 98% was observed, when considering the total amount of ionic silver added to the reaction was 500 μg Ag^+/mL for B1, B2, B5, B6, and 5000 μg Ag^+/mL for B3, B4, B7, and B8. While for the compounds that were synthesized using $Na_3C_6H_5O_7$ as reducing agent, the ionic silver remaining was higher and reached about 10% in those samples that were produced using initially 5000 μg Ag^+/mL in the reaction process (C3 and C4). C1 and C2 presented 61.1% and 33.3%, respectively, of ionic silver in samples as the total Ag^+ added in the reaction was 500 μg Ag^+/mL. For AgNP with no CaGP added to the reaction (Table 2) obtained by $Na_3C_6H_5O_7$ route (nanoAg($Na_3C_6H_5O_7$)) the Ag^+ concentration was 107.25 μg/mL, whereas for AgNP produced through $NaBH_4$ (nanoAg($NaBH_4$)) the Ag^+ concentration was 576.19 μg/mL.

3. Discussion

In the present study, both of the synthesis methods proposed using sodium citrate or sodium borohydride as reducing agents, led to the anchorage between the silver nanoparticles and calcium glycerophosphate (Figure 2). Besides, in general, the nanocomposites (AgCaGP) were effective against reference strains of *Candida albicans* and *Streptococcus mutans*. Notably, CaGP substantively increased the antimicrobial effectiveness of silver in the AgCaGP, reducing up to a quarter their minimum inhibitory concentration when compared to the respective AgNP controls (Tables 1 and 2).

Although the CaGP has been previously nanoparticulated before the Ag-CaGP synthesis, in our study it was not characterized as being in nanoparticulated form when associated with silver. It might be happen due to the poor solubility of calcium at pH = 7 [29], even when using the same dispersant (NH-PM), as preconized by Miranda et al., whom synthesized AgNP with hydroxyapatite. A pastier bulk was particularly noted in micrographics of Ag-CaGP when water was used as solvent instead of isopropanol (Figure 2c–f), regardless of the reducing agent that was used in the reaction. Although there has not been difference between micro and nanoparticulated-CaGP in the SEM images, its form influenced the amount of silver ions in the compounds (Table 1). In addition, our results showed the antimicrobial effectiveness against *C. albicans* and *S. mutans* for the samples of group C, and it could be explained by the highest amount of silver ions that are present in those compounds [4,30–35].

This expressive difference in the quantity of silver ions between groups B and C would be related to the characteristics of the reducing agents used, being sodium borohydride considered a stronger reducing agent than sodium citrate [33]. Although silver ions are effective to kill several pathogenic microorganisms, they are easily dispersed, which quickly decreases its local concentration to levels of low effectivity. Moreover, ambient light reduces ionic silver forming typical black spots on skin or on any surface of contact [36]. This process causes aesthetic problems and it has potential to injure healthy living tissues. Silver nanoparticles, contrary to ionic silver, induce the production of reactive oxygen species (ROS), which is the primary antimicrobial mechanism [37]. However, AgNP tend to form aggregates in the absence of any support, reducing their efficacy. Therefore, substrates decorated with immobilized AgNP exhibits enhanced antimicrobial activity for longer periods, reducing the undesirable secondary effects that are associated to free ionic silver [38]. Although the difficult to separate the impact of free ionic silver from the AgNP antimicrobial action, differences that were observed in minimum inhibitory concentrations (MIC), as shown in Table 1, for *C. albicans* and *S. mutans* suggests the influence of their respective metabolism on the efficacy of silver against each microorganism.

Furthermore, other factors may influence the antimicrobial potential of AgNP [34]. For instance, how the compound containing silver interacts with the microorganisms would dependent on the characteristics of the AgNP formed, as well as the chemical and physical changes that may occur when they are added to the medium of interest [33]. In general, for the synthesis of AgNP, $AgNO_3$ is used as source of silver, water or ethanol as solvent, and sodium borohydride or sodium citrate as reducing agent [39]. Fabricated under similar conditions, the AgNP would have a negative surface charge [33,40], and this fact is noteworthy to elucidate the lower effectiveness of the compounds against *S. mutans*. Bacteria have a negative outer membrane charge [41] and the electrostatic attraction may have been hampered, and hence the action of the AgNP associated or not with CaGP on the *S. mutans* was diminished. On the other hand, apart from fungi present a neutral surface charge [41] and might enhance the attraction of AgNP, the presence of phospholipid components, which contain phosphate groups, may have improved the antimicrobial activity of silver by targeting these sites [42,43]. Indeed, the control of AgNP reduced by sodium citrate showed a lower amount of ions (107.2 µg Ag^+/mL) than the control that was produced using sodium borohydride (576.2 µg Ag^+/mL), and it was more effective against *C. albicans*, suggesting an antifungal potential of AgNP by itself, which may have afforded the disruption of the *C. albicans* cell membrane by damaging the inner layers of the cell wall, increasing their permeabilization and then allowing for the passage of these particles to into the cell.

On the contrary, against *S. mutans* plaktonic cells, Ag^+ may have played a preponderant role, particularly in view of the MIC values that are found for $AgNO_3$ (21.2 µg/mL) when compared to those for AgNP, regardless of the reducing agent that is used in the reaction (250 and 125 µg/mL, respectively, for AgNP ($Na_3C_6H_5O_7$) and AgNP ($NaBH_4$)). Noteworthy was the effect that was produced against *S. mutans* when CaGP was associated with AgNP (Table 1). CaGP afforded an increment in the silver activity and it could be related to the acidogenic and acidic characteristic of *S. mutans*, acting CaGP probably as a buffer, and hence might have prevented the proliferation of the cells in the medium [44–46]. So that, the CaGP buffer activity and the highest amount of Ag^+ ions could account for the better effectiveness of the samples of C group against that gram positive bacteria tested.

4. Materials and Methods

4.1. Synthesis of Silver-Calcium Glycerophosphate (Ag/CaGP) Nanocomposites

Ag/CaGP nanocomposites were synthesized at the Interdisciplinary Laboratory of Electrochemistry and Ceramics of the Chemistry Department in Federal University of São Carlos. Initially, the commercial form of calcium glycerophosphate (80% β-isomer and 20% rac-α-isomer, CAS 58409-70-4, Sigma-Aldrich Chemical Co., St. Louis, MO, USA) was acquired and was nanoparticulated using a ball mill for 24 h at 120 rpm, obtaining nanoparticles of approximately 10 nm. Then, two chemistry methods were employed for the synthesis. The first method was employed using sodium borohydride as reducing agent ($NaBH_4$, Sigma-Aldrich Chemical Co., St. Louis, MO, USA) and was based on the methodology that was proposed by Miranda et al. [29]. The synthesis was carried out in an alcoholic medium (isopropanol) or deionized water. For this, suspensions containing 5 g of CaGP and silver nitrate ($AgNO_3$ Merck KGaA, Darmstadt, Hessen, Germany) at 0.85 or 0.085 g were prepared in the presence of 0.5 mL of a surfactant (ammonium salt of polymethacrylic acid (NH-PM), Polysciences Inc., Warrington, PA, USA) (Table 1). Then, $NaBH_4$ (0.015 g) was added to each suspension, which caused the reduction of Ag^+ to metallic silver nanoparticles in the presence of CaGP. The molar stoichiometric ratio between Ag^+ and $NaBH_4$ was 1:1.26, respectively. The second method was based on that proposed by Turkevich et al. [47] and Gorup et al. [48] (2011, p. 355). The reducing agent of $AgNO_3$ was sodium citrate ($Na_3C_6H_5O_7$, Merck KGaA) and the stoichiometric ratio of each compound was, respectively, 1:3. Thus, in a flask containing 100 mL of deionized H_2O 5 g of CaGP was added following of 0.5 mL NH-PM and 1.4 g of $Na_3C_6H_5O_7$. This mixture was kept under magnetic stirring and heating. After reaching 95 °C temperature, $AgNO_3$ was added and this suspension was maintained stirring for 30 min until occurring the color change, which qualitatively indicated the formation of AgNP. Controls containing only the reducing agents and surfactant, and AgNP produced by both reducing were also prepared.

4.2. Characterization of Ag-CaGP Nanocomposites

In order to demonstrate the presence of AgNP and CaGP in the compounds, the UV-Vis absorption spectroscopy was employed. The measure is based on the phenomenon of plasmon resonance band, as observed in metallic nanoparticles. Thus, UV-Vis spectra of Ag-CaGP nanocomposites were obtained from aqueous solutions poured out in a commercial quartz cuvette with 1 cm optical path using a spectrophotometer (Shimadzu MultSpec-1501 spectrophotometer; Shimadzu Corporation, Tokyo, Japan) at 300 to 800 nm. Water was used as blank.

After a drying step, the resulting powder, Ag-CaGP, was subjected to X-ray diffraction (XRD) phase characterization using Cu Kα radiation ($\lambda = 1.5406$ Å), generated at a voltage of 30 kV and a current of 30 mA with continuous sweep in the range of $5° < 2\theta < 80°$, at a scan rate of $2°/\text{min}$ (Diffractometer Rigaku DMax-2000PC, Rigaku Corporation, Tokyo, Japan). The particles morphology was also characterized by scanning electron microscopy (SEM) on a Zeiss Supra 35VP microscope (S-360 Microscope, Leo, Cambridge, MA, USA), with field emission gun electron effect (FEG-SEM) operating at 10 kV. A drop of each sample were added with a micropipette and deposited on silicon metal plate (111) and dried at 40 °C for 2 h. With this technique, we can identify in the synthesized biomaterials the presence of silver, oxygen, silicon, phosphate, and calcium, which were artificially colored (Figures 3 and 4).

4.3. Minimum Inhibitory Concentration (MIC)

The MIC values for each sample were determined through the microdilution method and followed the Clinical Laboratory Standards Institute guidelines (CLSI, documents M27-A2 and M07-A9). *Candida albicans* (ATCC 10231) was cultivated on Sabouraud Dextrose Agar (SDA, Difco, Le Pont de Claix, France) and *S. mutans* (ATCC 25175) on Brain Heart Infusion Agar (BHI, Difco, Le Pont de Claix, France). Inocula from 24 h cultures on the respective media were adjusted to a turbidity equivalent to a 0.5 McFarland standard in saline solution (0.85% NaCl). This suspension was diluted (1:5) in saline solution, and afterwards diluted (1:20) in RPMI 1640 or BHI. Initially, the Ag-CaGP nanocomposite was diluted in deionized water in a geometric progression, from 2 to 1024 times. Afterwards, each Ag-CaGP nanocomposite concentration obtained previously was diluted (1:5) in RPMI 1640 medium (Sigma-Aldrich) for *C. albicans* and in BHI for *S. mutans*. The final concentrations of Ag-CaGP nanocomposite in the dispersion ranged from 5 to 0.01 mg/mL. Each inoculum (100 μL) was added to the respective well of microtiter plates containing 100 μL of each specific concentration of the Ag-CaGP nanocomposite solution. The microtiter plates were incubated at 37 °C, and the MIC values were determined visually as the lowest concentration of Ag-CaGP with no microorganism growth after 48 h for *C. albicans* and 24 h for *S. mutans*. All of the assays were repeated in triplicate on three different occasions.

4.4. Determination of Ag⁺ Concentration

The evaluation of Ag^+ concentration in Ag-CaGP and AgNP, as obtained by both reducing agents, was determined by a specific electrode 9616 BNWP (Thermo Scientific, Beverly, MA, USA) coupled to an ion analyzer (Orion 720 A⁺, Thermo Scientific, Beverly, MA, USA). A 1000 μg/mL silver standard was prepared placing 1.57 g of dried $AgNO_3$ into a 1 L volumetric flask containing deionized water. This solution was stored in an opaque bottle in a dark location and diluted in deionized water at the moment of dosage in order to achieve the standard concentrations used. Thus, the combined electrode was calibrated with standards containing 6.25 to 100 μg Ag/mL at the same conditions of samples. A silver ionic strength adjuster solution (ISA, Cat. No. 940011) that provides a constant background ionic strength was used (1 mL of each sample/standard: 0.02 mL ISA).

5. Conclusions

In conclusion, the synthesis that is proposed in this study promoted the anchorage of AgNP with the CaGP, and the nanocomposites produced using sodium citrate as reducing agent were effective against both of the microorganisms tested. Also, the highlight of our study was that the addition of CaGP to AgNP expressively reduced the MIC values when it was compared with the MIC values of AgNP by itself. These promising results strongly encourage further studies with the purpose of producing biomaterials with antimicrobial and remineralizing functions in the near future, particularly in the dental field.

Supplementary Materials: The following are available online at http://www.mdpi.com/2079-6382/7/3/52/s1, Figure S1: UV–Visible spectrum of Ag-CaGP nanocomposites, Figure S2: X-ray diffraction pattern of Ag-CaGP nanocomposites, Figure S3: TEM images of B4 Ag-CaGP nanocomposite, Figure S4: SEM images and energy-dispersive X-ray sprectroscopy (EDS) mapping in 2D elements issuance Si Kα, O Kα, P Kα,Ca Kα and Ag Kα false color. Analysis of the distribution of silver nanoparticles in the Ag-CaGP nanocomposites B1, B2 and B3, Figure S5: SEM images and EDS mapping in 2D elements issuance Si Kα, O Kα, P Kα,Ca Kα and AgKα false color. Analysis of the distribution of silver in the Ag-CaGP nanocomposites B5, B6 and B7, Figure S6: SEM images and EDS mapping in 2D elements issuance Si Kα, O Kα, P Kα,Ca Kα and Ag Kα false color. Analysis of the distribution of silver nanoparticles in the Ag-CaGP nanocomposites C1, C2 and C3.

Author Contributions: G.L.F., D.B.B., A.C.B.D. conceived and designed the experiments; G.L.F., R.A.F., J.G.d.A., J.A.S.S. performed the experiments; G.L.F., D.B.B., A.C.B.D., D.R.M. analyzed the data; F.N.d.S.N., L.F.G., E.R.C. contributed reagents/materials/analysis tools; G.L.F., D.B.B., A.M.A.H. wrote the paper.

Funding: This study was supported by: São Paulo Research Foundation (FAPESP, Processes 2013/24200-9 and 2014/08648-2), Brazil; and Coordination for the Improvement of Higher Education Personnel (Capes, Process 88881.030445/2013-01), Brazil.

Conflicts of Interest: The authors declare no conflict of interest. The founding sponsors had no role in the design of the study in the role on the design of the study in the collection, analyses, or interpretation of data; in the writing of the manuscript, and in the decision to publish the results.

References

1. De Oliveira, J.F.; Cardoso, M.B. Partial aggregation of silver nanoparticles induced by capping and reducing agents competition. *Langmuir ACS J. Surf. Colloids* **2014**, *30*, 4879–4886. [CrossRef] [PubMed]
2. Zandparsa, R. Latest biomaterials and technology in dentistry. *Dent. Clin. N. Am.* **2014**, *58*, 113–134. [CrossRef] [PubMed]
3. Perez-Diaz, M.A.; Boegli, L.; James, G.; Velasquillo, C.; Sanchez-Sanchez, R.; Martinez-Martinez, R.E.; Martinez-Castanon, G.A.; Martinez-Gutierrez, F. Silver nanoparticles with antimicrobial activities against *Streptococcus mutans* and their cytotoxic effect. *Mater. Sci. Eng. C Mater. Biol. Appl.* **2015**, *55*, 360–366. [CrossRef] [PubMed]
4. Besinis, A.; De Peralta, T.; Handy, R.D. The antibacterial effects of silver, titanium dioxide and silica dioxide nanoparticles compared to the dental disinfectant chlorhexidine on *Streptococcus mutans* using a suite of bioassays. *Nanotoxicology* **2014**, *8*, 1–16. [CrossRef] [PubMed]
5. Li, Q.; Mahendra, S.; Lyon, D.Y.; Brunet, L.; Liga, M.V.; Li, D.; Alvarez, P.J. Antimicrobial nanomaterials for water disinfection and microbial control: Potential applications and implications. *Water Res.* **2008**, *42*, 4591–4602. [CrossRef] [PubMed]
6. Kim, K.J.; Sung, W.S.; Suh, B.K.; Moon, S.K.; Choi, J.S.; Kim, J.G.; Lee, D.G. Antifungal activity and mode of action of silver nano-particles on *Candida albicans*. *Biomet. Int. J. Role Met. Ions Biol. Biochem. Med.* **2009**, *22*, 235–242. [CrossRef] [PubMed]
7. Kim, J.S.; Kuk, E.; Yu, K.N.; Kim, J.H.; Park, S.J.; Lee, H.J.; Kim, S.H.; Park, Y.K.; Park, Y.H.; Hwang, C.Y.; et al. Antimicrobial effects of silver nanoparticles. *Nanomedicine* **2007**, *3*, 95–101. [CrossRef] [PubMed]
8. Monteiro, D.R.; Silva, S.; Negri, M.; Gorup, L.F.; de Camargo, E.R.; Oliveira, R.; Barbosa, D.B.; Henriques, M. Silver nanoparticles: Influence of stabilizing agent and diameter on antifungal activity against *Candida albicans* and *Candida glabrata* biofilms. *Lett. Appl. Microbiol.* **2012**, *54*, 383–391. [CrossRef] [PubMed]
9. Metwalli, K.H.; Khan, S.A.; Krom, B.P.; Jabra-Rizk, M.A. *Streptococcus mutans*, *Candida albicans*, and the human mouth: A sticky situation. *PLoS Pathog.* **2013**, *9*, e1003616. [CrossRef] [PubMed]
10. Lemos, J.A.; Quivey, R.G., Jr.; Koo, H.; Abranches, J. *Streptococcus mutans*: A new Gram-positive paradigm? *Microbiology* **2013**, *159*, 436–445. [CrossRef] [PubMed]
11. Falsetta, M.L.; Klein, M.I.; Lemos, J.A.; Silva, B.B.; Agidi, S.; Scott-Anne, K.K.; Koo, H. Novel antibiofilm chemotherapy targets exopolysaccharide synthesis and stress tolerance in *Streptococcus mutans* to modulate virulence expression in vivo. *Antimicrob. Agents Chemother.* **2012**, *56*, 6201–6211. [CrossRef] [PubMed]
12. Rouabhia, M.; Chmielewski, W. Diseases Associated with Oral Polymicrobial Biofilms. *Open Mycol. J.* **2012**, *6*, 27–32.
13. Barbieri, D.S.V.; Vicente, V.A.; Fraiz, F.C.; Lavoranti, O.J.; Svidzinski, T.I.E.; Pinheiro, R.L. Analysis of the in vitro adherence of *Streptococcus mutans* and *Candida albicans*. *Braz. J. Microbiol.* **2007**, *38*, 624–631. [CrossRef]

14. Jarosz, L.M.; Deng, D.M.; van der Mei, H.C.; Crielaard, W.; Krom, B.P. *Streptococcus mutans* competence-stimulating peptide inhibits *Candida albicans* hypha formation. *Eukaryot. Cell* **2009**, *8*, 1658–1664. [CrossRef] [PubMed]

15. Monteiro, D.R.; Gorup, L.F.; Silva, S.; Negri, M.; de Camargo, E.R.; Oliveira, R.; Barbosa, D.B.; Henriques, M. Silver colloidal nanoparticles: Antifungal effect against adhered cells and biofilms of *Candida albicans* and *Candida glabrata*. *Biofouling* **2011**, *27*, 711–719. [CrossRef] [PubMed]

16. Monteiro, D.R.; Gorup, L.F.; Takamiya, A.S.; Ruvollo-Filho, A.C.; de Camargo, E.R.; Barbosa, D.B. The growing importance of materials that prevent microbial adhesion: Antimicrobial effect of medical devices containing silver. *Int. J. Antimicrob. Agents* **2009**, *34*, 103–110. [CrossRef] [PubMed]

17. Lynch, R.J.; ten Cate, J.M. Effect of calcium glycerophosphate on demineralization in an in vitro biofilm model. *Caries Res.* **2006**, *40*, 142–147. [CrossRef] [PubMed]

18. Bowen, W.H. The cariostatic effect of calcium glycerophosphate in monkeys. *Caries Res.* **1972**, *6*, 43–51. [CrossRef] [PubMed]

19. Grenby, T.H. Trials of 3 organic phosphorus-containing compounds as protective agents against dental caries in rats. *J. Dent. Res.* **1973**, *52*, 454–461. [CrossRef] [PubMed]

20. Duke, S.A.; Rees, D.A.; Forward, G.C. Increased plaque calcium and phosphorus concentrations after using a calcium carbonate toothpaste containing calcium glycerophosphate and sodium monofluorophosphate. Pilot study. *Caries Res.* **1979**, *13*, 57–59. [CrossRef] [PubMed]

21. Nordbo, H.; Rolla, G. Desorption of salivary proteins from hydroxyapatite by phytic acid and glycerophosphate and the plaque-inhibiting effect of the two compounds in vivo. *J. Dent. Res.* **1972**, *51*, 800–811. [CrossRef] [PubMed]

22. Grenby, T.H.; Bull, J.M. Use of high-performance liquid chromatography techniques to study the protection of hydroxylapatite by fluoride and glycerophosphate against demineralization in vitro. *Caries Res.* **1980**, *14*, 221–232. [CrossRef] [PubMed]

23. Naylor, M.N.; Glass, R.L. A three-year clinical trial of a calcium carbonate dentifrice containing calcium glycerophosphate and sodium monofluorophosphate. *Caries Res.* **1979**, *13*, 39–46. [CrossRef] [PubMed]

24. Mainwaring, P.J.; Naylor, M.N. A four-year clinical study to determine the caries-inhibiting effect of calcium glycerophosphate and sodium fluoride in calcium carbonate base dentrifices containing sodium monofluorophosphate. *Caries Res.* **1983**, *17*, 267–276. [CrossRef] [PubMed]

25. do Amaral, J.G.; Sassaki, K.T.; Martinhon, C.C.; Delbem, A.C. Effect of low-fluoride dentifrices supplemented with calcium glycerophosphate on enamel demineralization in situ. *Am. J. Dent.* **2013**, *26*, 75–80. [PubMed]

26. Zaze, A.C.; Dias, A.P.; Amaral, J.G.; Miyasaki, M.L.; Sassaki, K.T.; Delbem, A.C. In situ evaluation of low-fluoride toothpastes associated to calcium glycerophosphate on enamel remineralization. *J. Dent.* **2014**, *42*, 1621–1625. [CrossRef] [PubMed]

27. Ghosh, S.K.; Pal, T. Interparticle coupling effect on the surface plasmon resonance of gold nanoparticles: From theory to applications. *Chem. Rev.* **2007**, *107*, 4797–4862. [CrossRef] [PubMed]

28. Inoue, M.; In, Y.; Ishida, T. Calcium binding to phospholipid: Structural study of calcium glycerophosphate. *J. Lipid Res.* **1992**, *33*, 985–994. [PubMed]

29. Miranda, M.; Fernandez, A.; Lopez-Esteban, S.; Malpartida, F.; Moya, J.S.; Torrecillas, R. Ceramic/metal biocidal nanocomposites for bone-related applications. *J. Mater. Sci. Mater. Med.* **2012**, *23*, 1655–1662. [CrossRef] [PubMed]

30. Pal, S.; Tak, Y.K.; Song, J.M. Does the antibacterial activity of silver nanoparticles depend on the shape of the nanoparticle? A study of the Gram-negative bacterium *Escherichia coli*. *Appl. Environ. Microbiol.* **2007**, *73*, 1712–1720. [CrossRef] [PubMed]

31. Bradford, A.; Handy, R.D.; Readman, J.W.; Atfield, A.; Mühling, M. Impact of silver nanoparticle contamination on the genetic diversity of natural bacterial assemblages in estuarine sediments. *Environ. Sci. Technol.* **2009**, *43*, 4530–4536. [CrossRef] [PubMed]

32. Duran, N.; Duran, M.; de Jesus, M.B.; Seabra, A.B.; Favaro, W.J.; Nakazato, G. Silver nanoparticles: A new view on mechanistic aspects on antimicrobial activity. *Nanomedicine* **2016**, *12*, 789–799. [CrossRef] [PubMed]

33. Le Ouay, B.; Stellacci, F. Antibacterial activity of silver nanoparticles: A surface science insight. *Nanotoday* **2015**, *10*, 339–354. [CrossRef]

34. Lok, C.N.; Zou, T.; Zhang, J.J.; Lin, I.W.; Che, C.M. Controlled-release systems for metal-based nanomedicine: Encapsulated/self-assembled nanoparticles of anticancer gold(III)/platinum(II) complexes and antimicrobial silver nanoparticles. *Adv. Mater.* **2014**, *26*, 5550–5557. [CrossRef] [PubMed]

35. Lok, C.N.; Ho, C.M.; Chen, R.; He, Q.Y.; Yu, W.Y.; Sun, H.; Tam, P.K.; Chiu, J.F.; Che, C.M. Proteomic analysis of the mode of antibacterial action of silver nanoparticles. *J. Proteom. Res.* **2006**, *5*, 916–924. [CrossRef] [PubMed]

36. Hou, W.C.; Stuart, B.; Howes, R.; Zepp, R.G. Sunlight-driven reduction of silver ions by natural organic matter: Formation and transformation of silver nanoparticles. *Environ. Sci. Technol.* **2013**, *47*, 7713–7721. [CrossRef] [PubMed]

37. Onodera, A.; Nishiumi, F.; Kakiguchi, K.; Tanaka, A.; Tanabe, N.; Honma, A.; Yayama, K.; Yoshioka, Y.; Nakahira, K.; Yonemura, S.; et al. Short-term changes in intracellular ROS localisation after the silver nanoparticles exposure depending on particle size. *Toxicol. Rep.* **2015**, *2*, 574–579. [CrossRef] [PubMed]

38. Agnihotri, S.; Mukherji, S.; Mukherji, S. Immobilized silver nanoparticles enhance contact killing and show highest efficacy: Elucidation of the mechanism of bactericidal action of silver. *Nanoscale* **2013**, *5*, 7328–7340. [CrossRef] [PubMed]

39. Tolaymat, T.M.; El Badawy, A.M.; Genaidy, A.; Scheckel, K.G.; Luxton, T.P.; Suidan, M. An evidence-based environmental perspective of manufactured silver nanoparticle in syntheses and applications: A systematic review and critical appraisal of peer-reviewed scientific papers. *Sci. Total Environ.* **2010**, *408*, 999–1006. [CrossRef] [PubMed]

40. Solomon, S.D.; Bahadory, M.; Jeyarajasingam, A.V.; Rutkowsky, S.A.; Boritz, C. Synthesis and Study of Silver Nanoparticles. *J. Chem. Educ.* **2007**, *84*, 322–325.

41. Malanovic, N.; Lohner, K. Gram-positive bacterial cell envelopes: The impact on the activity of antimicrobial peptides. *Biochim. Biophys. Acta* **2016**, *1858*, 936–946. [CrossRef] [PubMed]

42. Loffler, J.; Einsele, H.; Hebart, H.; Schumacher, U.; Hrastnik, C.; Daum, G. Phospholipid and sterol analysis of plasma membranes of azole-resistant *Candida albicans* strains. *FEMS Microbiol. Lett.* **2000**, *185*, 59–63. [CrossRef]

43. Lara, H.H.; Romero-Urbina, D.G.; Pierce, C.; Lopez-Ribot, J.L.; Arellano-Jimenez, M.J.; Jose-Yacaman, M. Effect of silver nanoparticles on *Candida albicans* biofilms: An ultrastructural study. *J. Nanobiotechnol.* **2015**, *13*, 91. [CrossRef] [PubMed]

44. Stoimenov, P.K.; Klinger, R.L.; Marchin, G.L.; Klabunde, K.J. Metal Oxide Nanoparticles as Bactericidal Agents. *Langmuir ACS J. Surf. Colloids* **2002**, *18*, 6679–6686. [CrossRef]

45. Hamouda, T.; Baker, J.R., Jr. Antimicrobial mechanism of action of surfactant lipid preparations in enteric Gram-negative bacilli. *J. Appl. Microbiol.* **2000**, *89*, 397–403. [CrossRef] [PubMed]

46. Lynch, R.J. Calcium glycerophosphate and caries: A review of the literature. *Int. Dent. J.* **2004**, *54*, 310–314. [CrossRef] [PubMed]

47. Turkevich, J.; Stevenson, P.C.; Hillier, J. A study of the nucleation and growth processes in the synthesis of colloidal gold. *Discuss. Faraday Soc.* **1951**, *11*, 55–75. [CrossRef]

48. Gorup, L.F.; Longo, E.; Leite, E.R.; Camargo, E.R. Moderating effect of ammonia on particle growth and stability of quasi-monodisperse silver nanoparticles synthesized by the Turkevich method. *J. Colloid Interface Sci.* **2011**, *360*, 355–358. [CrossRef] [PubMed]

Review

Antimicrobial Silver in Medicinal and Consumer Applications: A Patent Review of the Past Decade (2007–2017)

Wilson Sim [1], Ross T. Barnard [1,2], M.A.T. Blaskovich [3] and Zyta M. Ziora [3,*]

1   School of Chemistry & Molecular Biosciences, The University of Queensland, Brisbane, QLD 4072, Australia; wilson.sim@uq.net.au (W.S.); rossbarnard@uq.edu.au (R.T.B.)
2   ARC Training Centre for Biopharmaceutical Innovation, The University of Queensland, Brisbane, QLD 4072, Australia
3   Institute of Molecular Bioscience, The University of Queensland, Brisbane, QLD 4072, Australia; m.blaskovich@imb.uq.edu.au
*   Correspondence: z.ziora@imb.uq.edu.au; Tel.: +61-733-462-101

Received: 20 August 2018; Accepted: 16 October 2018; Published: 26 October 2018

Abstract: The use of silver to control infections was common in ancient civilizations. In recent years, this material has resurfaced as a therapeutic option due to the increasing prevalence of bacterial resistance to antimicrobials. This renewed interest has prompted researchers to investigate how the antimicrobial properties of silver might be enhanced, thus broadening the possibilities for antimicrobial applications. This review presents a compilation of patented products utilizing any forms of silver for its bactericidal actions in the decade 2007–2017. It analyses the trends in patent applications related to different forms of silver and their use for antimicrobial purposes. Based on the retrospective view of registered patents, statements of prognosis are also presented with a view to heightening awareness of potential industrial and health care applications.

Keywords: antibiotic resistance; antimicrobial activity; medicinal silver; patents; silver; silver nanoparticles; synergism

---

1. Introduction

Silver is a soft and shiny transition metal which is known to have the highest reflectivity of all metals [1]. Among its many useful properties, silver it recognized to have antimicrobial activity. Silver is known to be biologically active when it is dispersed into its monoatomic ionic state (Ag^+), when it is soluble in aqueous environments [2]. This is the same form which appears in ionic silver compounds such as silver nitrate and silver sulfadiazine, which have been frequently used to treat wounds [3]. Another form of silver is its native nanocrystalline form (Ag^0). The metallic (Ag^0) and ionic forms can also appear loosely associated with other elements such as oxygen or other metals and can form covalent bonds or coordination complexes [3].

To date, there are three known mechanisms by which silver acts on microbes. Firstly, silver cations can form pores and puncture the bacterial cell wall by reacting with the peptidoglycan component [4]. Secondly, silver ions can enter into the bacterial cell, both inhibiting cellular respiration and disrupting metabolic pathways resulting in generation of reactive oxygen species [5]. Lastly, once in the cell silver can also disrupt DNA and its replication cycle [6] (Figure 1). A recently published review includes more details about the bactericidal mechanisms of silver, along with methods of silver nanoparticle preparation [7]. Throughout history, silver has consistently been used to restrict the spread of human disease by incorporation into articles used in daily life. The earliest recorded use of silver for therapeutic purposes dates back to the Han Dynasty in China *circa.* 1500 B.C.E [8]. Silver vessels and plates were frequently used during the Phoenician, Macedonian, and Persian empires [9]. Families of the higher socioeconomic classes during the middle-ages were so acquainted with the usage of silver that they developed bluish skin discolorations known as *argyria*, an affliction which may have led to the term 'blue blood' to describe members of the aristocracy [10]. Modern medicine utilizes medical grade forms of silver, such as silver nitrate, silver sulfadiazine, and colloidal silver [11].

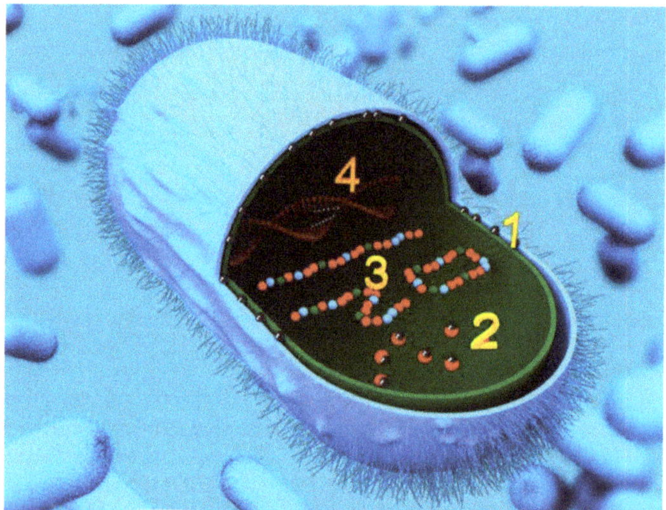

Figure 1. Silver's action on a bacterial cell. 1. Silver can perforate the peptidoglycan cell wall. 2. Silver inhibits the cell respiration cycle. 3. Metabolic pathways are also inhibited when in contact with silver. 4. Replication cycle of the cell is disrupted by silver particles via interaction with DNA.

The discovery of antibiotics in the early 20th century led to a cessation in the development of silver as an antimicrobial agent. However, the development of increasing levels of bacterial resistance to most antibiotics in recent years has led to reexamination of the potential of this ancient remedy [7,12] including studies with patients using colloidal silver and antibiotics [13]. This review aims to demonstrate the wide and ever-expanding applications of silver in medicine, health care, and other daily life activities, with a focus on the patents registered during the past decade. A similar patent review was published in Expert Opinion on Therapeutic Patents in 2005 [14], covering patents compiled from 2001–2004. The current review extends to the years 2007–2017. An analysis of the growth of patents describing antimicrobial silver applications is presented throughout this review, along with commentary of selected examples demonstrating some of the more interesting applications. Our analysis has separated these discussions of the use of silver into four general categories: Medical applications, personal care products, domestic household products, and agricultural/industrial applications.

2. Discussion

2.1. Antimicrobial Silver for Clinical and Medical Usage

This section presents a selection of some of the most interesting and unique patented products which utilize silver for their bactericidal action in the medical field, including therapies based on silver's antimicrobial properties. The product numbers referred to in this section correspond to those listed in Table 1.

Table 1. List of patented medical grade products and medically related products containing silver as an antimicrobial agent.

No.	Patent Title	Brief Product Description	Patent Number	Filing Date	Ref.
1	Ready to use medical device with instant antimicrobial effect	A medical apparatus packaging designed to activate a bioactive silver coating and other bactericidal elements upon opening of package.	US20150314103A1	5 November 2015	[15]
2	Use of silver-containing layers at implant surfaces	The method of coating medical implants with various forms of silver for infection prevention.	US20130344123A1	4 March 2013	[16]
3	Medical needles having antibacterial and painless function	Surgical needles coated with silver nanoparticles for the prevention of infection.	WO2006088288A1	5 January 2006	[17]
4	Fracture-setting nano-silver antibacterial coating	A method to coat invasive medical instruments with silver nanoparticles.	CN203128679U	3 April 2013	[18]
5	Antimicrobial closure element and closure element applier	The coating of the internal structure of a vascular portal device and the interior of its applicator.	US20080312686A1	9 June 2008	[19]
6	Bone implant and systems that controllably release silver	A specially designed bone implant which allows surgeons to control the release of silver ions.	WO2012064402A1	19 May 2012	[20]
7	Nanometer silver antibacterial biliary duct bracket and preparation method thereof	A biliary duct implant bracket made from plastic coated uniformly with silver nanoparticles to prevent biofilm formation at site of implant.	CN102485184A	3 December 2010	[21]
8a	Antimicrobial coatings on building surfaces	A coating with antimicrobial silver applied to the interior surface of a building's exterior wall.	US7641912B1	5 January 2010	[22]
8b			US8282951B2	9 October 2012	[23]
9a	Silver nanoparticle dispersion formulation	A topical gel to treat dermal infections with 1% w/w silver nanoparticles as active ingredient.	WO2007017901A2	16 February 2007	[24]
9a			EP3359166A1	15 August 2018	[25]
10	Antimicrobial silver hydrogel composition for the treatment of burns and wounds	An aqueous gel with a range of silver salt as its active ingredient made for the treatment of wounds specifically caused by burns.	WO20120282348A1	5 May 2011	[26]
11	Polysaccharide fibers for wound dressings	The method of coating wound dressing with a gel matrix where silver can be immobilized and applied to a wound to aid healing.	WO2013050794A1	5 December 2012	[27]
12	Antimicrobial, silver-containing wound dressing for continuous release	Wound dressing capable of releasing silver ions to aid healing upon contact with fluids from the wound.	US20070286895A1	24 August 2007	[28]
13	Nano-silver wound dressing	A wound dressing with enhanced antimicrobial properties for improved scarring.	US20070293799A1	9 December 2008	[29]
14a	Metal containing materials	Silver containing materials for treatment of bacterial conditions.	US8425880B1	23 April 2013	[30]
14b			US7255881B2	14 August 2007	[31]
15	Dental Uses of Silver Hydrosol	Silver suspended in aqueous gel used to reduce infection risks of dental procedures.	US9192626B2	24 November 2015	[32]
16	Antimicrobial silver nanoparticle additive for polymerizable dental materials	Denture material made with the addition of silver nanoparticles for additional antimicrobial effect.	US20070213460A1	13 September 2007	[33]
17	Silver ion coated products for dental and other body restoration objects	Silver coating with antimicrobial, antifouling and deodorant properties	US20180245278A1	30 August 2018	[34]

Surface coatings incorporating silver are a common application. One new approach is a method for producing ready packed medical apparatus which sterilizes itself upon the opening of the package, by creating a vapor that activates a silver-containing hydrophilic surface coating (Product No. 1). The antimicrobial properties of silver have been highly valued in medical application where implanted devices are coated with silver nanoparticles for the antimicrobial effects, but manufacturers need to be aware that this application is claimed in a patent application (Product No. 2) with a very broad claim 1: "An article that is implantable in an animal, the article comprising a microparticulate silver-containing antimicrobial layer stably adhered upon at least one surface of the article." However, the application does not appear to have progressed towards granting. Invasive surgical tools such as medical grade needles (Product No. 3) can also be coated with silver nanoparticles as described in its related patent (Product No. 4). Medical devices that are directly introduced into the human body that contain silver include vascular catheters (Product No. 5), bone implants (Product No. 6), and biliary duct brackets (Product No. 7). Another topical application of antimicrobial silver has incorporated it into coatings applied to the interior surface of a building's exterior wall (Products No. 8a and 8b).

Another general use is for topical treatments. Numerous topical gels with different formulations of silver have been patented. Silver was first used to treat burn wounds in the form of 0.5% silver nitrate solution and silver sulfadiazine cream in 1960 [35]. However, this was impractical as the dressings required rehydration every couple of hours. To overcome this limitation silver nanoparticle-based gels and silver salt-based gels have been developed (Products No. 9a, 9b and 10), with all approaches still considered novel.

Silver based wound dressings have greatly improved in efficacy compared to standard dressings, and more complex dressings have been developed. New knowledge in burn wound management led to the discovery of a method to immobilize silver nanoparticles on a gel-support matrix which is attached to a wound dressing (Product No. 11). A recently commercialized wound dressing allows a prolonged use of the dressing for up to 7 days or until saturation, without reapplication (Figure 2). It is made possible through its design, which slowly releases silver ions upon contact with wound exudates. Its highly absorbent padding is also coated with a layer of silicone which is aimed to reduce pain during removal and reapplication of the dressing. (Product No. 12). The use of silver with wound dressings is known to reduce scarring and such formulations are widely used (Product No.13). Silver-based wound dressings are available under brand names with different compositions, such as Mepilex® Ag, Acticoat™, Aquacel®, Flaminal®, Allevyn® Ag, and Biatain® Ag, SILVERCEL™. Other products containing a silver component, not specifically developed for wound healing, have been patented for treatment of bacterial infection (Products No. 14a and 14b).

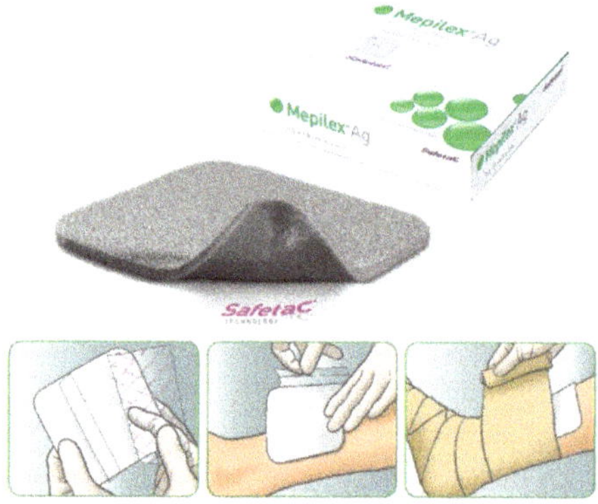

Figure 2. Mepilex® Ag with instructions for application. Images used with permission of Mölnlycke Health care, Sweden.

Silver has also been applied across the dental field. Silver has been the key component in dental amalgam fillings for more than one hundred years. However, its antimicrobial properties were not patented. Silver is used in the prevention of infection during and after dental surgery (Product No. 15). Dental support fixtures made out of silver and denture materials, and other body restoration objects, having silver nanoparticles as additives can reduce bacterial infections, especially during first few months of installation (Products No. 16 and 17).

2.2. Antimicrobial Silver in Personal Care Products

This section presents grooming products and devices which utilize silver for its sanitizing effects, summarized in Table 2. Hygiene and grooming products such as shavers (Product No. 18), toothbrushes (Product No. 19) and sanitary pads (Product No. 20) are frequently employed under adverse conditions where they encounter the bacteria microbiome, but are relied upon to be sanitary. One example of such usage is by the German public company "Beiersdorf AG" which has products incorporating silver for its added antimicrobial properties. They have applied silver over a wide range of products from shower gels and deodorants to first aid bandages (Figure 3).

Table 2. List of patented personal-care products containing silver as an antimicrobial agent.

No.	Patent Title	Brief Product Description	Patent Number	Filing Date	Ref.
18	Cosmetic and /or medical device for antimicrobial treatment of human skin with silver particles	Technology in which shaving devices can deposit silver ions unto skin in place of traditional antiseptic medium.	DE102012224176A1	26 June 2014	[36]
19	Antimicrobial thermoplastic polyurethane for toothbrush and preparation method for antimicrobial thermoplastic polyurethane	Addition of silver nanoparticles into plastic materials which are used to manufacture bristles of tooth brushes.	CN103254401A	28 April 2013	[37]
20	Sanitary towel capable of removing peculiar smell and manufacturing method thereof	Infusion of silver nanoparticles into fibers of sanitary pads which prevents the growth of odor causing microbes on menstrual discharges.	CN102961778A	21 November 2012	[38]
21	Solid oil cosmetics containing antimicrobial Ag zeolites and aluminum chlorohydrate	Deodorants and topical creams for the prevention of odor causing bacteria.	JP2013071914A	22 April 2013	[39]
22	Antimicrobial agents containing fine silver particle-carrying polypeptides and daikon radish fermentation products, and cosmetics containing them	The invention of a cosmetic lotion preservative consisting of silver nanoparticles.	JP2010059132A	18 March 2010	[40]
23	Functional cosmetic including nano silver	Addition of silver nanoparticles into manufactured cosmetics for its antimicrobial effect which aids in the prevention of acne and pimples.	KR20070119971A	21 December 2007	[41]
24	Antimicrobial contact lenses and methods for their production	Contact lenses manufactured from materials infused with silver nanoparticles for antimicrobial effects.	US20030044447A1	6 March 2003	[42]
25a	Silver-containing antimicrobial fabric	Textile material manufactured from fibers embedded with silver nanoparticles.	US20050037057A1	17 February 2005	[43]
25b			US7754625B2	13 July 2010	[44]
25c			WO2018160708	7 September 2018	[45]
26	Breast pump assemblies having an antimicrobial agent	Suction cup segment of device coated with silver ion exchange resin to prevent possible microbial contamination into breast milk.	US20080139998A1	12 June 2008	[46]
27	Nano-silver inorganic antibacterial nutritional hair dye	Hair colorant having additional silver nanoparticles as preservatives.	CN104224617A	27 August 2014	[47]

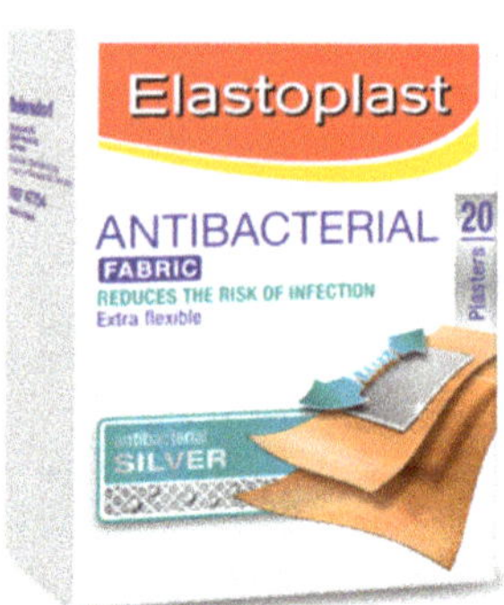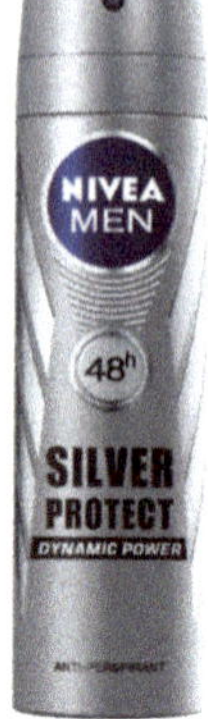

Figure 3. Commercial utilization of silver in consumer product lines, including Elastoplast (First aid bandages) and NIVEA (Shower gels and deodorants) for enhanced antimicrobial properties. *Images used with permission of Beiersdorf AG, Germany.*

Since using silver to treat skin infections is common, researchers in dermatology frequently resort to silver for treating conditions related to bacterial colonization, such as body odors (Product No. 21), acne outbreaks (Product No. 22), eczema and rash (Product No. 23).

A range of other personal health products have also added silver to improve their hygienic capacities, including contact lenses (Product No. 24), antimicrobial fabric garments (Products No. 25a, 25b and 25c), breast pump assemblies (Product No. 26), and hair dye (Product No. 27).

Most cosmetic products come in the form of cream, aqueous lotions, or hydrogel medium. It is observed that most manufacturers favor the incorporation of silver colloids into their products as they do not precipitate and separate, with the added benefit of acting as a preservative. Colloidal silver is defined as a mixture of silver ions and silver nanoparticles suspended in an aqueous medium. They are usually synthesized by electrolysis using a set of silver cathodes [48]. Colloidal silver was first used in 1891 by a surgeon named B.C Crede to sterilize wounds [9]. The use of silver grew in popularity between 1900 to the 1940s. Subsequently, antibiotics supplanted the use of silver [9]. Today, many products are offered not only as colloidal silver solutions, but also as personal devices suitable for home use, that synthesize colloidal silver. However, the commercialization of colloidal silver has been accompanied by inconsistencies in colloidal silver production and properties, as well as cases of unexpected side effects. Therefore, the Food and Drug Administration (FDA) has excluded any commercialized colloidal silver that claims health benefits without scientific evidence [49]. Similar action has been taken by the Therapeutic Goods Administration (TGA) in Australia [49] and the European Commission (EC) [50]. The commercial sales of colloidal silver are not banned, but claims of health benefits without scientific support are not permitted.

2.3. Antimicrobial Silver in Domestic Products

The antimicrobial applications of silver started in ancient times in domestic products like silver plates and pitchers [9]. With that in mind, there continue to be domestic applications of silver, particular for surface treatments (Table 3).

Silver is widely incorporated into surface coatings of electrical goods such as automated bathtubs (Product No. 28), laundry washing machines (Product No. 29), air purifiers with silver filters (Product No. 30) and refrigerators (Product No. 31), to produce 'bacteria-free' products. Application of silver nanoparticles to other household objects with frequent handling such as keyboards (Product No. 32), bath safety aids (Product No. 33), and bathroom safety handles (Product No. 34). Special stand-alone products such as containers for meat or water/wine/milk storage (Products No. 35a and 35b) are useful applications where bacterial contamination may present a health issue.

Table 3. List of patented home-use products containing silver as an antimicrobial agent.

No.	Patent Title	Brief Product Description	Patent Number	Filing Date	Ref.
28	Automatic cleaning system for bathtub or piping	A cleaning system attached to a Silver ion generator which flows through hot water inlet pipe which sanitizes bathtubs as a self-cleaning function.	JP2009268576A	19 November 2009	[51]
29	Clothes washing machine	Laundry washing machine consisting of a silver ion generator which will be released during each wash cycle.	US20080041117A1	21 February 2008	[52]
30	Air purifier, useful for neutralizing bad smells	An electronic air cleaning device which draws unclean air through an immobilized silver filter killing any airborne odor causing bacteria.	DE102007040742A1	3 March 2009	[53]
31	Antibiotic method for parts of refrigerator using antibiotic substance	Distribute silver ions within the fridge to slow the spoilage of food spoilage.	US7781497B2	24 March 2010	[54]
32	Submersible keyboard	A waterproof and washable keyboard with key caps made from plastic embedded with silver ions.	US20090262492A1	22 October 2009	[55]
33	Bactix silver-based antimicrobial additive in bath aids	Bath safety aids made from silver impregnated polymers for long lasting antimicrobial effects.	US20130029029A1	31 January 2013	[56]
34	Bacteria-resistant grab bar	Disability support bar paddings made out of silicone rubber impregnated with silver nanoparticles as antimicrobial additives.	US20100148395A1	17 December 2008	[57]
35a	Antimicrobial reusable plastic or glass container	Collapsible food storage containers mainly for meat or water/wine/ milk storage in kitchen composing of an antimicrobial silver fabric as its bottom inner layer.	US20070189932A1	10 February 2006	[58]
35b			WO2018137725A1	2 August 2018	[59]
36	Nano-silver antibacterial gloves	Domestic latex gloves impregnated with silver nanoparticles and other antimicrobial elements.	CN202738872U	20 February 2013	[60]
37	Natural silver disinfectant compositions	General surface cleaner containing soluble silver salt for added antimicrobial effect.	US20100143494A1	10 June 2010	[61]
38	Antiseptic solutions containing silver chelated with polypectate and EDTA	Laundry liquid having aqueous suspension of colloidal silver as additive for its antimicrobial properties.	US7311927B2	25 December 2007	[62]

Despite the many beneficial innovations in the use of silver as an antimicrobial agent, its application in cleaning products and disposable tools such as gloves (Product No. 36), disinfectant wipes (Product No. 37), and cleaning detergent (Product No. 38) may have negative environmental impacts. Cleaning products, once used, usually end up in sewage treatment systems, and eventually the environment. This is a concern for silver nanoparticles, as there are currently no effective methods for filtering out silver nanoparticles. The release of large amount of silver products into the environment may lead to disturbances of the microbiological ecosystem, and potentially lead to bacterial resistance to silver [63]. Consequently, alternative methods of sanitization should be considered such as the application of alcohol or bleach which are sufficient for domestic purposes, or employing 'fixed' silver containing surfaces that reduce the risk of environmental release.

Apart from being a threat to beneficial environmental bacteria, another issue to be addressed is the possible longer-term reduction of the potency of silver in killing microbes. Since the discovery of antibiotics, the efficacy of antibiotics has been compromised by over-prescription and over-usage, leading to the current antibiotic crisis. The presence of low levels of antibiotics in the environment fosters the generation of multiple drug resistant strains [64]. Silver is not immune to the generation of bacterial resistance, with several reports in recent years [65,66]. This history suggests a need for a systemic reassessment of the usage of silver in domestic products, so that it is not used too extravagantly, or released haphazardly.

2.4. Antimicrobial Silver in Agricultural and Industrial Products

Silver has also been used for a variety of agricultural and industrial products. In industry, large scale water purification can be made cost effective by using colloidal silver for purification as it is needed only in small quantities and can purify large quantities of water, though potential environmental risk needs to be considered [67]. For agriculture use, silver has been incorporated in nylon ropes that are used to tie down plants, cover them with netting, and for various other applications. These ropes normally decay after time due to bacterial biofilm formation, so the silver prevents this decomposition [68]. Agricultural use of silver products must be carefully assessed to avoid any impact on the microbial flora and symbiosis. The growth of healthy crop plants relies heavily upon the formation of symbiotic microbes around the roots such as nitrifying bacteria and mycorrhiza [69]. Studies have shown that the contact of bioactive silver to nitrifying bacteria impedes the formation of symbiotic channels [70].

Table 4 presents patented industry and agricultural related products utilizing silver as an antimicrobial agent. An example of agricultural use is Product No. 39 which uses Ag (I) and Ag (II) to treat infections in plants, while Products No. 40 and 41 with coating of a single rope strand with silver to prolong resistance to biofilm formation. Silver coatings can be beneficial in industrial machines which require a completely sterile environment to manufacture food or medical grade products, employing silver in the parts that come in direct contact with the products (Product No. 42). Machinery parts are usually designed for prolonged periods and incorporating silver particles into these materials provides an effective means of isolation and retention of silver so that it is not released into the environment easily.

Table 4. List of patented industry and agricultural related products utilizing silver as an antimicrobial agent.

No.	Patent Title	Brief Product Description	Patent Number	Filing Date	Ref.
39	Method and compositions for treating plant infections	Method of applying high valency silver to treat infection in plants of the *Rosaceae* family.	US20120219638A1	21 November 2011	[71]
40	Silver yarn, plied yarn silver yarn, functional fabric using same, and method for producing	Strong weather resistant rope for agricultural purposes made with polyester and silver-plated fiber yarn to prevent growth of biofilms.	CN102439205A	2 May 2012	[72]
41	Silver coated nylon fibers and associated methods of manufacture and use	The method and manufacture of industrialized antimicrobial fabric woven from nylon fibers impregnated with silver.	US20100166832A1	1 July 2010	[68]
42	Rolling apparatus having plastic parts containing antibacterial and antifungal silver (oxides)	Industry scale food grade rollers made from silver impregnated plastic for better hygiene and disease prevention.	JP2005201385A	28 July 2005	[73]
43	An antimicrobial food package	Food grade polymer containers with interiors coated with silver nanoparticles to prolong food freshness.	WO2014001541A1	28 June 2013	[74]
44	Rotationally molded plastic refuse container with microbial inhibiting inner surface and method	Industry scale plastic garbage container interiorly lined with silver nanoparticles for improved waste treatment.	US20080185311A1	7 August 2008	[75]
45a	Sustained silver release composition for water purification	Water filtration unit containing immobilized silver nanoparticles for water purification purposes.	WO2012140520A8	23 March 2012	[76]
45b			US20180186667A1	5 July 2018	[77]
46	Method for Producing Antimicrobial Agent Micro-Particle	An industry cleaning liquid having silver nanoparticles as its active ingredients.	JP2007161649A	28 June 2007	[78]

Polymers are extremely versatile and when impregnated with silver nanoparticles, they can be used for numerous applications, such as mass-produced food storage containers (Product No. 43) and industrial scale waste bins (Product No. 44). Sterility in the food and therapeutics industry is crucial, so the incorporation of silver into manufacturing equipment in contact with consumer products can be regarded as an appropriate usage. However, in the case of daily used food containers, frequent usage of silver may not be ideal as there is a risk of accumulation in the human body if the silver leaches, potentially leading to similar side effects as were observed in the middle ages when silver utensils were frequently used [9].

Products No. 45a and 45b describes a water filtration unit containing immobilized silver nanoparticles for water purification purposes. The invention and manufacture of industrial cleaning solutions containing silver (Product No. 46) is a potentially widespread application, as there is a need for instant effective sanitization to prevent bacterial transmission. However, precautions must be observed to prevent environmental release.

2.5. Overview of Patent Literature from 2007–2017

In the previous sections, applications of antibacterial silver in a variety of fields were discussed. This section presents an overview of silver-related patents on a global level for the purpose of understanding the trends, major applications as well as major contributors. Methods by which the data sets were obtained are reported in Section 3.

Tracking the number of patents disclosing antimicrobial applications of silver for each year over the past decade, as summarized in Figure 4, shows that there has been a steady upward trajectory in the number of silver patent applications in recent years. The increase may have reached a plateau in 2016/17, but it will be necessary to consider data from 2018 and onwards to confirm this hypothesis.

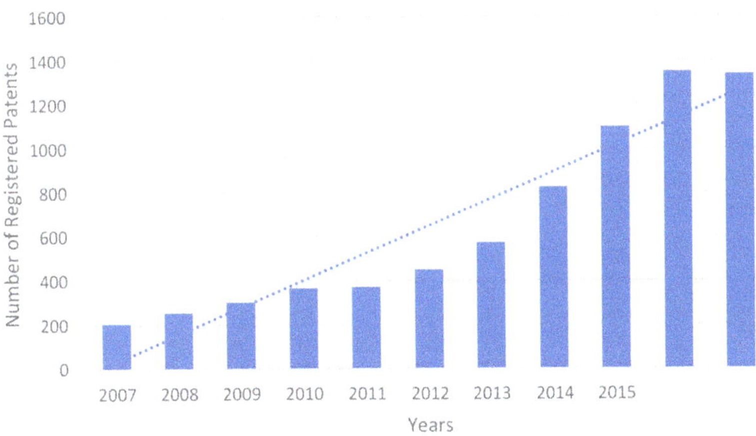

Figure 4. Trend analysis of yearly number of patent registrations involving antimicrobial silver applications over the past decade.

The data obtained from Figure 4 can be further dissected to reveal patents registered under each language, as shown in Figure 5. This can be linked to a deduction of the country of origin of these patents. This analysis demonstrates that patents claiming antimicrobial silver products are predominantly contributed by Asian countries, with China (55%), Korea (7%), and Japan (8%) comprising 70% of the chart. Patents registered in English compose 25% while another 5% are various European language patent registrations. It has been speculated that since the FDA and EMA have reduced influence in the Asian countries [49], this opens up opportunities in Asia for innovations with antimicrobial silver. Indeed, there is some basis for this speculation given the large number of silver related patents being registered in Asian languages, but additional research will be needed to establish the causes of this phenomenon. One of the potential reasons explaining the more substantial contribution of Asian countries in patented silver related innovations and consumer products is the fast-track approval pathway for new drugs in China, supported by consumer trials that are significantly cheaper than in other countries, which can be performed on higher number of participants. There is also a Chinese government strategy to commercialize non-Chinese ideas in China and financially support innovators who are willing to patent their ideas in China, with a preference to procure products whose IP is owned or registered in China. Finally, the database search may produce multiple results for a single patent that are particularly difficult to detect for Chinese applications because Chinese individuals' names typically have only three or two characters [79].

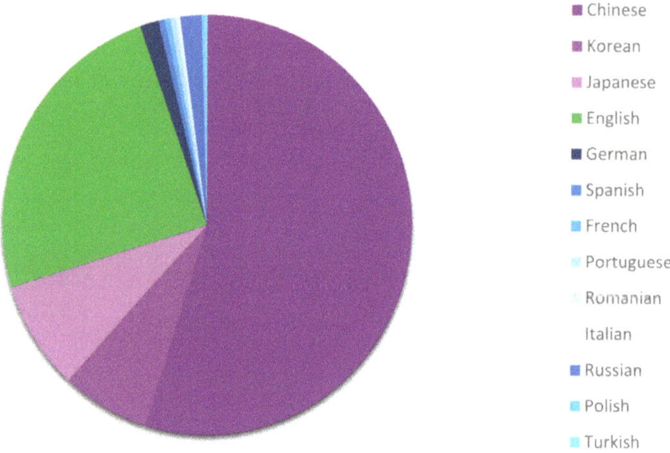

Figure 5. Analysis of language of patent registrations involving antimicrobial silver over the past decade with major contribution of Chinese (55%), English (25%), Korean (7%), Japanese (8%), and others (5%).

We have also assessed the application of silver according to its field of use. Based on the analysis from Figure 6, only about 20% of the patents in each year claimed medical uses of silver. This could mean that approximately about 80% of the patented silver applications are for other usages, such as domestic, agricultural and industrial usage. If this trend continues, there is potential for damage to the ecosystem if the non-medical uses result in release to the environment. This could lead to a worrisome situation where bacteria evolve resistance to silver, one of few promising alternatives to current classes of antibiotics.

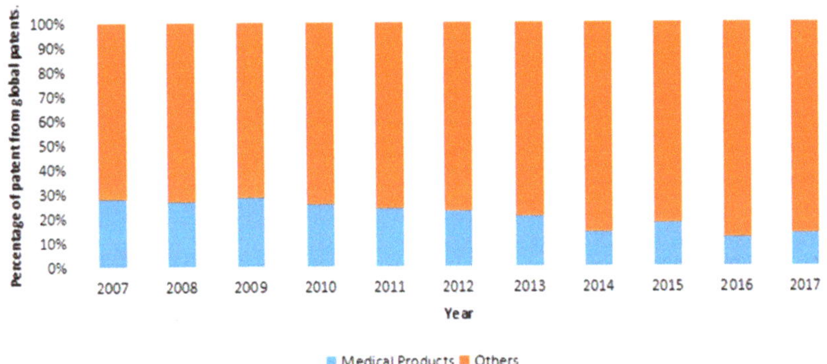

Figure 6. Total patents registered globally for silver used in the medical field compared to all other applications.

3. Materials and Methods

The patent analysis data and trend chart were generated through search results obtained from the scientific publication database, SciFinder® by the American Chemical Society. It was accessed through the University of Queensland Library portal. Results generated were accurate as of 15 June 2018.

3.1. Dataset 1—Application of Antibacterial Silver from the Global Perspective

The keywords used to generate this search were "antibacterial + silver" and "silver + medical" that narrowed down the number of silver related patents to those presenting the word "silver" in the title and describing only silver components. Therefore, patents with the general terms such as "metal nanoparticles" covering all nanomaterials with potential antibacterial applications like silver, platinum, gold, palladium, copper, zinc, and other metals, are not included here. Result limiters used were publication years (2007–2017) and document type (Patent). Search results yielded 5054 hits of exact words and concepts related to its words. No duplicates were found throughout the result set. The full result was analyzed by publication year and was sorted out by natural order to reflect results in yearly order. Obtained results were used to generate a bar chart as Figure 5 by using Microsoft Word chart sketching function.

3.2. Dataset 2—Application of Antibacterial Silver from a Regional Perspective

The keywords used to generate this search were as for Dataset 1, "antibacterial + silver". Result limiters used were publication years (2007–2017) and document type (Patent). Search results yielded 5054 hits of exact words and concepts related to its words. No duplicates were found throughout the result set. The full result was analyzed by language and was sorted out by frequency of region. Results obtained were used to generate the pie chart (Figure 6) by using the Microsoft Word chart sketching function.

3.3. Dataset 3—Application of Antibacterial Silver in the Medical Field

The keywords used to generate this search were "silver + medical". Result limiters used were publication years (2007–2017) and document type (Patent). Search results yielded 1310 hits of exact words and concept related to its words. No duplicates were found throughout the result set. The full result was analyzed by publication year and was sorted out by natural order to reflect results in yearly order. Results obtained were tabulated against dataset 1 to in order to obtain a 100% stacked bar chart presented at Figure 6 by using Microsoft Word chart sketching function.

4. Conclusions

The use of silver for its antimicrobial properties is increasing in numerous fields, including the medical, consumer, agricultural and industrial sectors. In just over 10 years, nearly 5000 new applications have been registered. The majority of the patents are from Asian countries, with Chinese language applications representing more than 50% of the global total, followed by Korean and Japanese language filings. Only about 20% of patents are registered in English.

While the potential benefits of silver are attracting increased attention, a number of publications have pointed out potential adverse effects from the overuse of silver, such as ecosystem disturbance [80], and bacterial resistance to silver [81]. Since our "armory" of antibiotics has been depleted by the rise in antimicrobial resistance, silver represents a new hope, but mindful use must be considered at an early stage to prevent a repetition of past mistakes.

We suggest that the application and commercialization of silver related products should be critically reassessed to avoid, or at least minimize, these adverse effects. In particular, while incorporation of silver products in an enclosed environment is justifiable, products which are expected to release silver into the environment should be avoided. There is ample evidence [82] that there can be adverse long-term effects from consumption or exposure to silver, so silver products should only be used in circumstances where (1) there is an absolute need for it, such as a medical intervention, and (2) in modes where silver is immobilized and containable.

Author Contributions: W.S. and Z.M.Z. designed and wrote the manuscript, Z.M.Z. coordinated the writing progress. All authors discussed the results and commented on the manuscript with the input from R.T.B. and M.A.T.B.

Funding: This research received no external funding.

Conflicts of Interest: The authors declare no conflict of interest.

References

1. National Institute of Standards and Technology. *CRC Handbook of Chemistry and Physics*, 81st ed.; Lide, D.R., Ed.; CRC Press: Boca Raton, FL, USA, 2000; p. 2556. ISBN 0-8493-0481-4.
2. Hoffman, R.K.; Surkiewicz, B.F.; Chambers, L.A.; Phillips, C.R. Bactericidal action of movidyn. *Ind. Eng. Chem.* **1953**, *45*, 2571–2573. [CrossRef]
3. Fong, J.; Wood, F. Nanocrystalline silver dressings in wound management: A review. *Int. J. Nanomed.* **2006**, *1*, 441–449. [CrossRef]
4. Jung, W.K.; Koo, H.C.; Kim, K.W.; Shin, S.; Kim, S.H.; Park, Y.H. Antibacterial activity and mechanism of action of the silver ion in staphylococcus aureus and escherichia coli. *Appl. Environ. Microbiol.* **2008**, *74*, 2171–2178. [CrossRef] [PubMed]
5. Morones-Ramirez, J.R.; Winkler, J.A.; Spina, C.S.; Collins, J.J. Silver enhances antibiotic activity against gram-negative bacteria. *Sci. Transl. Med.* **2013**, *5*, 190ra181. [CrossRef] [PubMed]
6. Yakabe, Y.; Sano, T.; Ushio, H.; Yasunaga, T. Kinetic studies of the interaction between silver ion and deoxyribonucleic acid. *Chem. Lett.* **1980**, *9*, 373–376. [CrossRef]
7. Möhler, J.S.; Sim, W.; Blaskovich, M.A.T.; Cooper, M.A.; Ziora, Z.M. Silver bullets: A new lustre on an old antimicrobial agent. *Biotechnol. Adv.* **2018**, *36*, 1391–1411. [CrossRef] [PubMed]
8. Yamada, K. The two phases of the formation of ancient medicine. In *The Origins of Acupuncture and Moxibustion, The Origins of Decoction*; International Research Center for Japanese Studies: Kyoto, Japan, 1998; p. 154.
9. Alexander, J.W. History of the medical use of silver. *Surg. Infect. (Larchmt)* **2009**, *10*, 289–292. [CrossRef] [PubMed]
10. Davies, O. *They Didn't Listen, They Didn't Know How*; AuthorHouse: Bloomington, IN, USA, 2013; p. 805.
11. Hill, W.R.; Pillsbury, D.M. *Argyria: The Pharmacology of Silver*; Williams & Wilkins Company: Philadelphia, PA, USA, 1939; p. 188.

12. Möhler, J.S.; Kolmar, T.; Synnatschke, K.; Hergert, M.; Wilson, L.A.; Ramu, S.; Elliott, A.G.; Blaskovich, M.A.T.; Sidjabat, H.E.; Paterson, D.L.; et al. Enhancement of antibiotic-activity through complexation with metal ions—Combined ITC, NMR, enzymatic and biological studies. *J. Inorg. Biochem.* **2017**, *167*, 134–141. [CrossRef] [PubMed]

13. Ooi, M.L.; Richter, K.; Bennett, C.; Macias-Valle, L.; Vreugde, S.; Psaltis, A.J.; Wormald, P.-J. Topical colloidal silver for the treatment of recalcitrant chronic rhinosinusitis. *Front. Microbiol.* **2018**, *9*. [CrossRef] [PubMed]

14. Melaiye, A.; Youngs, W.J. Silver and its application as an antimicrobial agent. *Expert. Opin. Ther. Pat.* **2005**, *15*, 125–130. [CrossRef]

15. Hannon, D.; Gilman, T.H. Ready to Use Medical Device with Instant Antimicrobial Effect. US20150314103A1, 5 November 2015.

16. Ostrum, R.; Hettinger, J.; Krchnavek, R.; Caputo, G.A. Use of Silver-Containing Layers at Implant Surfaces. US20130344123A1, 26 December 2013.

17. Yang, W.-D. Medical Needles Having Antibacterial and Painless Function. WO2006088288A1, 24 August 2006.

18. Li, Y. Fracture-Setting Nano-Silver Antibacterial Coating. CN203128679U, 14 August 2013.

19. Ellingwood, B.A. Antimicrobial Closure Element and Closure Element Applier. US20080312686A1, 18 December 2008.

20. Dehnad, H.; Chopko, B.; Chirico, P.; McCORMICK, R. Bone Implant and Systems that Controllably Releases Silver. WO2012064402A1, 19 May 2012.

21. Linghu, E.; Wang, K.; Wang, Y.; Yang, J. Nanometer Silver Antibacterial Biliary Duct Bracket and Preparation Method Thereof. CN102485184A, 6 June 2012.

22. Redler, B.M. Antimicrobial Coatings for Treatment of Surfaces in a Building Setting and Method of Applying Same. US7641912B1, 5 January 2010.

23. Redler, B.M. Antimicrobial Coatings for Treatment of Surfaces in a Building Setting and Method of Applying Same. US8282951B2, 9 October 2012.

24. Omray, P. Silver Nanoparticle Dispersion Formulation. WO2007017901A2, 16 February 2007.

25. Meledandri, C.J.; Schwass, D.R.; Cotton, G.C.; Duncan, W.J. Antimicrobial Gel Containing Silver Nanoparticles. EP3359166A1, 15 August 2018.

26. Yates, K.M.; Proctor, C.A.; Atchley, D.H. Antimicrobial Silver Hydrogel Composition for the Treatment of Burns and Wounds. WO2012151438A1, 8 November 2012.

27. Miraftab, M. Polysaccharide Fibres for Wound Dressings. WO2013050794A1, 11 April 2013.

28. Bowler, P.; Parsons, D.; Walker, M. Wound Dressing. US20070286895A1, 13 December 2007.

29. Ma, R.-H.; Yu, Y.-H. Nano-Silver Wound Dressing. US20070293799A1, 20 December 2007.

30. Lyczak, J.B.; Thompson, K.; Turner, K. Metal-Containing Materials for Treatment of Bacterial Conditions. US8425880B1, 23 April 2013.

31. Gillis, S.H.; Schechter, P.; Stiles, J.A.R. Metal-Containing Materials. US7255881B2, 14 August 2007.

32. Willoughby, A.J.M.; Moeller, W.D. Dental Uses of Silver Hydrosol. US9192626B2, 24 November 2015.

33. Ruppert, K.; Grundler, A.; Erdrich, A. Antimicrobial Nano Silver Additive for Polymerizable Dental Materials. US20070213460A1, 13 September 2007.

34. Ukegawa, S. Silver-Ion Coated Object Obtained by Microwave Irradiation and a Method for Coating a Silver-Ion Onto a Target Object. US20180245278A1, 30 August 2018.

35. Nherera, L.M.; Trueman, P.; Roberts, C.D.; Berg, L. A systematic review and meta-analysis of clinical outcomes associated with nanocrystalline silver use compared to alternative silver delivery systems in the management of superficial and deep partial thickness burns. *Burns* **2017**, *43*, 939–948. [CrossRef] [PubMed]

36. Banowski, B.; Garnich, F.; Simmering, R.; Device, I.E. Handset, for Performing Cosmetic and Medical Treatment for Human Skin, Has Application Surface Contacted with To-Be-Treated Skin Zone and Silver Portion, Which is Provided for Delivering Antimicrobial Silver Ions. DE102012224176A1, 26 June 2014.

37. Su, J.; Wang, D. Antimicrobial Thermoplastic Polyurethane for Toothbrush and Preparation Method for Antimicrobial Thermoplastic Polyurethane. CN103254401A, 4 March 2015.

38. Zhao, H. Sanitary Towel Capable of Removing Peculiar Smell and Manufacturing Method Thereof. CN102961778A, 13 March 2013.

39. Hasegawa, S. Solid Oily Cosmetic. JP2013071914A, 22 April 2013.

40. Miyata, S.; Kakihara, H.; Takahashi, F.; Nagaoka, H.; Kubota, T.; Ueda, G. Antimicrobial Agent. JP2010059132A, 18 March 2010.

41. Jang, H.C. Functional Cosmetic Including Nano Silver. KR20070119971A, 21 December 2007.

42. Zanini, D.; Alli, A.; Ford, J.; Steffen, R.; Vanderlaan, D.; Petisce, J. Antimicrobial Contact Lenses and Methods for their Production. US20030044447A1, 6 March 2003.

43. Schuette, R.; Kreider, J.; Goulet, R.; Wiencek, K.; Sturm, R.; Canada, T. Silver-Containing Antimicrobial Fabric. US20050037057A1, 17 February 2005.

44. Hendriks, E.P.; Trogolo, J.A. Wash-Durable and Color Stable Antimicrobial Treated Textiles. US7754625B2, 13 July 2010.

45. Hutt Pollard, E.A.; Morham, S.; Brown, D.E.; Kray, J.S. Systems and Processes for Treating Textiles with an Antimicrobial Agent. WO2018160708, 7 September 2018.

46. Silver, B.H. Breastpump Assemblies Having Silver-Containing Antimicrobial Compounds. US20080139998A1, 12 June 2008.

47. Wang, X. Nano-Silver Inorganic Antibacterial Nutritional Hair Dye. CN104224617A, 24 December 2014.

48. Panáček, A.; Kvítek, L.; Prucek, R.; Kolář, M.; Večeřová, R.; Pizúrová, N.; Sharma, V.K.; Nevěčná, T.J.; Zbořil, R. Silver colloid nanoparticles: Synthesis, characterization, and their antibacterial activity. *J. Phys. Chem. B.* **2006**, *110*, 16248–16253. [CrossRef] [PubMed]

49. Rulemaking History for OTC Colloidal Silver Drug Products. Available online: https://www.fda.gov/drugs/developmentapprovalprocess/developmentresources/over-the-counterotcdrugs/statusofotcrulemakings/ucm071111.htm (accessed on 27 September 2018).

50. Hartemann, P.; Hoet, P.; Proykova, A.; Fernandes, T.; Baun, A.; De Jong, W.; Filser, J.; Hensten, A.; Kneuer, C.; Maillard, J.-Y.; et al. Nanosilver: Safety, health and environmental effects and role in antimicrobial resistance. *Mater. Today* **2015**, *18*, 122–123. [CrossRef]

51. Sasaki, H. Automatic Bathtub Washing System. JP2009268576A, 19 November 2009.

52. Lee, Y.S. Clothes Washing Machine. US20080041117A1, 21 February 2008.

53. Nuernberger, C.; Nienaber, R.D. Air Purifier, Useful for Neutralizing Bad Smells, Preferably for Air Purification in Refrigerators, and for Controlling Bad Smells E.G. In Textiles and Vacuum Cleaners, Comprises a Silver Zeolite with a Nanoparticulate Metallic Silver. DE102007040742A1, 5 March 2009.

54. Kim, H.-K. Antibiotic Method for Parts of Refrigerator using Antibiotic Substance. US7781497B2, 24 August 2010.

55. Whitchurch, B.W.; Vaillancourt, D.; Jack, P.C.I.; Chen, W.; Huang, J. Submersible Keyboard. US20090262492A1, 22 October 2009.

56. Davis, W. Bactix Silver-Based Antimicrobial Additive in Bath Aids. US20130029029A1, 31 January 2013.

57. Gifford, S. Bacteria-Resistant Grab Bar. US20100148395A1, 17 June 2010.

58. Glenn, J.; Vogt, K.; Bridges, D. Antimicrobial Reusable Plastic Container. US20070189932A1, 16 August 2007.

59. Molnár, M. Vessel with Transparent Antimicrobial Silver Coating. WO2018137725A1, 2 August 2018.

60. Wang, X.; Gao, S. Nano-Silver Antibacterial Gloves. CN202738872U, 20 February 2013.

61. Scheuing, D.R.; Szekres, E.; Bromberg, S. Natural Silver Disinfectant Compositions. US20100143494A1, 10 June 2010.

62. Miner, E.O.; Eatough, C.N.; Miner, E.O.; Eatough, C.N. Antiseptic Solutions Containing Silver Chelated with Polypectate and Edta. US7311927B2, 25 December 2007.

63. Yu, S.-J.; Yin, Y.-G.; Liu, J.-F. Silver nanoparticles in the environment. *Environ. Sci. Process. Impacts* **2012**, *15*, 78–92. [CrossRef]

64. Stewart, P.S.; Costerton, J.W. Antibiotic resistance of bacteria in biofilms. *Lancet* **2001**, *358*, 135–138. [CrossRef]

65. Muller, M. Bacterial silver resistance gained by cooperative interspecies redox behavior. *Antimicrob. Agents Chemother.* **2018**, *62*, e00672. [CrossRef] [PubMed]

66. Elkrewi, E.; Randall, C.P.; Ooi, N.; Cottell, J.L.; O'Neill, A.J. Cryptic silver resistance is prevalent and readily activated in certain gram-negative pathogens. *J. Antimicrob. Chemother.* **2017**, *72*, 3043–3046. [CrossRef] [PubMed]

67. Oyanedel-Craver, V.A.; Smith, J.A. Sustainable Colloidal-Silver-Impregnated Ceramic Filter for Point-of-Use Water Treatment. *Environ. Sci. Technol.* **2008**, *42*, 927–933. [CrossRef] [PubMed]

68. Ingle, E.M.; Fisher, B.J.; Finney, J.W. Silver Coated Nylon Fibers and Associated Methods of Manufacture and Use. US2010166832A1, 1 July 2010.

69. Hayat, R.; Ali, S.; Amara, U.; Khalid, R.; Ahmed, I. Soil beneficial bacteria and their role in plant growth promotion: A review. *Ann. Microbiol.* **2010**, *60*, 579–598. [CrossRef]

70. Choi, O.; Hu, Z. Size dependent and reactive oxygen species related nanosilver toxicity to nitrifying bacteria. *Environ. Sci. Technol.* **2008**, *42*, 4583–4588. [CrossRef] [PubMed]

71. Olson, M.E.; Harding, M.W. Method and Compositions for Treating Plant Infections. US20120219638A1, 30 August 2012.

72. Song, Y.S.; Kim, M.H.; Won, M.H. Silver Yarn, Plied Yarn Silver Yarn, Functional Fabric Using Same, and Method for Producing Same. CN102439205A, 2 May 2012.

73. Yabe, S. Rolling Device. JP2005201385A, 28 July 2005.

74. Morris, M.; Kerry, J.; Cruz, M.; Cummins, E. An Antimicrobial Food Package. WO2014001541A1, 3 January 2014.

75. Maggio, R.A.; Pearson, R.C. Rotationally Molded Plastic Refuse Container with Microbial Inhibiting Inner Surface and Method. US20080185311A1, 7 August 2008.

76. Pradeep, T.; Chaudhary, A.; Sankar, M.U.; Rajarajan, G. Anshup Sustained Silver Release Composition for Water Purification. WO2012140520A8, 7 November 2013.

77. Pradeep, T.; Chaudhary, A.; Sankar, M.U.; Rajarajan, G. Sustained Silver Release Composition for Water Purification. US20180186667A1, 5 July 2018.

78. Takahashi, H.; Arakawa, H. Method for Producing Antimicrobial Agent Micro-Particle. JP2007161649A, 28 June 2007.

79. He, Z.-L.; Tong, T.W.; Zhang, Y.; He, W. A database linking chinese patents to china's census firms. *Sci Data* **2018**, *5*, 180042. [CrossRef] [PubMed]

80. Tlili, A.; Jabiol, J.; Behra, R.; Gil-Allué, C.; Gessner, M.O. Chronic exposure effects of silver nanoparticles on stream microbial decomposer communities and ecosystem functions. *Environ. Sci. Technol.* **2017**, *51*, 2447–2455. [CrossRef] [PubMed]

81. Gugala, N.; Lemire, J.; Chatfield-Reed, K.; Yan, Y.; Chua, G.; Turner, R. Using a chemical genetic screen to enhance our understanding of the antibacterial properties of silver. *Genes* **2018**, *9*, 344. [CrossRef] [PubMed]

82. Drake, P.L.; Hazelwood, K.J. Exposure-related health effects of silver and silver compounds: A review. *Ann. Occup. Hyg.* **2005**, *49*, 575–585. [PubMed]

MDPI

St. Alban-Anlage 66

4052 Basel

Switzerland

Tel. +41 61 683 77 34

Fax +41 61 302 89 18

www.mdpi.com

Antibiotics Editorial Office

E-mail: antibiotics@mdpi.com

www.mdpi.com/journal/antibiotics